The Years without Summer

Tracing A.D. 536 and its aftermath

Edited by

Joel D. Gunn

BAR International Series 872
2000

Published in 2016 by
BAR Publishing, Oxford

BAR International Series 872

The Years without Summer

ISBN 978 1 8417 1074 7

BAR Publishing is the trading name of British Archaeological Reports (Oxford) Ltd. British Archaeological Reports was first incorporated in 1974 to publish the BAR Series, International and British. In 1992 Hadrian Books Ltd became part of the BAR group. This volume was originally published by Archaeopress in conjunction with British Archaeological Reports (Oxford) Ltd / Hadrian Books Ltd, the Series principal publisher, in 2000. This present volume is published by BAR Publishing, 2016.

Printed in England

BAR
PUBLISHING

BAR titles are available from:

BAR Publishing
122 Banbury Rd, Oxford, OX2 7BP, UK
EMAIL info@barpublishing.com
PHONE +44 (0)1865 310431
FAX +44 (0)1865 316916
www.barpublishing.com

Dedicated to
Professor Richard Kaye,
who in my 1968 senior seminar on the
Early Middle Ages at the
University of Kansas
fused History and Archaeology for me.

CONTENTS

1. A.D. 536 AND ITS 300-YEAR AFTERMATH

Joel D. Gunn

Department of Anthropology, The University of North Carolina at Chapel Hill, Chapel Hill, North Carolina 27514

Introduction

In the fall season of A.D. 536 Cassiodorus sat at his writing table. He was a member of the Roman Senate and the most prolific European scholar of the middle first millennium A.D. Cassiodorus had returned to his estate, departing Rome because of relentless warfare, and because heaped on that trouble was a persistent dry fog that blued the sun; it cast no shadow at midday and had not for months. Without sunlight the grapes were bitter; the drought and cold of summer precluded wheat harvest and storage. Gaunt faces walked the streets of Rome and all of Europe; hunger was everywhere as well as its attendant dangers imposed by disease and violence. As Cassiodorus turned to his letter writing, he contemplated the once-beautiful, but now war-ravaged roadways of Roman Italy. Cassiodorus took up his pen to reassure the young subordinate in Rome who would stand in his place. After all, he wrote, last year was an unusually good harvest. We should be thankful for that blessing in the middle of all this shortage.

In the following years war and famine fed on each other and the unusually bountiful harvest of A.D. 535 was consumed without replacement. In A.D. 541 large portions of the human population began to die from hunger, disease, and warfare, so much so that the number of Europeans fell to its lowest between classical and modern times.

This book examines the first millennium A.D. worldwide context of Cassiodorus and the situation he and his contemporaries experienced. Is there sound evidence for such an event? The answer is "Yes." Tree-rings worldwide mark the evidence of it, as well as other lines of evidence, and I will return to the drama of strange first millennium climate in the second part of this introduction. First, I would like to examine the issues with regard to the environmental changes of the sixth century A.D., and how they relate to archaeology, history, ethnology, and the near future of the world's peoples.

Global Change and Local Cultures

Issues of Global Change and Archaeology/History

Although it is not clear Cassiodorus was aware that others from as far away as China were writing reports of similar problems, he was nevertheless evaluating, as the most important administrator of the Roman Senate, what to do locally about a sudden, worldwide change in environment. Problems that could be of similar scope face local and worldwide administrators today for other reasons. The global change study community has until recent years focused its efforts on understanding the effects of global warming on physical aspects of the environment (see for example Houghton et al. 1990). However, in recent years awareness of the need to study the biological, and to some extent the human side of global change has emerged. The IPCC (Intergovernmental Panel on Climate Change) published reports in 1995 and 1997 on growing concern for human and regional impacts of global change (see www.ipcc.com). In the latter publication they cited baseline data on extreme events as a key issue for future research. To this end the authors of this book contribute studies of the archaeological past when more extreme global conditions than at present challenged governments and peoples. There are, however, signs of recognition of the potential value of archaeology. This is most readily seen as shifts in emphasis of the programs at global change conferences that suggest that archaeological data will become more important in coming years. Frank Hole of the Yale University Department of Anthropology was a keynote speaker in a July 1996 Geneva conference on global change sponsored by IASA. This could be taken as an important turning point.

This introductory chapter to A.D. 536 and its aftermath explores two domains. The first is the potential of the emerging relationship between archaeology and earth system science, perhaps emblematized in Frank Hole's presentation. Earth system science is as new to the world scene as the space shuttle, and comparatively, archaeology and related historic science disciplines are as old as the hills whence archaeologists excavate artifacts. Nevertheless, it is not too early to forge fruitful relations between the two perspectives. The second part of this chapter introduces some background information on the first millennium A.D. Since the central focus of the book is the A.D. 536 event, its worldwide scale requires a global perspective in order to evaluate much of the meaning and context of the surrounding situation. Of particular interest to the authors are the variable effects that a worldwide event had on local cultures. Some of the history and methodology of the study of global-to-regional climate linkages are discussed with references to a broader literature.

At the many global change conferences held to promote interest in looming global scale problems, the regions treated in this book are frequently considered as being in special danger. On the east side of the Atlantic basin, Europe has become a center of concern because of the instability of the worldwide maritime conveyor system revealed by marine chemistry studies (Broecker 1995). Africa is already prominently featured on the global climate stage because of killing droughts in the Sahel. Asian populations and economies will be immensely important. The southeastern United States is mentioned as an important area of study because of its huge economic role in the US economy and its surprising vulnerability to global climate change. The lowlands of Mexico's Yucatan Peninsula are of interest because of a long record of difficulties encountered by complex civilizations, and also because they are vulnerable to sea level changes (Folan et al. 2000).

In this chapter I outline the book, but also attempted to design a program of participation for archaeologists in the global change research arena. It is at best an introductory attempt. However, as David Anderson discusses in his chapter, there is much that is promising, if formidable, about the responsibilities participating in the study of global change. I believe archaeologists can contribute to the understanding of global change because they study periods when local climates were influenced by global temperatures more extreme than any observed instrumentally in the last century. Archaeologists can prepare themselves for this role by reexamining the role of their discipline in a long and accomplished history of studying climate and human organization. Also, they need a program and a method of action.

As fate would have it, there has been a great convergence of growing archaeological knowledge of the remote past with the need to know about the past in the global change study community. Since mid-century knowledge of the human past has grown immensely. As a feat of scholarship and field work it is every bit as impressive as parallel accomplishments in space exploration and medical technology. Examples abound. The Maya, one of the most curious examples of collapsed civilization, have transcended the structures and ceramics of prehistory to speak plainly of their history through writing; scholars have learned to read their complex multi-sensory writing. Similarly the Hittite Empire, only briefly mentioned in biblical texts, has emerged from obscurity to stand fully equal to well-known New Kingdom Egypt. Human entry to the western hemisphere has been extended into the indefinite past when the great shift from ice age to interglacial must have carried human cultures through a rugged regimen of change. Plant domestication has grown from an isolated invention in the Near East to a worldwide emergence around 10,000 years ago at the end of the ice age.

As has been the case for two centuries, some of this growth of knowledge has been through university's academic archaeology. Great amounts of additional data have been generated since the passage of the US Moss-Bennett bill in 1975 through the efforts of so-called cultural resource managers. The institutional affiliation of cultural resource managers varies from country to country and includes universities (US), but also extends to private firms (US) and state bureaucracies (England, France), communes (China) and Native American tribes (US). Archaeologists discover evidence of former times and rescue it for the benefit of humankind. Because the sponsorship of cultural resource management has always been involved with politics and policy, cultural resource managers are aware of many of the ways and means by which academic knowledge and policy as implemented by legislation and government officials.

During the same time, not only has the concern for preserving knowledge of the past risen in the estimation of policy makers and the public, but policy makers also have been confronted by concerns for rapidly changing global climate. This awareness has risen from recognition of the linked nature of the world environment and how tampering with that environment could adversely influence the well-being of humanity. Archaeologists may be able to contribute to that debate because humans have experienced important impacts from the world environment in the remote past. Knowing of these encounters minimally adds to a list of known successful encounters with change, and could contributed to preemptory management scenarios directed toward mitigating future impacts and adapting to them where they cannot be prevented. The past impacts have been as important as severing the cultural thread between classical and modern times, some of the substance of this book. The availability of these knowledge sets in archaeology suggests a natural and comfortable wedding of archaeological information with other research into managing global climate change. Among all else that is going on in archaeology, I believe that this wedding, represented in a small but focused way in this book, is a dividend on the investment that has been made in the study of the past by the citizens of the world.

In recognition of the dividend and its potentially helpful outcome, I have organized the book to include a brief management summary at the beginning of each subsection. As is usual, the summary discusses the findings of the authors. Also, as usual in policy related reports, each summary contains a recommendation, in this case a brief discussion of how the section, concept and data of the chapters bear on issues of global change and their impacts on humans. One could hope that efforts such as this will help avoid unintended outcomes of our own actions as a culture by providing warnings of thing to come.

Relevant Archaeological Concepts

The archaeologists, historians and ethnographers who contributed to this book wish to identify overlaps between global climate change research questions and their own scholarship. As a starting point, an understanding of the archaeological concepts of region and time, and the inherent veracity of archaeological data could be useful in the global change studies.

Region. Understanding cultural change in regions and subregions is important for global change studies because there resides the root of social organization. Human social organization achieves its vital functions within aggregates of families in communities. It probably has for three million years or more. In modern societies, this is sometimes masked by large urban areas, and now even so-called megacities, but it is still the neighborhoods--made up of families and communities--that constitute the social fabric of a town, city, nation, and the world. The scales of these communities seem to remain similar throughout time and space. For example, among non-western societies, Birdsell (1973) notes that Australian desert tribes begin to fission when their numbers increase to more than 500 persons. This same figure is known in western modern educational circles to define the size of a high school, or subcommunity, in which the school principal knows everyone's name; to maintain that community intimacy and familiarity and accompanying social viability.

By working with archaeological sites as units of analysis, archaeologists start their study of cultural change at the same community-based scale of social fabric. Archaeological sites are the trace or foot print of past interactions between households, communities, and the environment. A lesson that can be learned, both in the archaeological and historical chapters of this book (Lilly and Webb chapter, see also Crumley 1994), and in the ethnological literature of living groups, is that societies change the relationships between households and communities during sudden environmental changes. These are regional social phenomena that re-rank heterarchically to address new conditions.

Time. The trace of A.D. 536^{cal} (see **Date Notation** text box for superscript) is important as an example of sudden change set in time remote enough to provide historical perspective. Because it is temporally apart from the present, both its antecedent conditions and long term aftermath can be studied. Thus, while A.D. 536^{cal} is pivotal to the concept of this book, the book is a study soundly set in the context of the first millennium A.D.: little can be understood about the impact of A.D. 536^{cal} without the informative company of the centuries before and after. Because it is remote in time, archaeological methods and historical scholarship are inherently important to its study.

In an even broader perspective, many of the more dramatic examples of human experience in the face of changing world and regional climates can only be accessed by archaeological methods. Without archaeology, most of the history of the world fades into prehistory. There are two reasons for this. First, without archaeology, everything social before 4,500 years ago disappears because writing emerged at that time as a sufficiently refined art for people to reveal their intentions and relationships. This is important to current global change issues because the three thousand years prior to 4,500 years ago was the most extreme instance of global warming in the Holocene, the Altithermal or Middle Holocene global climate episode.

Second, after 4,500 years ago, literacy spreads from its hearthstone in Mesopotamia like a cone in time and space ("cones" according to some). Only in recent centuries does it reach the most distant cultures of the Earth; in non-literate regions of the world, archaeology, and in some cases oral tradition, remain the only avenue of access to regional culture history until virtually this century. Even in regions where historic records exist, much about human adaptation to climate is incomprehensible unless supplemented by archaeologically recovered material culture. The design of this book was specifically set to include case studies from within and without the cone of history. It is important to recall also that in spite of the prominence of megacities on the world political stage, the greater part of the world's population remains rural and basically Neolithic in outlook.

Date Notation

Dates are superscripted with

"cal" for calendar,

" c14" for radiocarbon age, and

"cor" for corrected radiocarbon.

Compounding world cultural regions and archaeological time periods illuminates the scope of cultural diversity, which includes the catalog of worldwide human answers to coping with the environment. The number of local human responses to global climate is potentially quite large. Although opinions vary, one group of ethnologists/ethnohistorians found that the world harbors about 180 regional varieties of culture (Murdock and White 1968). Archaeologists generally divide the time of the last 13,000 years, the Holocene, into more than a dozen periods. These periods correspond in some general fashion to the climatic episodes of the Holocene (Denton and Karlén 1973; Wendland and Bryson 1974). Thus, in theory there are more than 12 x 180 = 2,160 climates and cultures to be found and studied in the worldwide Holocene. If those who make world policy are to accommodate human cultural diversity in planning, a substantial range of these regions and times need to be sampled to understand the responses of world cultures to global climates. Of that number, this book represents about a dozen regions in one time period, which is one half of one percent of the total.

Veracity. Archaeologists struggle with biases inherent in the archaeological site record and in their own training and outlooks (Nassaney and Sassaman 1995). Even so, a certain relentless truthfulness resides in pot sherds from agriculturists' dwellings, whether on the floodplain of the Yadkin River in North Carolina or from an excavation on the seven hills overlooking the Tiber River at Rome. Selection of temper inclusions and twist of cordage in exterior decoration reveals direction and ethnicity of origins as surely as an alphabet. Both of those neighborhoods faced the changes of A.D. 536^{cal}. The physical remains of past cultures keep re-presenting themselves for re-examination and re-interpretation. The interpretations do not slide easily from one intellectual fad to the next without reflecting on their sherdy touchstones. The archaeological record provides an alternative cross-check to be laid against other historical, mythological, ethnohistorical, and ethnological sources on regional cultures and environments.

World civilization is or is trying to become, a planning society, and the global change study community is a part of that milieu. As planning for world cooperation becomes an important element of international and intercultural action, controverting ethnocentric biases and cultivating knowledge of the total human experience will become infinitely more important in international and intercultural relations. Plans should be designed to recognize and preserve cultural diversity, just as biological diversity should be preserved. Using the spatial, temporal and cultural dimensions of archaeological knowledge and methods, the authors of this book offer important contributions to that knowledge and how to obtain it.

Archaeologist's Role in Global Change Research

The question raised among by the contributing authors is what they might contribute to understanding global changes. We suppose this contribution lies at least in part in the realm of extreme conditions not observed in the twentieth century, and their impact on human social organization. The globe-circling case studies to follow illustrate differing personal and population reactions to the events surrounding A.D. 536^{cal} in different regions of the world. By doing so, the authors address issues of regional variation of global change. Concerns over variations in regional manifestations of global change and their potential effects on stakeholders are an emerging topic in the global change community.

Social adaptations by ancient stakeholders are often the domain of effort unique to archaeologists. Students of the earth system cannot be certain at this time that global warming will not precipitate climatic transformations as vast of magnitude and broad of expanse as those experienced in the sixth century A.D., or perhaps even equivalent to the end of the last ice age 10,000-12,000 years ago. Wallace Broecker (1995) warns in a *Scientific American* article that he foresees potential significant changes as soon as a century in the future; he gathers his conclusions from the study of global warming at the end of the last ice age. Archaeologists investigate social phenomena at this time depth and in ways otherwise inaccessible. By studying social phenomena associated with the end of the ice age over 10 millennia ago, archaeologists illuminate the social aspect of that great shift in global climate. In fact, the fates of Paleolithic cave painters in Europe and Paleoindian elephant hunters in the Americas have been an enduring fascination for over 200 years. Publication on the topic began with John Frere in 1797, now the two hundredth anniversary of his work.

Archaeologist's methods of understanding the past lie primarily in material culture, the material objects people use in everyday life to adjust to their circumstances. By studying material culture, archaeologists' methods discover how people adjusted other aspects of their social and physical environment. To be sure, these methods are limited in some cultural dimensions compared to ethnologists or historians. Archaeologists cannot demonstrate that the sun cast no shadow for 18 months after A.D. 536^{cal} as historians can. On the other hand, archaeological methods reveal other than personal aspects of culture that frequently serve as complementary source of insight to the findings of the more cognitively rooted historical and cultural disciplines. Historians can determine that in the decade after A.D. 536^{cal} Gregory of Tours (see Young chapter) rebuilt burned churches; what can archaeologists add to that picture? They could detect materials and styles that indicate the conditions under which the churches were rebuilt. Were materials obtained locally under economic stress, or imported from afar reflecting relative affluence?

In the case of Gregory of Tours, where both history and archaeology reveal the nature of his culture, the issues of complementary roles of the disciplines are clear. How, on the other hand, are the non-material aspects of Paleolithic cave painter's lives to be discovered in the absence of history? Answers generally emerge from a delicate balance of material culture studies, ethnohistory and ethnology to provide general guidance, and simulations of comparable social organizations. By careful combinations of methods, studies in recent years have revealed a surprisingly sophisticated cognitive world in the Paleolithic; it included tallying the passage of months and other enumerations, perhaps keeping track of food gathering activities. But this is not really a surprise once one has stood at the foot of Lascaux's walls.

Such reconstructions are precisely the complement of the scenarios of future civilizations that will be required of the global change community if it is to evolve understandings of future change events. In those cases, however, the scenarios will be yet one step further from reality because they will require simulated rather than observed climate context for social interactions.

The question is, then, how do archaeologists fit into the interdisciplinary structure that constitutes, as near as academically possible, the whole of human history, including its future? Does their part comprehend information useful to understanding sudden global change and the human consequences thereof, especially in the future? Because of the complementary methods of collecting data, we believe there are roles that archaeologist can play in the modeling and reconstruction of societies responding to environmental changes.

The organization of the book arises from this concern. The global scale of the A.D. 536^{cal} atmospheric disruption provides a venue from which to search from region to region for the differing perspectives of cultures toward the same event. It also transpired during a time when part of the world was literate and part was not. As such, records from the literate part of the world, within the cone of history, can be used to broaden understanding of archaeological data on sudden, dramatic change. The incident opens the usually close-knit web of culture and environment to inspection, and at the same time we are challenged by the great reaches of the world surface that are accessible only through the archaeological record.

Toward an Event-Sensitive Archaeology

Given current concerns for sudden global change, it is clear that for archaeologists to participate in global change studies, their methods and outlooks need to be event sensitive. At the same time, if A.D. 536^{cal} exists as a milestone in history, it has to be dealt with in a productive manner simply for the study of the past.

One of the compelling issues of the volume, the question that glues history and archaeology together, is the ill-defined relationship between archaeological data and global events. Although archaeologists usually think in terms of century time scales, they have utilized near-event scale phenomena since the 1930s when the concept of horizon styles was offered as a dating technique (Trigger 1989:192; Willey and Phillips 1958). Horizon styles are constituted of rapidly spreading, short-lived items of material culture. When the spread of these traits can be traced across more than one region, they provide a means of cross-dating other cultural phenomena in other affected regions. An example of a horizon style is the use of copper ornaments obtained from the Great Lakes during the Hopewell period in the eastern United States. Characteristic Hopewell artifacts appeared in Illinois around 100 B.C. and, apparently carried by a religious cult, spread rapidly over most of eastern North America (Fiedel 1987:235-244). The Hopewell domain of influence included colonies and copies as remote as Yellowstone Park, Maryland's Eastern Shore, and south to Florida. It encompassed all of the southeastern regions studied in this book, and local variants figure in most author's deliberations since the radiocarbon dated end of Hopewell and its related manifestations at around A.D. 400 approximates the A.D. 536 event in calendar years.

In the chapters of this book we attempt to apply a modified concept of horizon style to the global spatial scale. The vehicle of this application is a historically and tree-ring documented global scale event of over a decade in duration, A.D. 536^{cal}. To be sure, the issues are very much more complicated than the authors of the horizon style concept might have imagined. The concept not only has to be adjusted for worldwide scale; it also has to account for a greatly evolved conceptualization and valuing of the earth system, and it must adjust to new ideas about cultural and environmental interactions such as complex and chaotic systems theories. Finally, and perhaps most challenging, the concept of horizon style must undertake the work of making archaeology, and its sister historical science disciplines, of utility in understanding the impacts of global change on the human condition, and on the regional ecosystems in which humans participate. As we shall see, the region-to-region variations can be immense.

The use of A.D. 536 as a horizon style marker substantially broadens the concept of horizon style. Events are not artifacts, but presumably they are generators of artifacts as styles adjust to new circumstances. To accommodate this reversal of roles, I suggest extending the horizon style concept in the sense of a snap-to-grid scheme. Snap-to-grid is an option in computer drawing programs that aligns drawn objects along an invisible underlying grid: drawn objects are attracted to an invisible underlying grid as though by a magnet. In reading about A.D. 536^{cal}, it seemed to me that A.D. 536^{cal} provided a compelling underlying time grid, which at least allows consideration of temporally refined causes even though the precise proof of the relationship is at the time wholly or partly inferential. Mathis, for example, in his chapter on the North Carolina coast, has insufficient radiocarbon dates to pinpoint the immigration of Algonkian speakers into the region to A.D. 536^{cal}. However, the knowledge of a potentially potent climate influence at approximately the correct time, and the ethnographic understanding that people sometimes move south during colder episodes, such as that found by Jones in her chapter on the opposite side of the ocean, minimally suggests a hypothesis, a focus for a research program in North Carolina: did the conditions around A.D. 536^{cal}, contribute to this immigration? Though not in the usual sense, the A.D. 536^{cal} event provides a type of horizon style of worldwide scope, a snap-to-hypothesis.

In addition to the economic advantages of a focused research program, an event-sensitive, but not event-bound, archaeology provides a laudable environment for research because it surpasses the limitations of gradualist evolution and comfortably joins archaeological research with the more plausible paradigm of change by punctuated equilibrium. Punctuated equilibrium allows cultures and biological organisms sudden changes precipitated by outside influences on internal system functions, but also extended periods of stasis.

Toward and Earth System Sensitive Archaeology

Having traveled the related topics of archaeological methodology and the relationships of archaeological data to worldwide events, we come back to the question of earth system context for the A.D. 536 event. As noted above, the earth system has come to be regarded as a tightly woven fabric of interacting components: water, air, land, biota, and human. The often mentioned analogy between fabric and social and environmental interactions is apt and can be extended to reveal the utility of studying the social implications of A.D. 536^{cal}. Usually the fabric of culture and the earth system is such that their interactions are seamless. This is an acknowledged property of complex or emergent systems; their behavior is so obscured by feedback and dialectical relationships that responses cannot be anticipated. It cannot be predicted, but it can be understood (Kauffman 1995). The value of looking at the A.D. 536 event, and other similar circumstances, is that the fabric is stressed, parted, perhaps even rent, in such a way as to allow insights into how climate and culture interact or intertwine. When we have a package of these partings and resulting insights, then we have information to take to the global climate community for scenario building. Many of the chapters offer useful insights on social responses to sudden change and Jones' chapter can be used as an example. In the context of sudden global change, the social fabric of Roman British society and its links to the Near East opened the door to plague from Asia; through this intermediary effect, the unusual, abrupt transition from Roman Celtic towns to Saxon dispersed hamlets transpired in southern England. Such lengthy chains of interaction should be expected when regional social systems are joined to a global ecology.

The greatest challenge offered by the earth system concept lies wholly in the future. Beyond the usual concept of the earth system as a biological entity, the concept that historical scientists should accommodate in addressing global change issues is Lovelock's (1987) geophysiology. Lovelock uses global temperature as an index of the status of the earth system. Global temperature as an index of earth system conditions has been the underlying premise of my work in

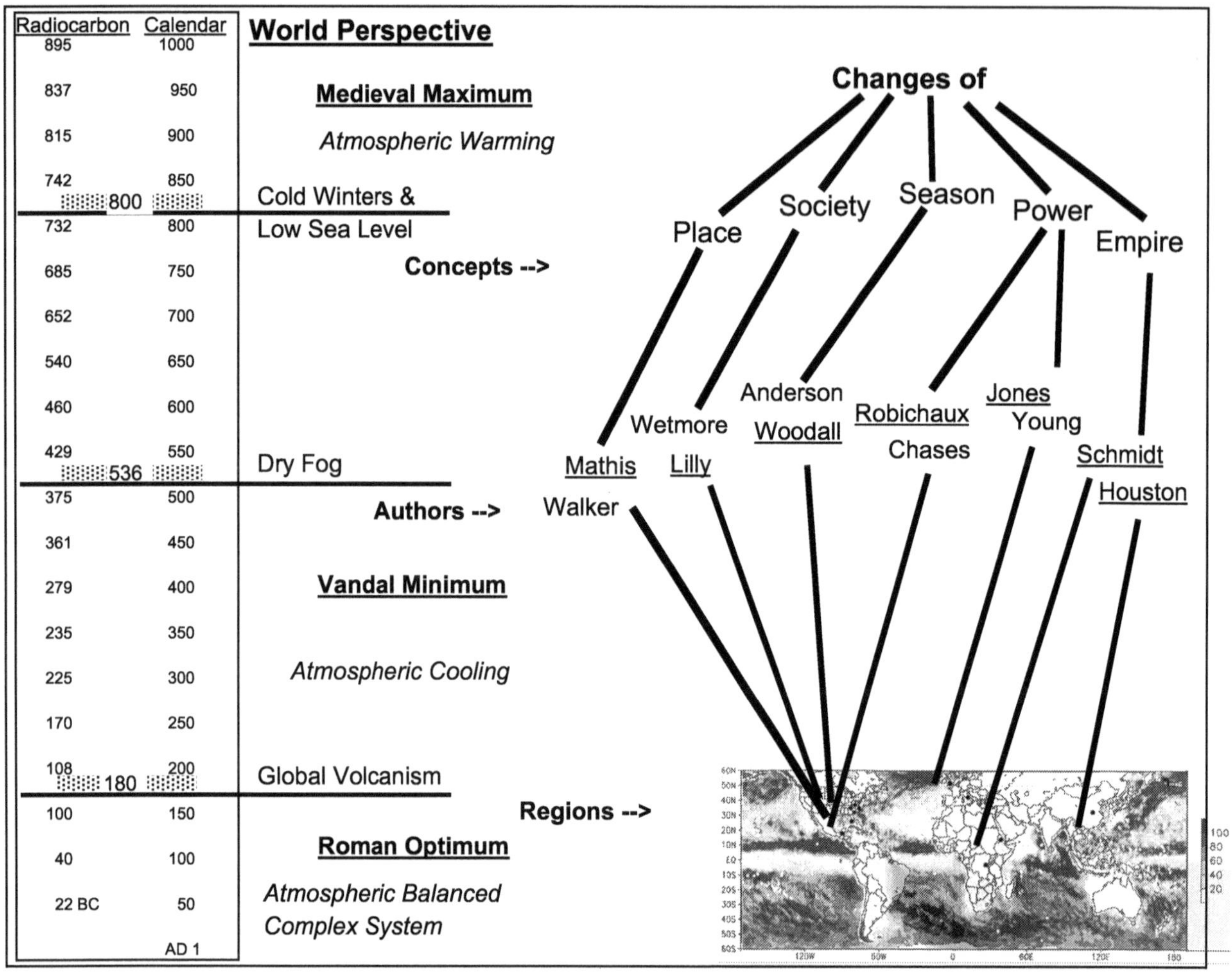

Figure 1.1. Time Line and World Regional Distributions in the Book. (Inset world map is cloud liquid water satellite image for August 1999 from *Climate Diagnostics Bulletin*.)

the study of global climate, and is what I am referring to when I say that Tanner's lowered sea levels indicate lowered global temperatures (see Tanner chapter); as the global temperature diminished, ocean water is confined in glaciers. Broecker (1995) explains some dangerous thresholds relative to global temperature warming. However, geophysiology seems to imply much more than merely watching thresholds. If the earth system is to be treated as a physiological being, then it, like the human body, has homeostatic processes that maintain critical variables within acceptable boundaries for its survival. If these mechanisms fail to maintain global temperature at a level acceptable for existing species, then problems ensue of great magnitude, such as the mass species extinctions now so much a topic of interest (Donnovan 1989). From a preventative perspective, we should be concerned with global temperature, but also how hard the earth system physiology is working to maintain the correct range of global temperature. How hard is the system working to sequester excess carbon dioxide that will, if unchecked, overheat the atmosphere and turn it into a scorched Earth like Venus? Is it near its capacity to do so? One might suspect that the answers to these questions reside with microbiologists (see for example Volk 1998), but human social systems are often the canaries of the earth system. Will the study of the A.D. 536 event and other climate disturbances help uncover the signs of earth system overwork as read in the archaeological record through stress manifest in social systems?

The First Millennium A.D.: An Unsettled Time in History

Global Climate Change in the First Millennium A.D

The previous section provided an overview for the study of global impacts on regional cultures. I now turn to the specifics of environmental context of the A.D. 536 event, the conditions of the first millennium A.D. In the first part of the chapter we examined Senator Cassiodorus' observations early during the episode. It has been known among climatologists for some time that an atmospheric event of significance occurred around A.D. 536[cal]. Historical accounts report as much 18 months during which the sun did not cast a shadow at noon (see Young chapter). The cause was a persistent "dry fog." Records indicate that crop failure, famine and disease stalked Eurasia and Africa. The apparent magnitude of the event increases as research in tree rings and historical records reveal its dimensions: 15 years of reduced tree growth (the severest such period in a 6,000 year record

[Baillie 1994, 1995]), marked decline of populations to the nadir between classical and modern times (Crumley and Marquardt 1987:252), and transitional moment from classical to modern historical time in Europe. In the Americas, there is a broadly recognized suite of changes in eastern North American cultures around the sixth century (Nassaney and Cobb 1991). The Maya Early-Late Classic Hiatus in Mesoamerica also corresponds in time (see Robichaux chapter). Is there a plausible link between these cultural conditions, and could the atmospheric event of the early sixth century be it?

The authors of this book attempt to discover the global scale cultural ramifications of the sixth century conditions by focusing on its detection in subregional records, whether they be historic, archaeological or paleoenvironmental. "Records" is interpreted broadly as information obtained in the historical sciences, which include geology, archaeology, history and related disciplines.

It is clear from these presentations that the world of the first millennium A.D. was no stranger to rapid global change, neither in A.D. 536 [cal], nor in other centuries. The first millennium A.D. has been recognized for some time as an unruly period of world climate (Figure 1.1). One of the more spectacular cold events of the first millennium was the freezing of the Nile River A.D. 829 [cal] Lamb (1977:427). In broader perspective, the freezing of the Nile River was one season in a series of unusually cold winters between A.D. 760 and 840 [cal].

The freezing of the Nile bookends the first millennium time of changes with the third century A.D. Annales School scholars such as Leroy Ladurie (1971; Bryson and Murray 1977; Crumley 1993) have long been fascinated by the disappearance of viticulture from northern Europe after the second century A.D. and following the Roman Empire Golden Age. The lapse of high latitude viticulture was a fragment of the greater disruption of life in the Roman Empire that included a collapse of administrative control in the western Empire, the "Third Century Crisis," and the emergence of the mysterious "Gallic Empire," famine and disease (perhaps the first appearance of Bubonic plague in Europe), and subsequent emergence of a substantially different administrative structure and attitude toward things Roman (Berry 1987:458-461; Crumley 1994; Gunn, Crumley

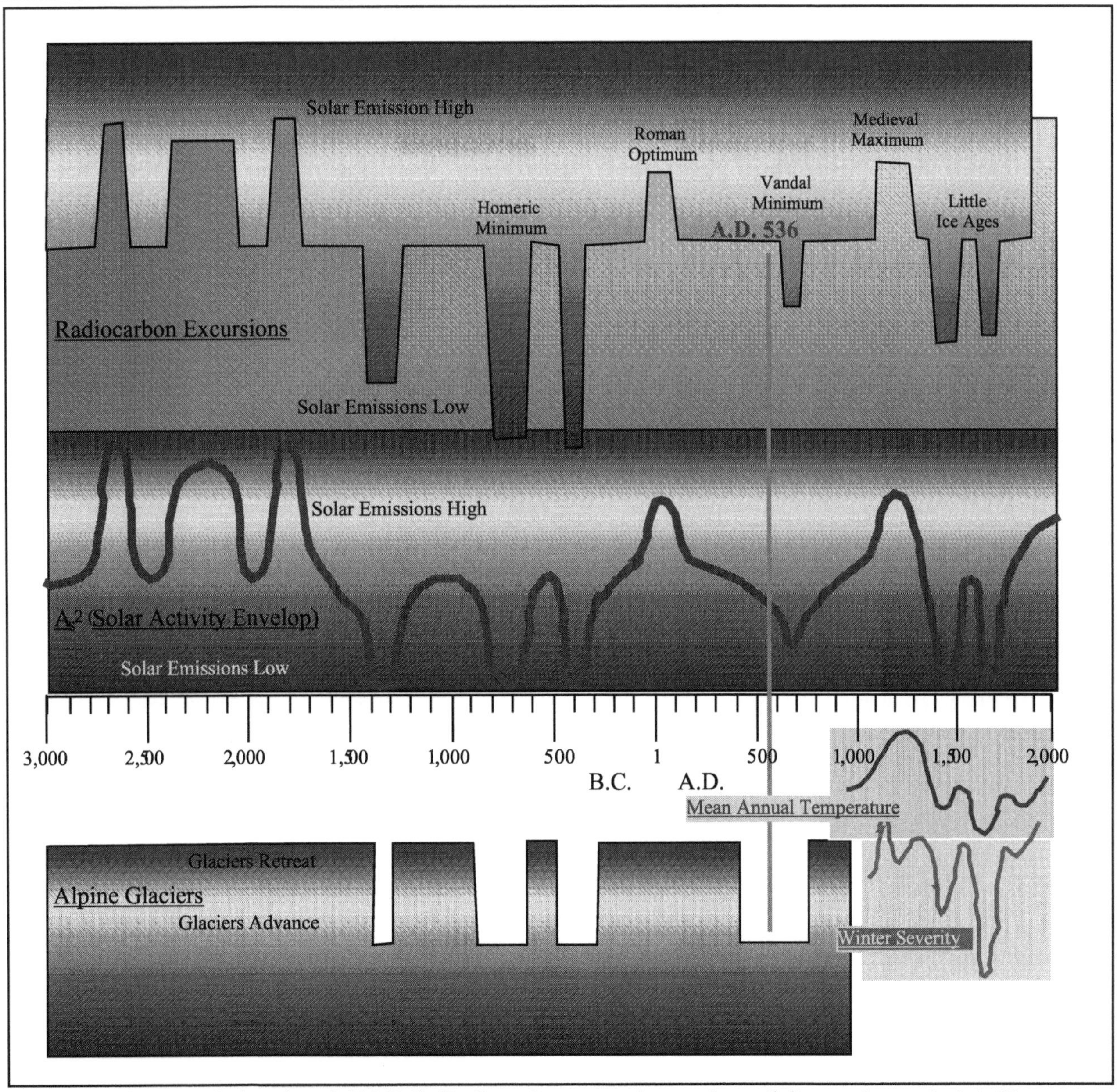

Figure 1.2. Solar Emission since 3,000 B.C. Radiocarbon excursions, solar activity envelope calculated by Landscheidt, and European climate (adapted from Landschiedt 1987). Note especially that Alpine glaciers are well-advanced around A.D. 500.

et al. 1995). The third century, in fact, coincided with the greatest detectable episode of volcanism in the Holocene (Bryson 1994; Bryson and Goodman 1980). These and many other notable weather anomalies led Reid Bryson (1988, 1994) to designate the middle centuries of the first millennium as the "Vandal Minimum."

Sandwiched between these two specters of the first millennium, and strangely undetected until recently, is A.D. 536^{cal} and its surrounding circumstances. Lines of evidence other than historical show that the precursor conditions to a global cold were present, especially low solar emissions[1], produced cold and unstable global climate (Gunn 1997). Among the empirical evidence of advanced global cooling in the middle first millennium is extensive expansion of alpine glaciers (Denton and Karlén 1973; see Figure 1.2). A similar and better documented coincidence of low solar emissions and active volcanism can be seen in the seventeenth century (Eddy 1977, 1994). There is no reason to doubt that the sixth century could have been the nadir of the Vandal Minimum.

Why is this third grim face of the first millennium so dimly seen? The magnitude of the event is brought into focus by Baillie's (1995) work with tree rings (Figure 1.3); after A.D. 536^{cal} he detected 15 years of the slowest growth in a 6,000 year long record of Irish oak tree rings. He believes that the episode was so serious that normal processes of historical transmission were disrupted (Baillie 1995, 1999). What is known of the occasion in most parts of the world was largely transmitted through traditional oral history rather than written words.

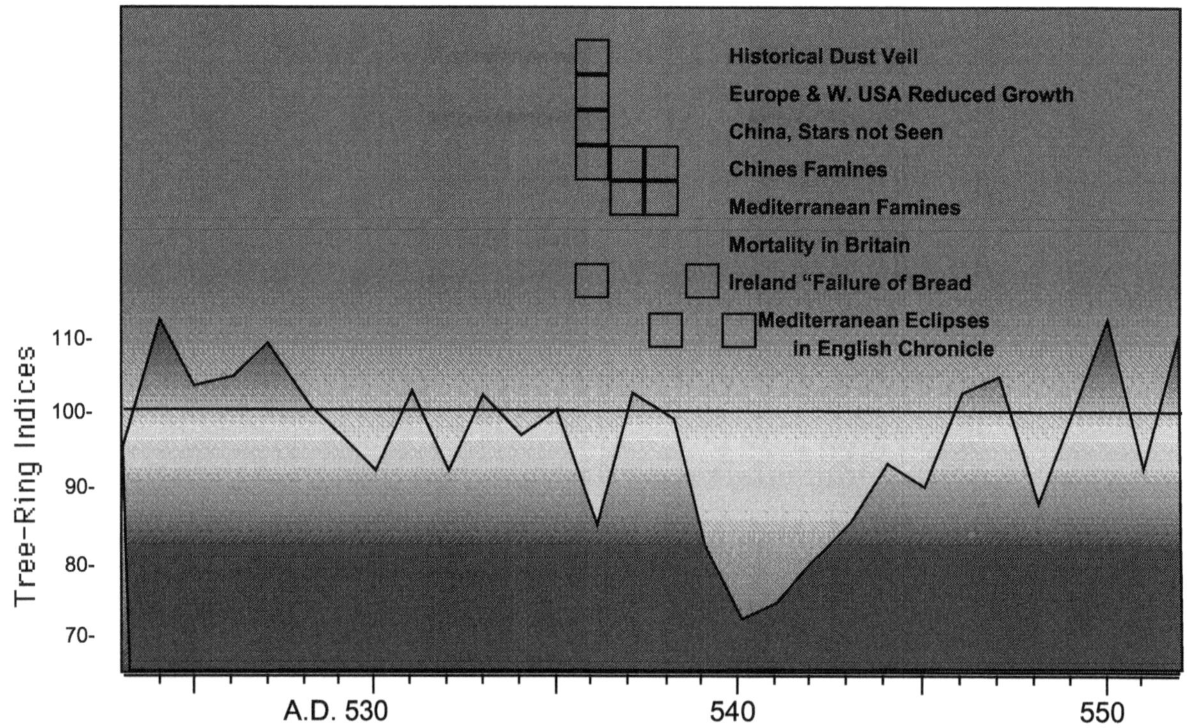

Figure 1.3. Tree Ring Growth Shows Reduced Ring Width from 3 to 8 Years following A.D. 536 (Baillie 1994). The tree rings are from Irish oaks and 15 other tree ring sites in Europe. Similar depressions in tree ring growth occurred in the United States in bristlcone pine and foxtail pine (Scuderi 1990).

and that there indeed were expectable effects of global cooling around the North Atlantic basin. (Figure 1.2; Landscheidt 1987 [see especially p. 438 discussion of Eddy 1977]). Reinforced by atmospheric dust from geophysical and/or astronomical sources, the combination of low solar emissions and blocking of incoming radiation would have

I refer to the period as the "years without summer" because of the extended number of years affected. In Baillie's publications, he argues that the impacts on European cultures were profound and the environmental consequences were varied and worldwide as well. He, for example cites freeze rings on high altitude trees in the Southwestern United States (Baillie 1994). The effects on trees were not universally

[1] For models of solar emissions and global cooling see Schneider and Mass (1975), Gilliland (1982), Gunn (1991); Gunn and Grzymala-Busse (1994), and Lean et al. (1992). For paleoclimatic studies of solar forcing of climate see Willett, Fairbridge and Sanders, Schove, and Landscheidt chapters in Rampino et al. (1987).

negative. In the lower altitudes of the Southwest, trees experienced a period of increased growth because cooler conditions relieved some of the drought stress normally imposed by the dry climate of the Southwest (J. Dean, 1995, personal communication).

The causes of the A.D. 536^{cal} atmospheric phenomena are not yet known. An extended period of volcanic eruptions could have been responsible (Stothers 1984; Keys 1999). Based on some negative evidence, Baillie (1994, 1995, 1999) suggests that a near miss or hit by a meteor or comet could have caused the atmospheric disturbance. An aerial burst of a meteor in 1908 over Tunguska, Siberia provides an analogous description of such events (Baillie 1995; Gribbin and Gribbin 1997; Rasmussen et al. 1995). Light and weather effects in 1908 appear to correspond to some of those described in mythology for A.D. 536^{cal}. Some of them were spectacular. For example, on July 2, 1908, following the explosion in the upper atmosphere, a Danish newspaper reported that "At eleven o'clock, it was almost daylight, and at twelve o'clock letters could still be discerned in a book or newspaper." (Rasmussen et al. 1995:634). This was because great amounts of dust in the upper atmosphere bent the sun's light around the curvature of the earth. Similar conditions are reported for A.D. 536^{cal} (Baillie 1995). There was also a brief episode of global cooling after 1908. The global cooling has been credited to the effects of a volcanic eruption in the Caribbean, but could have been in part or whole due to the Tunguska impact. Similar but more extreme atmospheric cooling followed the A.D. 536 event (Baillie 1994).

Although not extensively studied, comet debris could be an influence on global climate. For example, cumulative tree ring graph for eastern North Carolina shows a decline in spring moisture following A.D. 700 (Figure 1.4). Since drier and cooler conditions in eastern North America generally correlate with cooler global temperatures, the contemporary approach of the Tempel-Tuttle comet inside the earth's orbit around A.D. 710^{cal} could have been a contributing factor. Like volcanoes, comets can leave dust in the upper atmosphere, and in space as well, when they pass inside the earth's orbit, and have the potential to block incoming solar radiation (Rao 1996).

On the cultural side, some of the authors in this volume report that comets are dark portents in various cultures. Because of their oft-noted astronomical alertness, if comets were repeatedly associated with negative consequences, ancient people would have rapidly identified the link. This is particularly likely if a comet left dust that filtered into the atmosphere for several years, as the Chixilub asteroid was thought to have done 65 million years ago. We await with interest the resolution of this question.

Cross-Structuring of the Book

Although separating the book into parts and chapters provides a broad outline, the chapters in the volume are cross-structured to make its subject material accessible to the reader something like an encyclopedia or other reference work. A road map to this underlying structure is provided in Figure 1.1. Within parts, chapters are paired into "concept sets" that address similar environments or similar regional questions relating to changes in place, society, season, power, and empire.

To give a consistent point of reference in time and space, the first illustration in each chapter is a map with a time line on the left. Reference to the map shows the location of the region under study on a world map of precipitation for August of 1999, showing the dry summer state of the Bermuda-Azores High. The figure also shows a regional map with place names discussed in the chapter. The time line marks important events in the region during the first millennium A.D. These are often given as archaeological phases, time periods during which a persistent culture can be recognized from the duration of artifact types and styles of ceramics and other material remains. In regions where history is available, the time line shows recognized historical periods, dynasties, or other helpful time indicators. The volume is divided into two parts.

Part I presents the historical evidence from literate societies demonstrating that A.D. 536^{cal} was a watershed moment in the world history. A search for a plausible boundary between Classical and Modern times has been a subject of debate for much of the twentieth century. Young argues that A.D. 536^{cal} was the break line between classical and modern times. The other five chapters in Part I profoundly support this contention.

Part II is initiated by two chapters designed to set the stage for understanding the A.D. 536^{cal} global cooling and the generally cooler, often unstable, global climate of the first millennium A.D. The objective of these chapters is to understand global temperature and global change in a time framework compatible with the human scale of interaction and social studies. William Tanner's chapter provides an approximately 50-year time scale analysis of sea level (resembling global temperature) during the period of study. The issue of time scales is further pursued in a chapter on archaeological time through radiocarbon dating and how that relates to the normal sense of annual time.

The remaining six chapters in Part II address the question of how archaeologists with less resolved time scales outside the cone of history can make useful contributions to understanding precisely timed environmental events. Case studies from six subregions of southeastern United States are presented. The answer is found more in the aftermath, the 300 years following, than in the A.D. 536 event itself. However, rapid culture changes and population movements are implicated in more than one region. Especially important are comparisons with historically observed parallel changes, or parallel events (Gunn 1994b), in other regions of the world.

The parts are divided into themes, or *concept sets*. Each concept set is a pair of chapters that explores a geographic and environmental issue. The concept sets emerged from the regional time transects scanned by the authors through the centuries of the first millennium A.D. (see Figure 1.1). Studies of subregional variation, both in climate and cultural impacts, are greatly facilitated by pairing the contributions. Six chapters of three pairs present studies of the Atlantic slope (Anderson, Lilly and Webb, Mathis, Walker, Wetmore et al., Woodall). To gain the perspective of other parts of the world, chapters from Insular Europe (Jones) and Italy and Burgundy (Young) provide a view of subregional variation there. Another pair addresses the Maya lowlands of Mesoamerica (Chase and Chase, Robichaux). Two additional chapters open fronts in Africa (Schmidt) and Asia (Houston).

Each concept set emphasizes the best data a region offers students of global change, either historical and/or archaeological. In Eurasia, Africa and Mesoamerica, historical records reveal the scope of A.D. 536 ^{cal} consequences among empires, dynasties, and economic and military balances of power (see Figure 1.1). In central Italy, the personal ambitions of Justinian, Emperor of the Eastern Roman Empire, clashed with the consequences of the A.D. 536 event. In other areas such as southern Britain, history and archaeology are combined with mythology and poetry to fathom the consequences. This is also true of the Mesoamerican studies of the regional states of Tikal and Caracol. The Lilly and Webb chapter is paired with Wetmore et al. as they address multiple instances of cultures in highly dissected terrain. They are just across an artificial administrative border from each other. This methodology amounts to subdividing a statistical sample to see if results can be replicated. I should emphasize that the studies are replicated; I did not anticipate that the results would be replicated. Contradictory opinions were tolerated, even encouraged. The only requirement was that the authors come to thoughtful conclusions.

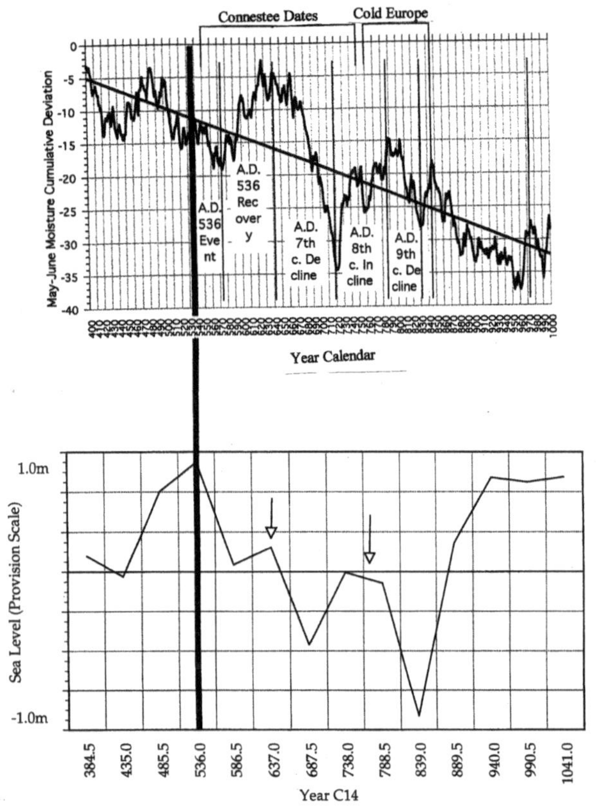

Figure 1.4. Visual Correlation of Spring Moisture from Black River North Carolina Tree Rings and Sea Level.

Linking the chapters also reduces the organizational complexity of the book from 12 regions to six concept sets (see Figure 1.1). This facilitates the process of abstracting principles from experience by providing multiple examinations of data and varied conclusions. Other authors make use of major contributions of knowledge from archaeologists and paleoclimatologists. In the chapters focused in the southeastern United States, discussions are dominated by changes of season of precipitation and temperature, organization of society, and relocation of place. In regions where the coverages of historians and archaeologists converge, much is gained. An interesting example is the chapters set in coastal regions on both sides of the north Atlantic ocean: they indicate southward movements of populations that are most intense at higher latitudes (Jones and Mathis).

The audience of first concern for the book is practitioners of the historical, and specifically, the archaeological sciences. However, an effort was made to make it accessible to non-archaeologist, and non-regional specialists within archaeology. Broader readability was implemented first by explaining or avoiding some of the more esoteric concepts that archaeologists use to refine analyses of various artifact classes; this information can be found in the many site and survey reports cited in the chapters.

To avoid controversy over correlation and causation, a distinction was made between stage I and stage II research chapters. In the stage I chapters, emphasis is on identifying correlations between global climate changes and regional cultural and climate phenomena. Stage I research is the necessary antecedent to more sophisticated stage II global-local linking models. Such works are generally offered in subregions where there are little existing data to support the analysis of complex global-local linkages. I regard the value of the stage I chapters for non-specialists and specialists alike as examples of how archaeologists think and approach the resolution of problems informed by some new piece of information, in this case A.D. 536 ^{cal}, and how they structure their activities to acquire data-based answers. Although the distinctions are not always clear cut, the chapters by Mathis, Lilly and Webb, Houston, and Schmidt are largely exploratory, stage I investigations.

The stage II, or linking model chapters, are directed toward establishing global-local links with reference to large existing data bases. Examples are Woodall, Walker, Wetmore et al., Anderson, Jones, Young, Chase and Chase, and Robichaux.

In arranging the chapters, I have attempted as much as possible to have a chapter representing each approach in each concept set. In most cases, the first chapter in each set is the stage I (correlation) chapter and the second is the stage II (linking) chapter. The correlation chapters are generally easier to read and serves as an introduction to the issues in the region represented by the concept set. The second chapter, the linking chapter, explores regional questions more deeply and more exhaustively, usually with more sophisticated methods. The linking chapters will be more satisfying to those who are ready to understand the difficult problems of global-local change studies. In Figure 1.1, the correlation and linking chapters are mapped onto other dimensions of the book to provide a road map to subject matter that will be of interest to readers with many perspectives and depths of interest. The authors of the correlation chapters are underlined. The concept sets provide both guidance to regions (e.g., Lilly and Webb, and Wetmore et al. in the Southern Appalachians), and the issues they most prominently address (e.g., changes in social structure and settlement pattern).

Regional Cultural Variations following A.D. 536^{cal}

The character and contribution of the chapters are discussed in introductory pieces to the concept sets. However, a brief overview will help to presage the findings. The central question for the authors concerns variation in the impact of global change from place to place. The moment of change is A.D. 536^{cal}, although there was an equally important 300 year aftermath. Part I presents the clear and precise historical record of the event supported in some instances by archaeological information. The focus of these records is the circum-Mediterranean, but Africa, the Near East, and the Maya lowlands also contribute. Part II experiments with a purely geological and archaeological outlook on the problem by turning to a cluster of studies on the Atlantic Sea Board of southeastern United States.

The scope of the long term change is examined by geologist William Tanner who studies sea level change: sea level, we assume, changes with global temperature. The sea level measurements show downward tendency for 300 years following A.D. 536^{cal}. Almost precisely 300 years after A.D. 536^{cal} and at the nadir of sea level decline is the remarkable freezing of the Nile River mentioned earlier. This is the most extreme manifestation of first millennium cold climate I know. Tanner's sea level change measurements are supplemented by eye witness accounts from the coastal lowlands of western Europe (Jones chapter) and archaeological discoveries along the North Carolina and Florida coasts (Mathis and Walker chapters). Following these sea level discussions, four colleagues and I who have accumulated large radiocarbon date data sets supply a chapter on archaeological time. We raise the question of the practical resolution of radiocarbon dates, which could be more refined than the commonly used statistical error functions suggest.

Variation in impact within regions is examined by pairing studies into concept sets in similar environments. In the southeastern United States, examples are drawn in pairs from the coast, piedmont and mountains. The selections from other parts of the world are also paired where possible and provide historical information and a global outlook on the situation. As nearly as the authors can determine from existing evidence, the results are starkly different from region to region. In the southeastern US coasts concept set, in one example the old population fades and a new one appears (North Carolina). In the other, the old population flourishes as never before or since (Florida).

The overarching question is always: Why do cultures change as global temperatures change? Thus, the search for local linkages to global climate is always lurking in the background no matter how directly it is addressed by each author. Some of the authors, such as Woodall and Walker, have given the issue of global-local linkages direct attention in the past, and thus can propose explicit models in their chapters. Others are just beginning their regional investigations. Their chapters represent the initial step in identifying linkages, which must always be to look for correlations between global events and local activities.

Especially from the archaeologist's point of view, the A.D. 536 event is no more important than the subsequent 300 year global cooling episode. Also, global warming of substantial proportions during the tenth century A.D. is a part of the warming-cooling change over time (see Figure 1.4); eventually another research effort will be focused on the post-A.D. 900 warm episode. The broader question concerns global change, not just global warming.

How is the cloak of history extended to strictly prehistoric region-time domains? The solution obviously has to do with bold inferences. Engaging the logical system of alternative models helps assure that we know what we think we know. Walker's chapter has a good feeling of finality because a combination of circumstances makes it unlikely that alternative models other than rapid cooling and warming can be effective in her study sites. It is a methodology of multiple cross checking to a point when there can be little question. Is cross checking across disciplines the key to the analytical needs of this era? Is it the only strategy?

In terms of an interregional model of the A.D. 536 event based on this set of case studies, rather clear patterns emerge through the contradictions and inverse relationships. In the US Southeast, China, and Europe, the more northerly regions, there were notable changes at around A.D. 200^{c14}. This coincides with the Holocene's grandest episode of volcanism. Similar conditions hold in Europe. A.D. 200^{c14} is the end of the Roman Empire Golden Age. The western Roman Empire collapsed and had to be reconstituted by the Diocletain Reforms late in the third century; it was reconstituted on an entirely different basis: military administration rather the social cooperation. Something similar transpired in China; A.D. 536^{cal} was a trouble among ongoing difficulties at higher latitudes.

At lower latitudes, another set of outcomes emerges from the author's discussions. On the Gulf Coast of the United States (Walker), and maybe in all coastal areas, including the North Carolina coast (Mathis), and the Savannah River of coastal South Carolina and Georgia (Anderson), immediate disruptions might have transpired, but conditions actually improved in the aftermath; lower sea levels and cooler, perhaps more temperate, climate facilitated population growth. The history of the Maya Lowlands provides additional insight into this low latitude insight. Troubles in the Maya Lowlands could have been due to counterbalancing complex and inflexible foreign ruling hierarchies unable to cope with sudden changes, with the more flexible local social organization. This argument emerges from the findings of Caracol's period of ascendance over Tikal after A.D. 536 ^{cal}. During that time Caracol imposed more democratic domination of the economy, possibly more hierarchical but locally governed. As such Caracol was better able to cope with the situation than Tikal, distantly associated with Teotihuacan in Highland Mexico (Chase and Chase, Robichaux). Under conditions like those following the A.D. 536[cal] cold world conditions, year-to-year weather would have been inherently less stable because of the thermodynamic characteristics of a dry atmosphere (see Gunn 1997). A study of the eastern Maya lowlands (Fialko 1996, 1997; Fialko et al. 1998) shows that linkages are similar there to the ones identified in the southwestern Maya lowlands (Gunn et al. 1994, 1995). This helps explain Caracol's more than 100 years of ascendance. Not only were the inhabitants of the city able to sustain through the event, but they were actually advantaged by the subsequent more than 100 years of instability because of local management. The following Late Classic is a reassertion of hyperhierarchy and the attendant liabilities of that system.

Between the interior basins and the coasts of the Southeast are the question-lands. Anderson suggests that these regions are wholly up for grabs in terms of their response to the A.D. 536 [cal] disruption. Lilly and Webb, and Wetmore et al., both present chapters from the highland Southeast, yet one pattern or another emerges because of very local preconditions. Bryson has noted that the meteorological models used to predict weather and simulate climate are unreliable in mountains; most weather forecast maps are virtually blanks over dissected terrain because it atmospheric processes are so difficult to model. While there is no ability to predict weather in the mountains, it can be predicted on plains on plateaus much more reliably. Thus, in the mountains, every canyon is a microhabitat. Relative to weather forecasting, canyons could be a meteorological nemesis, but for long term climate studies an archaeological boon because of the way the two disciplines are practiced, one by top down modeling of the atmosphere and the other bottom up modeling of a landscape location. Archaeologist with their ground-up metrology can provide relatively even coverage regardless of terrain.

Objectives and History of the Research

Whatever the cause(s) of the A.D. 536 event, it is not the primary concern of this book. That is an issue for scientists with telescopes, ice cores, tree rings, and other relevant methods and data. We do feel that sufficient information has accumulated on the circumstances of the sixth century to justify a first approximation examination of the information available, and to assess the importance of that time period relative to others with respect to global change problems.

By "first approximation," I am referring to the scope of the book relative to the overall, complex problem of establishing linkages between global scale phenomena and local environmental and cultural change. This book is preceded by two decades of efforts to establish useful linkages between global and local climate parameters and cultural changes. Participants in this effort refer to the project as the "Atlantic Basin design" because it was conceived as a broad enough project to encompass transoceanic relations between cultural change chronologies. The transoceanic design largely eliminates the problem of spurious correlations between cultural and climate change due to intimate contacts between subject cultures. In ethnology this is referred to as Galton's Problem.

From the point of view of archaeologists, we initially understood global-local relationships to be useful as a means of addressing local environmental problems by using otherwise global scale proxy measures of climate to refine and extend local paleoclimate chronologies (Folan et al. 1983; Gunn 1979). However, beginning in the mid-1980s, as the global change controversy heated up, we also began to think of the modeling of global-local functions as potentially useful for forecasting the impact of future global changes on regional ecologies (Gunn and Crumley 1991), and understanding the impacts of climatic extremes on human populations in the remote past. The latter are also perhaps useful for understanding the impacts of anthropogenic influences of global conditions such as greenhouse warming on local societies.

As discussed above, the discovery of global-local linkages is both an extensive and arduous task. It is extensive because every region in the world has to be studied and defined as a specific set of global temperature-to-local hydrology functions (Gunn and Crumley 1991; Gunn et al. 1994, 1995). It is arduous because it not only is based on evaluation of local hydrology relative to global temperatures, but also requires information about local food production customs through literature review and/or interviews with contemporary agriculturists to understand their particular solutions to inter-annual environmental variations, and archaeological and historical studies of how those adaptations have changed in the past.

The discovery of global-local linkages requires multiple stages of effort. First is the identification of potential relationships between local and global conditions. Stage I investigations are usually accomplished through simple visual and verbal correlation of local biological and cultural phenomena with very generalized global scale conditions as summed in the idea of episodes of similar year-to-year climates (Eddy 1977; Gunn 1994a, 1997). As an example, during the last two decades the Atlantic Basin design's research group efforts have concentrated in two case study areas. One is Campeche, a state in the southwestern Yucatan Peninsula of Mexico, and the other is Burgundy, a region in east-central France. The importance of studying Europe and Mexico in parallel were first published in an original stage I article on Mexico (Gunn and Adams 1981) in which it was noted that cultural florescences, primarily as reflected in architectural construction, seemed to occur in opposite, or out of phase, episodes, i.e., while Dark Age Europe languished, the Maya Lowlands were in a cultural florescence, and when Europe was at its apogee in the High Middle Ages, the Maya Lowlands lay ruined in collapse. Understanding the why of these out-of-phase correlations would provide important insights into variations in regional social forms, and environment, and the interactions of global, regional, and local climate change processes. Basically, global warming moves tropical, subtropical and temperate zones to higher latitudes. However, studies proved most productive when related to large semi-permanent atmospheric cells (Figure 1.5) such as the Bermuda-Azores High, the Siberian High, and the Icelandic Low (Crumley

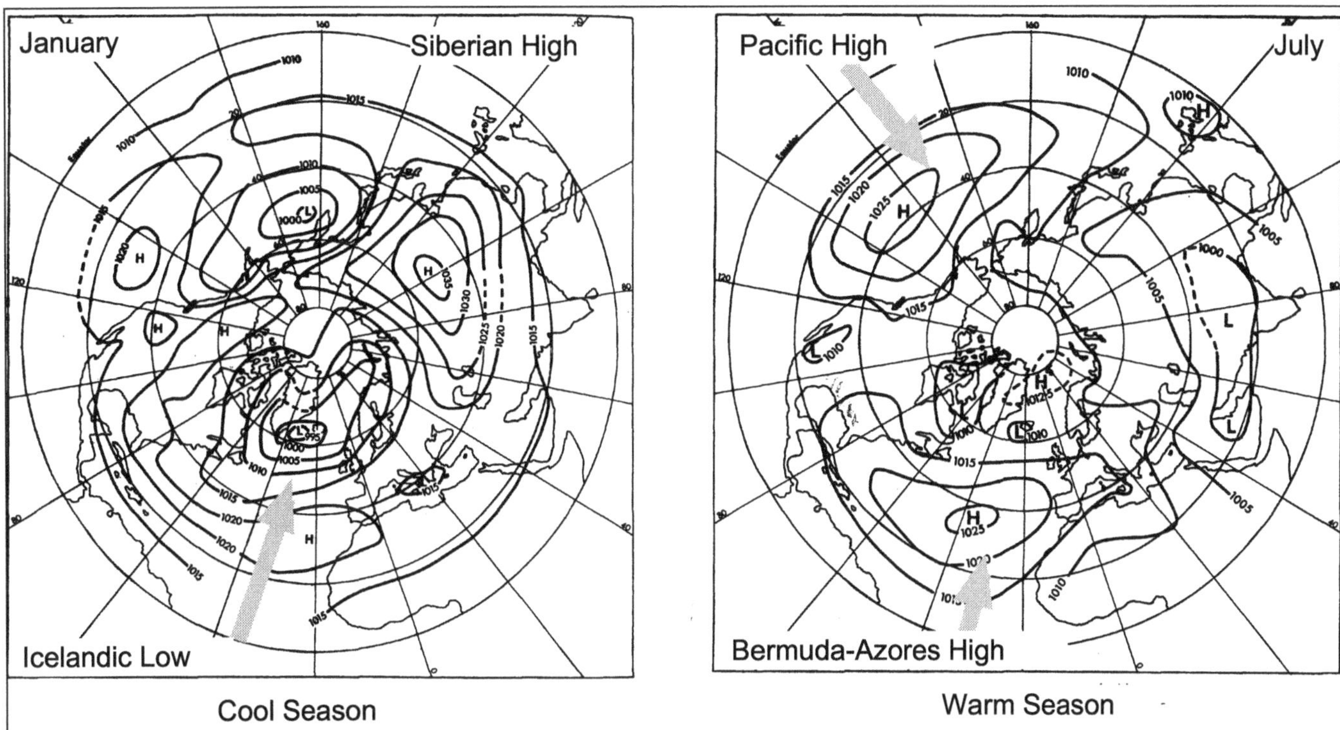

Figure 1.5. Worldwide System of Semipermanent High and Low Pressure Cells. Each world region is influenced by a combination of these semi permanent cells as they are moved around by global warming/cooling. (Adapted from Barry and Chorley 1992).

1993; Folan et al. 1983, Gill 1994 1998; Gunn and Crumley 1991, Gunn, Crumley et al. 1995). For Burgundy, stage I correlations were offered by Crumley and Marquardt (1987, 1990) in the context of a broad landscape analysis of the study area and further refined in subsequent publications (Crumley 1993, 1995). In both case-study areas, the correlation hypotheses were posed in terms of changes of seasonal concentrations of moisture and ecotonal shifts at different global temperatures as governed by the movement of semi permanent cells.

The stage I correlations are important for understanding local chronologies in global context. For example, it has become clear in the last few years, contrary to our original model, that the Maya Preclassic (450 B.C. to A.D. 250) corresponded in architectural vigor to that of Roman Europe; that is to say that the two areas were not always out of phase as originally thought. However, we also discovered that this period was a middle global temperature period (Gunn et al. 1995; Gunn 1997). The three-part global temperature model allowed for an intervening global temperature state, an "optimum", in which both high and low latitude cultures might find conditions favorable for large population aggregates and complex hierarchical social organizations capable of large public works. Our research has also shown that even though a region may be enjoying favorable climate circumstances for its type of cultural adaptation, all of its subregions will not participate in the conditions equally. In such a circumstance, some subregions can be disadvantaged by simply being between to other prosperous subregions in a favorable time (Gunn and Folan 2000). Anderson's (this volume) studies show similar findings in the southeastern United States.

Second stage global-local studies establish statistically verifiable linkages between more refined measures of global climate and local hydrology, and biological and cultural landscapes, or biocultures. This usually requires a battery of methods and disciplines working in concert within a region. The methods used in a region vary with local conditions and available data. In both Burgundy and Campeche, river discharge data were available to measure the impact of changing global temperatures on local seasonal distributions of moisture (Gunn and Crumley 1991; Gunn et al. 1994, 1995).

The third stage involves extensive intra-regional testing of the models developed during stage II. This process has already begun in the Yucatan Peninsula with Fialko et al's (1998) study of the central Petén, which shows that elements of the Gunn et al. (1994, 1995) model are applicable there, and is currently being extended into the Guatemalan Highlands. Eventually intra-regional studies should yield variations and serendipitous elaborations of the original models. Since establishing the global-local linkages requires a great deal of effort and preferably some consensus among regional specialists, it is best that the regional research community be kept abreast of developments. A means to accomplish this is through the publication of articles in peer reviewed journals, which was the approach taken in the

Atlantic Basin project. Most of the articles cited in the previous paragraphs were published in *Climate Change*, *Ecological Applications*, *World Archaeology*, *Southeastern Archaeologist*, *Human Ecology*, *Geoarchaeology*, *Latin American Antiquity*, *Journal of Field Archaeology*, *Current Anthropology*, and *Earth Surface Processes and Landforms*, and circulated to interested parties. Anyone interested in the more detailed concerns of global-local linkages should take time to review this literature. Venues other than journals could be utilized, such as the annual roundtables used so effectively by the Maya epigraphy translation study community, or e-mail based communications methods.

The history of the Atlantic Basin design research brings us to the objective of the A.D. 536 book. One obvious next step is to extend the Atlantic Basin design by incorporating additional case studies on the Atlantic rim. This is done by calling on a number of regional specialist from the Atlantic Slope of the southeastern United States and western European history to examine the environmental and cultural change characteristics of their regions. The issues are focused on the first millennium A.D. to constrain the scope of the effort and to take advantage of the many rapid global changes during that period. The A.D. 536 event provides a particularly poignant point of departure for the highly variable Vandal Minimum episode. The global scope of the focus on A.D. 536 offers an opportunity to cross-correlate prehistoric and historic records as historic record keeping processes were established both in Afro-Eurasia and the Maya lowlands at the time. The efforts of some of the authors are necessarily of stage I dimensions although Anderson and Woodall have proceeded to stage II linking efforts. Robichaux and Chase and Chase are part of the established stage III effort in the Yucatan Peninsula. Jones' and Young's studies might also be considered contributing to the stage III studies in Burgundy in a contextual sort of way, although neither study is centered in Burgundy. Walker's study of the Florida coast is richly connected to the Tanner sea level curve by mutually supported research designs; thus Walker's chapter is a stage II effort with regard to sea levels.

The outcome of the A.D. 536 studies presented in this book could be viewed simply as an effort to gain insight into how archaeologists can refine their time scales. Anderson discusses this issue in his chapter. However, we are also interested in the additional insights that archaeology lends to historical perspectives, and how historical perspectives and the comparison of historic and prehistoric social processes expand understanding of reactions to global scale climate change.

Conclusion: Lost Corridors

For me, the editor, reading and working with the chapters presented in this book was something like walking through the vast rediscovered tomb of Ramses II, which was prepared for the burial of his 52 sons. Each chapter is like exploring corridor strangely forgotten. I have been stimulated by the phenomenon of important things forgotten ever since I read the Stommels' (1979) *Scientific American* article on 1816, the "year without summer," incidentally to which the title of the book alludes. It was caused by the eruption of Mount Tambora in Sumatra, the largest volcanic eruption in the Holocene. In that year, the economy of the United States, still a rural agrarian economy, stopped; crops froze and never matured in June, July and August as far south as North Carolina and Missouri. Farmers in New England still refer to it, even though they are not aware of the event to which it refers, in the expression "1800 and froze to death." Joseph Smith was a 13 year old boy at the time, and one of the well known tenants of the Mormon Church he founded is that one should keep enough food in one's home to survive a year. How do people forget causes that are so important, even when acting on the experience? The practice remains, but the cause is again forgotten.

A similar forgetting phenomenon can be observed in the worldwide pandemic of 1918. A *Smithsonian Magazine* article discusses the pandemic in which millions of people all over the world died of a virulent form of influenza. I had known for some time that many people, almost everyone I know, had something unfortunate happen in their family that year: mostly infant children and pregnant women dying. Much of modern medical practice and public health policy stems from lessons learned in that devastating year. Again, it is interesting that the practices endure, but why did people forget the cause?

In the chapters of this book, the authors struggle to extract bits of information from history and archaeology on an important moment in the past. Every chapter is a hallway opened with difficulty and uncertainty in a tomb of strangely forgotten information. The moment was so important that the forces of history pivot on its weight. So-called modern times as distinguished from the Classical era were wrought in its forge. How was it forgotten? What can be learned from its revalation?

ACKNOWLEDGMENTS: I would like to thank Hugh Robichaux who jarred me into realization of the importance of the A.D. 536 event. Numerous persons including the Carole L. Crumley, David G. Anderson, Meredith Morris-Babb, Richardson Gill, Thomas McGovern, and several anonymous reviewers generously provided advice and counsel on the writing of this chapter.

References

Baillie, M. G. L. (1994) Dendrochronology Raises Questions about the Nature of the AD 536 Dust-veil Event. *The Holocene* 4:212-217.

Baillie, M. G. L. (1995) Patrick, Comets, and Christianity. *Emania* 13:69-78.

Baillie, M. G. L. (1999) Exodus to Arthur. B. T. Batsford Ltd., London.

Barry, Roger G., and Richard J. Chorley (1992) *Atmosphere, Weather & Climate*. Sixth Edition, Routledge, New York.

Berry, Walter E. (1987) Southern Burgundy in Late Antiquity and the Middle Ages. *In Regional Dynamics: Burgundian Landscapes in Historical Perspective*, edited by C. Crumley and M. Marquardt, pp. 447-609, Academic Press, Orlando.

Birdsell, Joseph B. (1973) A Basic Demographic Unit. *Current Anthropology* 4:337-356.

Broecker, Wallace S. (1995) Chaotic Climate. *Scientific American* 273:62-69.

Bryson, Reid A. (1988) Civilization and Rapid Climate Change. *Environmental Conservation* 15:7-15.

Bryson, Reid A. (1994) Orbital History, Volcanism, and Major Climate Changes: On Integrating Climate Change and Culture Change. *Human Ecology* 22:115-158.

Bryson, Reid A. and Thomas J. Murray (1977) *Climates Of Hunger: Mankind and the World's Changing Weather.* The University of Wisconsin Press.

Bryson, R. A. and B. M. Goodman (1980) Volcanic Activity and Climatic Changes. *Science* 207:1041-1044.

Crumley, Carole L. (1993) Analyzing Historic Ecotonal Shifts. *Ecological Applications* 3:377-384.

Crumley, Carole L. (1994) The Ecology of Conquest: Contrasting Agropastoral and Agricultural Societies' Adaptation to Climatic Change. In *Historical Ecology: Cultural Knowledge and Changing Landscapes*, C. Crumley, editor, pp184-201, School of American Research Advanced Seminar Series, Santa Fe.

Crumley, Carole L., and William H. Marquardt, eds. (1987) *Regional Dynamics: Burgundian Landscapes in Historical Perspective.* Academic Press, San Diego.

Crumley, Carole L., and William H. Marquardt (1990) Landscape: A Unifying Concept in Regional Analysis. In *Interpreting Space: GIS and Archaeology*, K. Allen, S. Green, and E. Zubrow, eds., pp 73-79, Taylor & Francis, London.

Denton, George and Wibjorn Karlén (1973) Holocene Climatic Variations-Their Pattern and Possible Cause. *Quaternary Research* 3:155-205.

Donovan, Stephen K. (1989*) Mass Extinctions: Processes and Evidence.* Columbia University Press, New York.

Eddy, John A. (1977) The Case of the Missing Sunspots. *Scientific American* 236:80-92

Eddy, John A. (1994) Solar History and Human Affairs. *Human Ecology* 22:23-36.

Fialko Coxeman, Vilma (1996) Yaxha and Nakum: Jerarquias y Patrones de Asentamiento en Sus Espacios Intersitios. *Mayab*, Numero 10:15-24.

Fialko Coxeman, Vilma (1997) Organizacion Territorial y de Asentamientos Mayas en los Intersitios de Yaxha y Nakum. *Los Investigadores de la Cultura Maya* 5:249-261, Universidad Autonoma de Campeche, Campeche.

Fialko, Vilma, William J. Folan, Joel D. Gunn, and Ma. del Rosario Domínguez Carrasco (1998) Investigations in the Intersite Areas Between Yaxha-Nakum-Tikal. Paper presented at the 63rd Annual Meeting of the Society for American Archaeology, Seattle.

Folan, William J. (1981) Comments on The Late Postclassic Eastern Frontier of Mesoamerica: Cultural Innovation along the Periphery by J. Fox. *Current Anthropology* 22:336-337.

Folan, William J., Joel Gunn, Jack D. Eaton, and Robert W. Patch (1983) Paleoclimatological Patterning in Southern Mesoamerica. *Journal of Field Archaeology* 10:453-468.

Fiedel, Stuart J. (1987) *Prehistory of the Americas.* Cambridge University Press, Cambridge.

Folan, William J., Betty Faust, Wolfgan Lutz, and Joel D. Gunn (2000) Social and Environmental Factors in the Classic Maya Collapse. In IIASA research report *Population, Development, and Environment on the Yucatan Peninsula: From Ancient Maya to 2030*, edited by W. Lutz, in press.

Gill, R. B. (1994) *The Great Maya Droughts.* Unpublished Dissertation, The University of Texas at Austin.

Gill, Richardson B. (1999) The Great Maya Droughts: Water, Life, and Death. Albuquerque: University of New Mexico Press.

Gilliland, R. L. (1982) Solar, Volcanic, and CO2 Forcing of Recent Climatic Changes. *Climatic Change* 4:111-131.

Gribbin, John, and Mary Gribbin (1997) *Fire on Earth.* St. Martin's Griffin, New York

Gunn, Joel (1979) Occupation Frequency Simulation on a Broad Ecotone. In *Transformations: Mathematical Approaches to Culture Change*, edited by C. Renfrew and K. Cooke, pp 257-274. Academic Press, New York.

Gunn, Joel (1991) Influences of Various Forcing Variables on Global Energy Balance During The Period of Intensive Instrumental Observation (1958-1987) and Their Implications for Paleoclimate. *Climatic Change* 19:393-420.

Gunn, Joel D. (1994a) Introduction: A Perspective from the Humanities - Science Boundary. Global Climate—Human Life: Physical Contexts of Historical Ecology. *Human Ecology* 22:1-22, Special Issue.

Gunn, Joel D. (1994b) Global Climate and Regional Biocultural Diversity. In *Historical Ecology*, edited by C. Crumley, pp 67-97. School of American Research, Santa Fe.

Gunn, Joel D. (1997) A Framework for the Middle-Late Holocene Transition: Astronomical and Geophysical Conditions. *Southeastern Archaeologist* 16:134-151.

Gunn, Joel D., and R. E. W. Adams (1981) Climatic Change, Culture, and Civilization in North America. *World Archaeology* 13:85-100.

Gunn, Joel, and Carole L. Crumley (1991) Global Energy Balance and Regional Hydrology: A Burgundian Case Study. *Earth Surface Processes and Landforms* 16:579-592.

Gunn, J., W. Folan, and H. Robichaux (1994) Un Analisis Informativo sobre La Descarga del Sistema del Rio Candelaria en Campeche, Mexico: Reflexiones acerca de Los Paleoclimas que Afectaron a Los Antiguos Sistemas Mayas en Los Sitios de Calakmul y El Mirador. In *Campeche Maya Colonial*, edited by W. Folan, pp. 174-197, Universidad Autonoma de Campeche, Campeche Mexico.

Gunn, Joel, and Jerzy Grzymala-Busse (1994) Global Temperature Stability by Rule Induction: An Interdisciplinary Bridge. *Human Ecology* 22:59-82.

Gunn, J., W. Folan, and H. Robichaux (1995) A Landscape Analysis of the Candelaria Watershed in Mexico: Insights into Paleoclimates Affecting Upland Horticulture in the Southern Yucatan Peninsula Semi-Karst. *Geoarchaeology: An International Journal* 10:3-42.

Gunn, Joel, C. Crumley, E. Jones, and B. Young (1995) Landscape Analysis of Western Europe during the Early Middle Ages. Submitted for review.

Houghton, J. T., G. J. Jenkins, and J. J. Ephraums (1990) *Climatic Change: the IPCC Scientific Assessment.* Cambridge University Press, Cambridge.

Kauffman, Stuart (1995) At Home in the Universe: The Search for Laws of Self-Organization and Complexity. Oxford University Press, New York.

Keys, David (1999) Catastrophe: An Investigation into the Origins of the Modern World. Ballatine Books, New York.

Lamb, H. H. (1977) Climate: Present, Past and Future: Climatic History and the Future. Vol 2, Methuen and Co. Ltd., London.

Ladurie, E. L. (1971) Times of Feast, Times of Famine: A History of Climate Since the Year 1000. Translated by B. Bray, Doubleday, New York.

Landscheidt, Theodor (1987) Long-Range Forecasts of Solar Cycles and Climate Change. In *Climate: History, Periodicity, and Predictability.*, edited by M. Rampino, pp. 421-445, Van Nostrand Reinhold Company, New York.

Lean, Judith, Andrew Skumanich, and Oran White (1992) Estimating the Sun's Radiative Output during the Maunder Minimum. *Geophysical Research Letters* 19:1591-1594.

Lovelock, James (1989) Geophysiology. Transactions of the Royal Society of Edinburgh: Earth Sciences 80:169-175.

Nassaney, Michael S., and Charles R. Cobb (editors) (1991) *Stability, Transformation, and Variation: The Late Woodland Southeast.* Plenum Press, New York.

Nassaney, M., and K. Sassaman (1995) Native American Interactions: Multiscalar Analyses and Interpretations in the Eastern Woodlands,. University of Tennessee Press, Knoxville.

Murdock, George P., and Douglas R. White (1968) Standard Cross-Cultural Sample. *Ethnology* 8:329-369.

Rampino, Michael R., John E. Sanders, Walter S. Newman, L. K. Konigsson (editors) (1987) *Climate: History, Periodicity, and Predictability.* Van Nostrand Reinhold Company, New York.

Rao, Joe (1996) The Leonids' Last Hurrah? *Sky & Telescope* 11:74.

Rasmussen, Kaare L., Henrik B. Clausen, and Gregory W. Kallemeyn (1995) No Iridium Anomaly after the 1908 Tunguska Impact: Evidence from a Greenland Ice Core. *Meteoritics* 30:634-638.

Scuderi, L. A. (1990) Tree-ring Evidence for Climatically Effective Volcanic Eruptions. *Quaternary Research* 34:67-85.

Schneider, S. H. and C. Mass (1975) Volcanic Dust, Sunspots, and Temperature Trends. *Science* 190(4216)741-746.

Stommel, Henry, and Elizabeth Stommel (1979) The Year Without a Summer. *Scientific American* 240:176-186.

Stothers, R. B. (1984) Mystery cloud of AD 536. *Nature* 307(5949):344-345.

Trewartha, Glenn T. (1961) *The Earth's Problem Climates.* University of Wisconsin Press, Madison.

Trigger, Bruce G. (1989) *A History of Archaeological Thought.* Cambridge University Press, Cambridge.

Volk, Tyler (1998) Gaia's Body: Toward a Physiology of the Earth. Copernicus, New York

Wendland, W. M., and Reid A. Bryson (1974) Dating Climatic Episodes of the Holocene. *Quaternary Research* 4:9-24.

Willey, Gordon R., and Philip Phillips (1958) *Method and Theory in American Archaeology.* University of Chicago Press, Chicago.

I. CHANGES IN POWER: DRY FOG IN EUROPE

-

The dry fog concept set, along with the chapter on China, brings forth direct eyewitness evidence of the proximate causes of the A.D. 536[cal] disturbance. As Baillie (1995, and others, see chapter 1) reports, widespread evidence implies an extended period of obscured solar radiation. The obscuring element is referred to in the scientific literature as "dry fog," or "aerosols," fine dust in the atmosphere that in some regions is reported to have hovered for a year and a half. This is well beyond the residence time of atmospheric dust from one volcano of any size. Stothers (1985) estimates from eye witness accounts of day length that the sun's strength was reduced to one tenth of normal. Empires and dynasties were broken and made across the eastern hemisphere with unusual rapidity in the years following A.D. 536[cal]. As we shall see in the next set of chapters, the pattern appears in the western hemisphere in the Maya lowlands as well.

The two chapters in the Dry Fog concept set focus on Italy, France, and Britain, although effect are also evident in Scandinavia and the eastern Mediterranean. In his chapter, Bailey Young resurrects ancient texts of the Classical Mediterranean world to examine the content, cultural context and phenomenology of eye witness accounts. Edward Gibbon published his celebrated history of *The Decline and Fall of the Roman Empire* over two hundred years ago. Since then, scholars have sought to understand the complex web of causes that transformed a predominantly urban and Mediterranean ancient civilization into a predominantly rural medieval world centered on northwestern Europe. In the wake of studies by Henri Pirenne and others in and following the 1930s, scholarly attention shifted from an emphasis on superficial political events to deeper underlying changes in economic, cultural, and mental processes. Young suggests here that the crucial period for the transition in the old Roman west occurred not during the barbarian invasions of the fifth century, when Gibbon placed it, or with the advance of Islam after the late seventh century, as Pirenne proposed, but during the sixth century. Political events such as the devastating Justinian's war, just reaching Italy in A.D. 536[cal], and the expansion of Frankish Merovingian power in Gaul, certainly played a part, but the impact of climatic and environmentally disastrous events on public health and nutrition were significant factors as well.

The written documents that are the grist of Young's discussions span the Mediterranean shores and the Near East. However, the writings of Roman Senator Cassiodorus provide an extensive corpus from the viewpoint of one person whose potential is yet to be realized. He lived and governed through the event. Slightly post-dating Cassiodorus, Bishop Gregory of Tours (A.D. 538-594[cal]), the major narrative source for life in sixth-century Gaul, observed astronomical events, climate, natural prodigies and disasters, and epidemic sickness in the decades following A.D. 536[cal]. Although Gregory's five decades of observations require an extensive data base from other periods to compare frequencies and magnitudes of events to verify their novelty, many questions are raised by his observations. In broad time-space perspective, could a "global climatic event" attested in the Mediterranean in A.D. 536[cal] by Cassiodorus have played a role and been a sign of deeper environmental changes that, although obscure in Gregory at first reading, forced northern Europeans to adapt to a changed world? Did that changed world feed the eventual relocation of power in western Europe from the Roman-era Mediterranean to the North Sea of Charlemagne?

Elizabeth Jones' chapter examines the North Sea region from the perspective of the British Isles. Methodologically the severe climate following A.D. 536[cal] provides a focal point for integrating archaeological, historical, literary, folkloric, linguistic, and environmental evidence from Dark Age Britain. Britain presents an opportunity to observe the effects of the climatic disturbance on two very different cultures living side by side, the Roman-Britains and Anglo-Saxons. Theoretically Britain would seem to have been somewhat buffered from the severest effects of the disturbance because of its position in the Atlantic and its southerly latitude compared to harder-hit Scandinavia. By comparison, the effects were so severe and extended in Scandinavia that Britain received a wave of emigration from there. However, secondary effects, of famine and epidemics, seem to have had more far-reaching impacts on local Roman British society (see also chapters by Young, Schmidt, and Houston for evidence of epidemics). The contrasting settlement patterns of Roman-Britains and Anglo-Saxons, along with their non-overlapping spheres of social interaction explain why the climatic disruption, in conjunction with ongoing social, political and economic processes, affected these two groups in dramatically different ways. Additionally, the A.D. 536 event sheds light on two long-debated problems of cultural memory and transmission—the problem of the apparent disappearance of Roman-Britain culture from the archaeological record following the Anglo-Saxon invasion, and the problem of an anomalous date often given for King Arthur's death. Complimenting several other chapters, Jones finds evidence for sea level changes in eye witness accounts and archaeological information.

Two issues in Jones' and Young's chapters deserve special highlighting as pertains to the broader view of Europe in A.D. 536[cal] and its aftermath. First, the twin giants of King Arthur and Emperor Justinian provide a sense of individual motives and the ramifications of individual ambitions in the era. Second, archaeological and historical methods relative to the overall objective of this book, the integration of the two for climate change studies, are particularly evident in this land where history and archaeology co-reside.

Arthur and Justinian

In and odd confluence of events and persons, the central political figures of the A.D. 536 event are the legendary King Arthur of Roman-Britain and Justinian of Constantinople, Emperor of the Roman Empire. Both figures are artifacts of the Roman Empire, then divided, and they were posted at opposite ends of its vast 2700 mile length. Sorting through the archaeological and historical evidence, our authors find that Justinian was alive and willfully involved in one of the great conflagrations of the pre-industrial era. Arthur, also a Roman Emperor of sorts, stands as a distant ghostly presence apparently directing his troops by a curious brand of post-mortem remote control. He nevertheless represents a tradition of Roman-Britain tampering in Roman politics typical of the era and complimentary power base to that of Justinian's Eastern Empire. It bespeaks a western empire that foreshadows the accomplishments of Charlemagne 250 years later.

Combined with an environmental disruption of substantial, perhaps unprecedented proportions clearly documented by tree rings, are social factors such as the cataclysm of a war precipitated by Justinian, an ambitious young ruler who insisted on flying in the face of a trend toward regional control in the western Roman Empire; that trend already had progressed over 300 years by the sixth century. Justinian began his war to reconcile the Empire in A.D. 533^{cal}. At the beginning of the A.D. 536^{cal} confluence of events and personalities, Constantinople was a proud city of a half million persons and Rome was only marginally changed from its imperial splendor. In the years following A.D. 536^{cal}, Rome was brought to ruin, and the population of Constantinople reduced by half. In the end it is clear that both Justinian and the A.D. 536 event share responsibility for the destruction of Italy, but Constantinople suffered as well in the same cascade of processes. Were it not for a broader context of disruption, one would expect the powerful perpetrator to remain relatively unscathed.

Eyes at the other end of Europe also turned toward Rome, at least in the century before A.D. 536. Elizabeth Jones reports that in the British Isles the ominous portents of Italy are repeated if not amplified as the Roman-Britains vanished mysteriously in the years following A.D. 536^{cal}. Jones' research follows the fate of a civilization as its society and population disintegrate leaving a situation probably as mysterious to the Saxons who immediately replaced them, as it is to us today. Woven through the tale are archaeological illuminations of the shadowy Arthur of Camelot whose life and legend has become the mythological underpinning of post-Antique culture in Insular Europe and southwest Europe in general. Though cloaked for us in the garb of the later High Middle Ages by writers of that time, the late Roman elite lurks in Arthurian folklore. Like Justinian, these shadowy figures, now conflated in legend, turned their aspirations toward Italy and in doing so provided a cultural memory of Roman times.

Perhaps most significant for the issues under consideration is both Arthur and Justinian presided over essentially healthy pre-A.D. 536^{cal} societies in Italy and Britain, maintaining with some success, if not in full glory, the lifeways and ideals of the Roman Empire Golden Age, then 300 years past. When contrasted with post-A.D.536^{cal} conditions, the suddenness and completeness of the transition from classical to medieval culture becomes striking. The social and cultural contrast adds substantial insight to the generally accepted view that Europe's population fell precipitously to its most diminished numbers between classical and modern times in the sixth century A.D (Crumley and Marquardt 1987:252).

Cadbury Castle, William Stukeley's drawing of the ramparts of Cadbury Castle as they appeared from the north in 1723. The apparent number of ramparts is exaggerated because the drawing does not distinguish between defensive works and agricultural terraces; but the steepness of the ramparts is faithfully depicted." (adapted from "By South Cadbury is that Camelot... Excavations at Cadbury Castle 1966-70", by Leslie Alcock, Thames and Hudson, London)

Archaeology and History in Western Europe

The startling juxtaposition of archaeology and history in the Jones and Young chapters fully displays the potential of the two approaches to interact in an event-oriented archaeology. In Italy and other points around the Mediterranean basin, writers of antiquity provided eye witness accounts of the conditions associated with the A.D. 536 event. In a way, the contrasts between eye witness accounts in Young's research on Italy and France, and the largely inferred conditions reported by Jones in the British Isles, highlight the differences between what historical figures saw, felt, and attempted to do about a problem as reported in their writings, and the alternative perspectives of folklorists, archaeologists and paleoclimatologists viewed through social and technological ramifications and tree rings.

Especially notable are the research questions that arise from the combinations of information. Among them are issues of individual action or agency. While literacy embellished the edges of European culture during the early Middle Ages, written records are relatively rare, and those that do exist, present problems of interpretation; inquires such as those in this book often require the reinterpretation of original sources. In this circumstance, archaeology and history can be joined in a systematic fashion to systematize these re-interpretations of events. Nevertheless, the historical records, more or less sparse, do provide a quality of understanding that is difficult to achieve with purely archaeological sources. This is ably and prolifically demonstrated both in Young's and Jones' discussions as noted above.

For prehistorians working without the assistance of historical color as in the southeastern United States, these efforts to integrate historical and archaeological data can offer a rich methodological analogy for defining an approach to event–oriented prehistory. Adding the environmental perspective of the sixth century to the otherwise political and military character of the transition suggests that Antiquity could have ended with a bang rather than a whimper. But clearly "Arthur" and Justinian did play significant roles. In a similar fashion, parallel cultural events in southeastern United States (migrations, diminished populations), or anywhere, require a theoretical component of individual involvement to be understood in some modeled sense.

Combined with post-contact personages and Native American legends, the two sources of information open the question of individual actions in the Americas during the A.D. 536 event. None of the authors of this volume engage this problem, but Redman's (1977) concept of the analytical individual could be useful in this context. The important point to be made is that individual agent's actions will compose some portion of what is observed in the archaeological record, and therefore they must comprise some portion of the explanation if there is to be a good fit between reconstruction and processes. The presence of agents in models also opens such reconstructions to evaluation in the domain of self-organizing complex systems theory.

What can archeologists do to further illuminate Europe's reaction to the A.D. 536 event? Cassiodorus' writing in Rome, and various other scholars of his time, noted the immediate consequences of the dry fog, while Gregory of Tours was a witness to the century-scale aftermath. He records unusual cold in A.D. 548^{cal} and notes rebuilding churches that were burned in A.D. 573^{cal} without further comment on causes. A reading of Gregory raises tantalizing questions. Is there internal evidence within Gregory, or alternative lines of evidence, to suggest unusual church burnings? Could this be unusual or is it just something that normally happened with early churches? Do archaeologists have the answers to these questions in their report archives somewhere that have gone unnoticed before the question of A.D. 536^{cal} was raised?

An interpretive nemisis of both archaeology and paleoclimatology is underestimating the magnitude of short term, un-witnessed (unrecorded) events. Historical writings more often than not do not suffer from this limitation, but to the contrary offer exaggerated and/or stylized accounts. They therefore require contexts for comparison. Gregory's writings do not have at this time a quantitative comparison to other times in the manner of the Annales school of historical methodology. Such comparisons would afford a sense of the severity of the conditions in the half century after A.D. 536^{cal}. As we see in these chapters, there were plagues, famines, and other disasters. Elizabeth Jones' dissertation on Burgundian demography will supply a comparison with the Little Ice Age, presumably a somewhat parallel climate episode. Future research should include a comparison of Jones' Burgundian demography and climate with Gregory that will provide insights into the relative severity of the sixth century aftermath. This can be further cross-referenced with the American prehistoric contexts.

Finally, I might suggest that archaeology serves a particular utility in western culture. We, i.e., western culture, almost always distance ourselves from past traumatic events. In many cases we totally eliminate them from everyday consciousness (see chapter 1). This seems not to be the case in at least one other culture I have experience first hand, the present day Maya, who constantly rehearse and remember traumas distant in time. Western cultures need to consciously remedy their historical perspectives by re-interring their buried cultural memories of cultural traumas and incorporate the implications gained there from into climatic policy.

Global Change Assessment. As discussed in Chapter 1, each concept set is evaluated in terms of the understanding it lends to questions of global changes and their impacts on humans. Two issues addressed by the Young and Jones chapters may prove valuable.

First, Jones' and Young's chapters taken together provide a prospective vista of regional variation in Europe that begs further examination. Although Gregory of Tours seems to report a troubled Gaul, the consequences are apparently not as extreme as those in Italy and Britain. One might be tempted to distinguish Italy and France in terms of consequences of Justinian's war that laid Italy waste. However, Jones' analysis of Roman British unaffected by the war suggests consequences of near equal proportions relative to the scale of Roman-Britain society. An interesting question is, was central France an enclave sheltered by either social or environmental conditions from extreme ramifications of sixth century disturbance? Young's chapter seems to indicate relative continuity in France, while Jones proffers discontinuity *and* continuity; the Roman-Britains discontinued and the Anglo-Saxons continued. The discontinuity seems to be attributable to cultural traditions of living in towns and cultural links to highly valued luxury goods from the more urbanized eastern Mediterranean where plague was rampant. Jones offers a theory of selective plague incidents based on trade routes and settlement patterns that does much to explain some scholarly mysteries. This is complex but very well explained. It is the opposite of what would have been expected given that the Roman Britishs were the closest of all Europeans to the ocean and therefore should have been buffered by its warmth.

In Italy, the situation also turns grim with decades of relentless warfare. Only with a full examination of the information, some of it only recently available, can the full breadth of regional variations in the middle first millennium disruptions in Europe be understood. The apparent relative stability in France and the conflicting influence of empire war in Italy open a number of questions concerning the variable contributions of politics, health issues, and human agendas in the face of global climate change. Perhaps a key element is that east central France could have been stabilized by its natural control of the Rhone-Rhine-Seine-Loire trading corridors. Another interesting issue is Scandinavia where the consequences were so far reaching as to precipitate major emigrations. This is a question that could profitably engage the attention of future scholarship and archaeological investigation relative to global change.

Second, Jones offers the engaging metaphor of climate as a funnel that channelizes existing social trends. She develops a parsimonious explanation of the transitions between Roman British and Saxon Britain by considering a broad range of disciplinary perspectives including the environmental instability associated with the A.D. 536 event, and secondary effects including famine and plague. All too often global change studies are posed as uncomplicated interactions between climate and society (see Crumley 1994 examples of the complex interactions between climate and society). Jones' funnel metaphor adds another tool to the arsenal for interpretation of complex systems that include human societies.

Baillie, M. G. L. (1995) Patrick, Comets, and Christianity. *Emania* 13:69-78.

Crumley, Carole L. (1994) The Ecology of Conquest: Contrasting Agropastoral and Agricultural Societies' Adaptation to Climatic Change. In *Historical Ecology: Cultural Knowledge and Changing Landscapes*, C. Crumley, editor, pp184-201, School of American Research Advanced Seminar Series, Santa Fe.

Crumley, Carole L., and William H. Marquardt, eds. (1987) *Regional Dynamics: Burgundian Landscapes in Historical Perspective*. Academic Press, San Diego.

Redman, Charles L. (1977) The "Analytical Individual" and Prehistoric Style Variability. In The Individual in Prehistory: Studies of Variability in Style in Prehistoric Technologies, edited by J. Hill and J. Gunn, pp 41-54, Academic Press, New York.

Stothers, R. B. (1984) Mystery cloud of AD 536. *Nature* 307(5949):344-345.

2. CLIMATE, ARCHAEOLOGY, HISTORY, AND THE ARTHURIAN TRADITION: A MULTIPLE-SOURCE STUDY OF TWO DARK-AGE PUZZLES

Elizabeth Jones

Department of Anthropology, University of North Carolina, Chapel Hill, North Carolina 27514

Resolving Conflicting Theories by Finding a Broader Perspective

[Britain; interaction, interdisciplinary integration, multisourcing; history, archaeology, literature, linguistics, folklore, environment, climate; Early Middle Ages, sixth century A.D.]

I have long been interested in the problem of integrating archaeological, environmental, historical, literary, linguistic, folkloric, and other types of data or source materials. Since it is now generally recognized that there are no simple causal explanations, and that phenomena are the result of a complex interaction of factors and processes, an inter-disciplinary, multiple-sourced study seems to offer the ideal approach for generating explanations that reflect the complexity of past phenomena. There is a difficulty, however, in integrating different types of data and investigative strategies. For example, a literary study and a climatic study may seem to represent entirely different worlds, rather than merely different facets of a single complex reality that made up a past era. I was delighted, therefore, to discover that focusing on the climatic *event* beginning in the year A.D. 536[cal] seemed to provide a perspective from which to effectively integrate a broad range of historical *processes* ranging from biological to social, as well as to pull together loose ends of some unresolved problems I had encountered in several different disciplines of medieval study relating to Dark Age Britain.

This chapter will focus on how the recognition of the climatic event of A.D. 536[cal] not only provides plausible explanations for two highly-debated problems in Dark Age studies, but also how it ties these seemingly unrelated studies from different disciplines together, providing an overall context within which they may be understood. Accordingly, I will suggest several hypotheses based on the A.D. 536 event which gather elements from the research of others. These hypotheses seem plausible to me because they meet criteria of validity that are common to most all disciplines, whether they be the "hard sciences" or the humanities. That is, these hypotheses lay out simple (parsimonious) explanations that replace several previously defined, separate, often convoluted explanations. Furthermore, these hypotheses reveal connections between types of data that were heretofore unsuspected (serendipity).

Britain Before A.D. 536[cal]

[Britain, Mediterranean, North Sea, Gaul; history, archaeology, climatology, tree rings, ice cores; migration, urbanism, trade, ethnic interaction, sociopolitical processes; Late Antiquity, Early Middle Ages, A.D. 410-560[cal]]

I will start by briefly laying out the context and background of Dark Age Britain, leading up to the year A.D. 536[cal] (Figure 2.1). By the third century A.D., Britain had been thoroughly incorporated—culturally, economically, and militarily—into the Roman Empire, but as the empire slowly began to disintegrate, Roman troops were withdrawn from Britain to defend areas on the Continent. By the year A.D. 410[cal] the Britons were entirely on their own (Davis 1982:6-7). Defensive measures were needed against the groups who had been raiding Britain for several centuries: the Picts from the area which is now Scotland; the Scots from Ireland (who not only raided, but invaded and established settlements in Wales and in Scotland, later named after them); and the group known as the Anglo-Saxons (which included not only Angles and Saxons, but also Jutes, Frisians, Franks, and other Scandinavian peoples—all from around the North Sea). These peoples later coalesced into the "English." When using the term "Anglo-Saxon" I am referring to this heterogeneous group of North Sea peoples.

The "Britons," on the other hand, are the Celtic peoples of Britain who were in possession of the southern portion of island before the arrival of either the Romans or the "Anglo-Saxons." At that point in time, the Britons all spoke a common Celtic language, and were considered one people, although they were organized along tribal divisions with tribal "kings" or leaders. The Britons had become highly Romanized. Urbanism, supported by rural villa agriculture, was firmly established, and the country was inter-connected by a vast network of Roman roads. The Latin language was used not only by the elite classes, but also, as the archaeological evidence of inscriptions and graffiti indicates, by tradesmen such as potters and brick makers (Wacher 1995:409). The civil administration seems to have continued after A.D. 410^{cal} relatively smoothly with tribal leaders replacing elected officials (Wacher 1995:409; Davies 1993:48). To be sure, there were changes as Roman influence waned, such as the end of a money-based economy. The towns shrunk in size as their economic functions withered, and as the large-scale villa system which had provided their agricultural sustenance slowly died out (Wacher 1995). Yet the Britons were still involved in the cultural life of the Continent, as shown by the Pelagian heresy, which spread to the Continent from Britain. St. Germanus of Auxerre came to Britain to root out the unorthodoxy in A.D. 429^{cal} and again in A.D. 440^{cal} (Wacher 1995:409). Records show that when he visited the important Roman regional center of Verulamium in A.D. 429^{cal} (see Figure 2.1) it was still functioning normally as a Roman

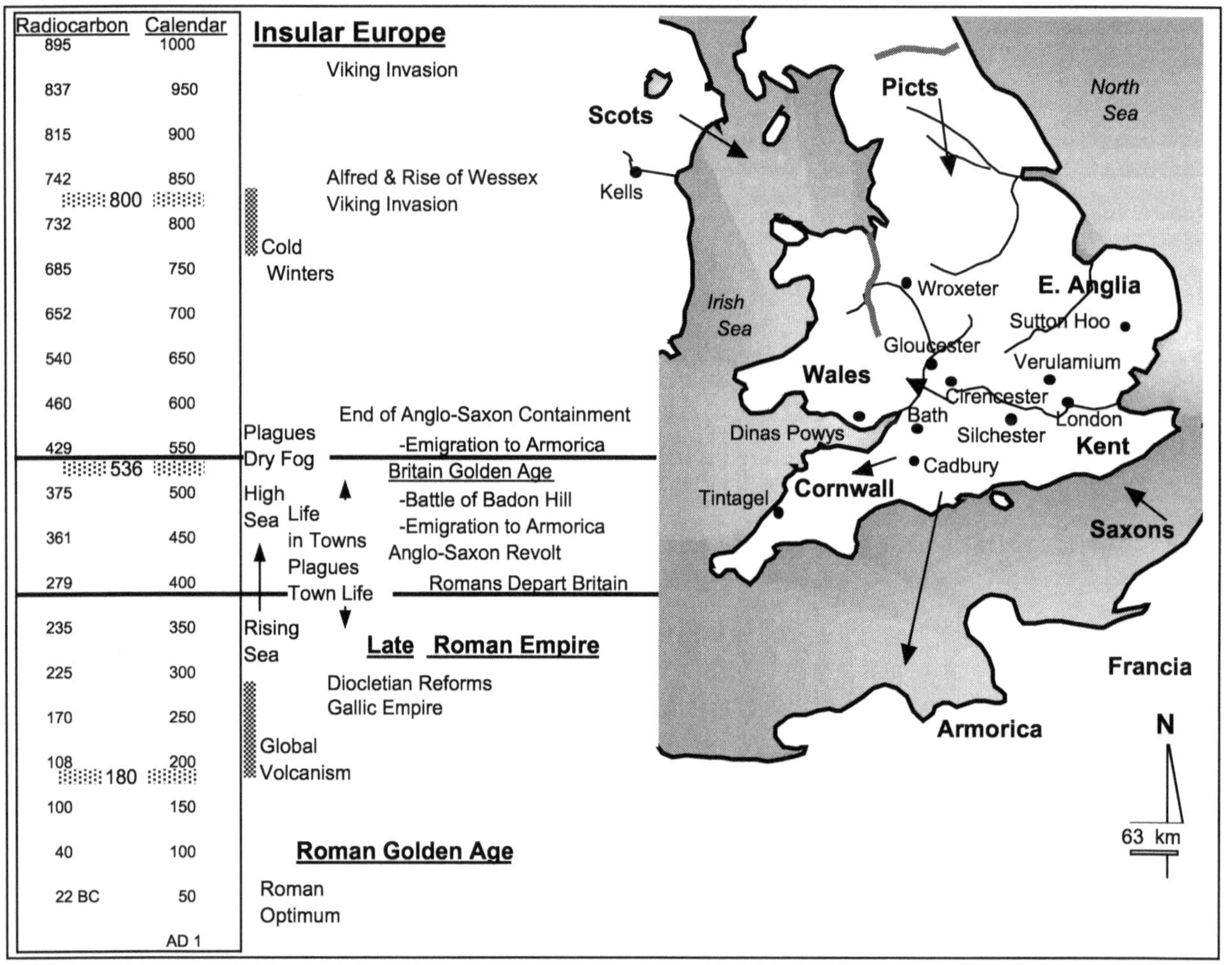

town (Davis 1982:42).

Trouble came to the former province when, around the year A.D. 425^{cal}, a ruler known from documentary sources as "Vortigern," a Celtic title meaning "overlord," came to power in a position that seems to have replaced the Roman Governor. To gain assistance in fighting off troublesome groups of raiders and invaders, he emulated the Continental practice of recruiting Saxon groups to serve as *Foederati* or "Federates." During the earlier Roman administration, Anglo-Saxons had also settled in Britain as part of the Roman military organization. By contrast, these later *Foederati* were allowed a semi-independent status, and were given both land and provisions in exchange for guarding Britain from its attackers. They were largely settled in Kent and southeastern Britain (Davis 1982:7-8, 105).

The revolt of the *Foederati,* that took place some time between A.D. 449 and 456^{cal} was described by the Briton Gildas writing around A.D. 540^{cal} (Campbell 1982:23, 26). Gildas is one of only two extant contemporary sources on Britain for this period, the other being the Byzantine historian Procopius who wrote around A.D. 560^{cal}. It seems that the mercenaries were too efficient for their own good; after the danger from raiders was under control, some Britons resented the burden of having to pay and provision them, and wanted them dismissed. The *Foederati* not surprisingly revolted and greatly expanded their territory through constant warfare. They were reinforced by new Anglo-Saxon immigrants who swelled the numbers of the original *Foederati,* especially in East Anglia. In response, there was a migration of the British to the regions of Wales and Cornwall. This period (about A.D. 458-460^{cal}) also saw a wave of emigration to Armorica, later called Brittany, in France, which was evidently stimulated by the Anglo-Saxon pressure and displacement, in addition to Irish raiding in Devon and Cornwall (Davies 1993:57-58; Morris 1996:251; Pearce 1978:167). The archaeological record shows that many British towns were abandoned at this time, and that organized civic town life appears to have come to an end (Snyder 1996; Wacher 1995). John Wacher (1995:421) makes an important distinction between the Roman-style, civic-governed "town life," which existed prior to the mid-fifth century, and "life in towns," which continued in some communities through the mid-sixth century in an unorganized squatter fashion. This later "life in towns" perpetuated, nevertheless, a congregated settlement pattern, and a degree of continued participation in trade (see sixth-century evidence from Verulamium, London, Bath, Cirencester, Gloucester, Silchester, Wroxeter, etc. in Snyder 1996).

In the A.D. 460s^{cal} the Britons finally rallied and turned the tide against the Anglo-Saxons. They started a long, slow recovery that culminated around the year A.D. 500^{cal} in a decisive battle at Badon Hill. This battle put a temporary end to the Anglo-Saxon threat. Many of the Anglo-Saxons were forced to flee Britain as both Frankish historical records and archaeological evidence on the Continent indicate (Davies 1993:58; Whittock 1986:51-53). Those Anglo-Saxons who remained were firmly controlled and circumscribed in reduced areas of settlement that can be detected archaeologically by the distribution of sites from the period. The divergence of styles in brooches and pottery in the separated Anglo-Saxon areas of Britain after about A.D. 500^{cal} shows that contact between the different Anglo-Saxon groups was virtually cut off (Whittock 1986:54). That the Anglo-Saxons were also isolated from the Continent is shown by the divergence of insular and Continental styles (Davis 1982:61). Another indication that the Anglo-Saxons were prevented from trading is that they dug up for reuse, from graves found in abandoned Roman-period settlements, centuries-old Romano-British pottery, coins, and brooches—the pennanular ones being the most common (White 1991:146). The items were then reburied in Anglo-Saxon burials. Other items such as Rhennish glassware (or claw-beakers) appear in burials 100 years after their manufacture, having been conserved for so long presumably due to lack of replacements (White 1991:146).

The Britons, in marked contrast to the Anglo-Saxons, especially in the newer settlements in Wales and Cornwall, remained in frequent contact with the Continent. They traded directly with the eastern Mediterranean, as is evidenced by the distribution of imported pottery (Snyder 1996; Campbell and Lane 1993; Fulford 1989). These new settlements were not towns, but seem to be headquarters for regional authorities and their entourage, who collected tribute from the interior and exchanged it for luxury goods from the Continent. Some settlements seem to have been involved in local exchange as well (Gilchrist 1988). Fortified hillfort sites such as Dinas Powys (Alcock 1987), Tintagel (Thomas 1993), and South-Cadbury (Campbell and Lane 1993) (see Figure 2.1) were centers of importation. Other sites such as Longbury Bank in Wales were unfortified, but still produced evidence of the Mediterranean trade (Campbell and Lane 1993). Even a few eastern inland towns such as Verulamium and London show some evidence of the Mediterranean imports (Snyder 1996:16-18). All major centers of importation, however, were located near the coast, except for South Cadbury (Campbell and Lane 1993:68), which is unique in this period, both for being an major inland center of importation, and for the massive scale of the improvements and fortifications erected in the fifth and sixth centuries. This site has traditionally been identified with the Arthurian Camelot. The coastal location of the rest of these establishments may reflect a breakdown of the interior Roman road and transport system (Campbell and Lane 1993:68).

Imports during this period included amphora or B-ware, containing wine, oil, and other commodities, as well as fine serving/dining dishes or A-ware directly from the eastern Mediterranean, presumably in exchange for tin (Campbell and Lane 1993:62; Fulford 1989; Graham-Campbell 1980:114-115). These were high status objects and had the aura of "Romaness" about them which may have lent a certain legitimacy to the ruler's position as inheritor of imperial authority.

The peace lasted two generations, and indicates some kind of united, organized, state-wide governmental effort on the part of the Britons to maintain it almost without exception throughout the island (Davis 1982:42). Gildas refers to the period of peace after Badon Hill as "years of good stable government" and a kind of golden age when harmony and justice were maintained throughout the land (Morris 1996:117). The *Anglo-Saxon Chronicle* of the ninth century which tends to record only Saxon victories, lists no major Anglo-Saxon activity from about A.D. 500^{cal} until after mid-century (Whittock 1986:53). Gildas comments that "in his day the English were not waging war upon the Britons" (Davies 1993:59).

This, then is the situation on the eve of the A.D. 536^{cal} climatic event which resulted in a dry fog that is recorded in historical sources from Byzantium to China, and caused a severe cold period that is globally registered in tree rings from Scandinavia, Ireland, and in the western United States (Baillie 1994). The worst of the cold occurred in the year A.D. 540-541^{cal} (Baillie 1994:212). The period of reduced growth in trees across northern Europe lasted about ten years, from A.D. 536 until 545^{cal} (Baillie 1995:71).

The cause of this sudden and protracted cold period is difficult to explain. Atmospheric evidence from the ice cores at both the north and south poles is ambiguous and subject to various interpretations. Historic sources are equally vague concerning the cause of the event. Baillie (1999) proposes that the event was caused by a comet passing close to the earth, thereby introducing dust and debris into the atmosphere along with larger pieces whose direct impact with the earth was strong enough to produce tectonic effects of volcanic eruptions and ocean outgassing. Keyes (1999), on the other hand, attributes the cause solely to the effects of volcanic activity. The exact cause of the event is not critical to the arguments presented in this chapter. The tree-ring evidence that the prolonged and extreme cold period did occur is incontrovertible. What is of immediate concern here is the influence of the event on human history.

Thus, whatever the cause, one would expect the event to have severe, if not catastrophic, effects on some plant, animal, and human populations. Due to the scarcity of historical records, especially in Britain, the question then follows: how might this event show up in the *archaeological record*, and how might it be recorded in *cultural memory*?

Britain After A.D. 536^cal

[Britain, North Sea, Mediterranean, Gaul; migration, Anglo-Saxon takeover, archaeology, history, linguistics, trade patterns; Late Antiquity, Early Middle Ages, Roman/Saxon transition, A.D. 440-560^cal]

In Britain, there are extensive changes that all appear around the mid-sixth century (because of the various dating methods, the dates given here must be seen as falling within a range of ten to 20 years). Gildas describes the period around A.D. 540^cal as a time when spiritual corruption and political chaos set in—a marked contrast with the immediately preceding golden age (Morris 1996:201-207). In doing so Gildas is echoing the feeling of post-A.D. 536^cal decline also expressed in Gaul by his contemporary, Gregory of Tours (see Young this volume). Other sources tell us that the Anglo-Saxon offensive started anew after mid-century, and resulted in the largely unchecked take-over of England (Morris 1996:225-248; Whittock 1986:201-202) (Figure 2.2A).

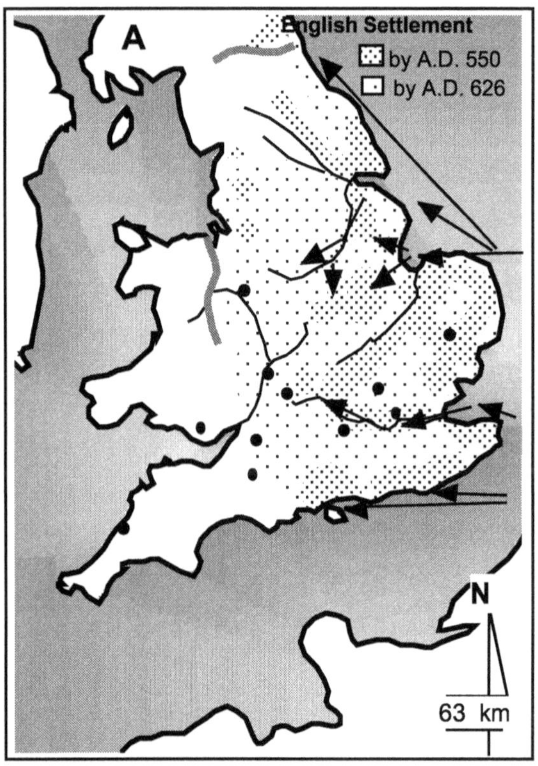

Figue *.2. Settlement and Artifacts Distributions in Dark-Age Britain. (A) English settlement by A.D. 550 and 626^cal (Adapted from Gilbert 1968:7).

Archaeological evidence reveals the end of the Anglo-Saxon containment at that same time, around A.D. 550^cal. The Anglo-Saxons ceased reusing buried Roman goods (White 1991:146), and free trade among the formerly separated groups was established (Davis 1982:62). The Anglo-Saxon areas of Kent and Sussex started trading with the Frankish areas in Gaul for the first time, breaking their earlier pattern of an exclusively North Sea economy (Whittock 1986:225). There was an explosion in Anglo-Saxon settlement expansion, and new waves of immigrants appeared (Davis 1982:62).

Occuring at the same time was another massive emigration of Britons to Armorica (around 540-560^cal). It was large enough to swamp the local culture and political structure, transforming Armorica into Brittany or "lesser Britain" (Morris, 1996: 254; Pearce 1978:167). "Linguistic studies show that Brittany was substantially settled by South West Britons who arrived in the middle of the sixth century" (Pearce 1978:157).

Finally, the British importation of products from the eastern Mediterranean abruptly ceased at mid-sixth century, and came to be replaced in the later sixth and seventh centuries by trade with western Gaul characterized by D-ware—wheel-made grey bowls and mortaria, E-ware—crude buff, yellow or orange stoneware pitchers, jars, beakers and bowls, and imported glass (Campbell and Lane 1993:67). In essence, the elite wares, containing exotic goods were replaced by utilitarian kitchen wares.

Roman Britain—Genocide or Assimilation?

[Britain; history, archaeology, linguistics; cultural continuity, genocide, assimilation, migration, linguistic borrowing, religion, Slavery; Early Middle Ages, A.D. 536-600^cal]

At this point I will address two problems which have been hotly debated by Dark-Age scholars for many years, and show how the analysis of these two problems in light of the A.D. 536^cal climatic event can help explain the multitude of changes we see in the archaeological record at mid-sixth century.

Scholars have long argued over the problem of what happened to the British populations in the areas taken over by the Anglo-Saxons (see Alcock's discussion 1987:267-281). There are two popular positions on this question—one espousing genocide and emigration, the other proposing assimilation. The evidence appears contradictory. On the one hand, the historical reports on both sides, that of the Briton Gildas as well as the *Anglo-Saxon Chronicle* and English historian Bede (writing around A.D. 730^cal), speak of the flight and annihilation of the Britons (Davies 1993:77; Alcock 1987:267; Davis 1982:9).

Supporting the historical record, is the fact that there was almost no linguistic borrowing of Celtic words into the English language anywhere, and in the majority of regions, especially South East England, very few Celtic place-names exist (Campbell 1982:38; Hills 1980:86). In the British areas that held out the longest (South West and Northern England), some place names and Celtic family names do survive (Davies 1993:59). This is in contrast with the situation on the Continent where the Germanic tribes adopted the Latin-derived speech, and the situation with the Irish, who though always at odds with the Britons, freely exchanged loan words, culture and even religion.

Additionally, there was practically no spread of Christianity from the Britons to the Anglo-Saxons, as one might expect would take place from the mingling of the populations (Davies 1993:77; Morris 1996:223). Without exception, the early British kings were all Christian, the early Anglo-Saxon kings were all pagan (Campbell 1982:22). The Saints' Lives of this period make no mention of any interaction with the Anglo-Saxons, yet record that the British were actively converting Irish, Armoricans, and Galicians in Spain (Davies 1993:77; Morris 1996:223).

Furthermore, the word *wealh*, which became the English word for Britons—"Welsh" originally meant "slave" in the West Saxon dialect, and records on the Continent show that the Anglo-Saxons in the periods of expansion sold great numbers of British slaves as far away as Rome (Hills 1980:86).

Finally, extremely little evidence of a continuation or adoption of any elements of British material culture is present in the archaeological record in the areas of Anglo-Saxon settlement (Campbell 1982:29).

On the other hand, those that argue for assimilation rather than genocide, counter that there is no evidence of the violent destruction and burning of towns, and that, although a few villas were burned, most, like the towns, were simply abandoned (Wacher 1995:411; Graham-Campbell 1980:113).

They point out that while Britons cannot be identified archaeologically in the areas taken over by Anglo-Saxon settlement, they also seem to be largely missing archaeologically from the areas known to have remained in British hands until late in the sixth century or early seventh century. To account for this phenomenon they assume that British material culture may have become so impoverished as to be largely "invisible." In other words, because trade and manufacture were so diminished, articles would be used until they disintegrated or became unrecognizable (for example, a shovel being reduced to a shapeless, corroded blob of iron) (Davis 1982:47).

Additionally, because dating is largely based on typologies rather than absolute measures, they suggest that many British or "sub-Roman" sites are mistaken for late Roman period sites, and that there was little recognizable change in the material culture stylistically (for example, British sub-Roman cemeteries would be indistinguishable from late Roman period ones) (Hills 1980:86). This is borne out by the evidence from Caerwent where a British burial contained a Roman coin dated to around A.D. 335-348^{cal} and an apparently late-Roman bracelet, but was radiocarbon dated to A.D. 540-770^{cal} (Snyder 1996:32).

They add that the historical evidence reveals clauses in the law codes of Kent and Wessex that deal with the rights of Britons (Davies 1993:68).

While admitting that some slaughter probably did take place (on both sides), they maintain that in an area as populous as Britain is estimated to have been, it would, on practical grounds, have been impossible for the less-numerous Anglo-Saxons to have killed the majority of Britons (Hills 1990:86). A related point is that there is an absence of the mass-battlefield burials that were associated with the earlier Roman conquest (Wacher 1995:411).

The Differential Effects of Climate: The Selective Epidemic Hypothesis

[Britain, Mediterranean, Scandinavia; history, archaeology, settlement pattern, linguistics, literature, folklore; epidemics, plague, emigration, rats, lice, abandonment; Early Middle Ages, A.D. 443-556^{cal}]

If we integrate the climatic disturbance of A.D. 536^{cal} and its aftermath into the argument, a third line of explanation emerges. It is well-established that drastic, relatively sudden changes in average global temperature—whether warmer or colder—cause imbalances in populations of bacteria, viruses, insects, rodents, and other animals that result in plagues and other epidemics, which are then spread to human populations (Linden 1996:56). As it happens, there were major outbreaks of epidemics around A.D. 443 and 540^{cal} immediately prior to *both* periods of Anglo- Saxon expansion in Britain, beginning about A.D. 449 and 550^{cal} respectively. John Wacher (1995), among others (e.g. Davies 1993; Whittock 1986; Davis 1982), has advanced a theory that the decimation of the British populations by plagues depopulated the countryside, and weakened the British ability to resist the Anglo-Saxons. Wacher and others have not, however, tied their theory to climate change.

Independent historical sources mention a world-wide plague about A.D. 443^{cal} in western Europe and A.D. 445-446^{cal} in Constantinople, coming just before the first revolt of the Saxon *Foederati* (A.D. 449-456^{cal}), followed by another outbreak that spread from Asia Minor in A.D. 455^{cal} (Wacher 1995:412). At this same time a great impetus to the emigration of the Anglo-Saxons from their coastal homelands was the rising sea levels, which inundated farms on the coast, and even some inland areas (Campbell 1982:30; Hills 1980:85). During second and third centuries rising sea levels had forced them to elevate their villages on mounds. In the fifth century further elevation became too labor intensive and coastal lowlands were abandoned. The global warming indicated by the rising sea levels that dispersed the Anglo-Saxons would be related to the global epidemics of the period as well.

Similarly in the sixth century, the Justinian plague appeared around A.D. 540^{cal} in the Middle East, and bubonic-like epidemics were recorded soon after in both Ireland and Wales (Davies 1993:69; Morris 1996: 222; Whittock 1986:203). Once again, epidemics immediately preceded this second explosion of Anglo-Saxon expansion. And as before, the sudden climate change that spawned the plague also gave impetus to the invaders. Tree ring evidence shows that the cold effects of the dry fog were more pronounced and longer-lasting in Scandinavia than in the British Isles (Baillie 1994:213; 1995:71). The majority of invaders in the sixth-century wave came from more northerly areas than in the fifth century—from Denmark and even from Sweden, as evidenced by the Sutton Hoo burial (Campbell 1982:27, 31,

34) and the overall distribution of saucer and cruciform brooches (Figure 2.2B).

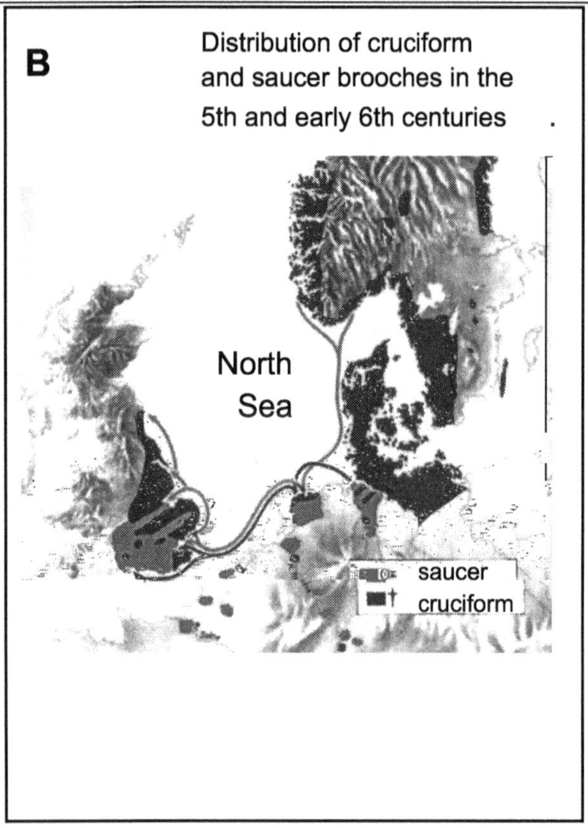

Figue 2.2. Settlement and Artifacts Distributions in Dark-Age Britain. (B) Distribution of Cruciform and Saucer brooches fifth and early sixth centuries (Adapted from Falkus and Gillingham 1987:31).

Although many scholars acknowledge the epidemics, few regard them as much of a causal factor (e.g. Morris 1996:224). Yet more recent historical examples, such as the European invasion of North America, show that aggression combined with high mortality rates from disease can easily decimate a population. Some forms of plague such as the pneumonic form have mortality rates of up to 98 percent. Other diseases such as small pox can have a similarly high death rate in populations with no genetic immunity (Whittock 1986:204-5). Whatever the actual disease may have been, the Justinian plague is believed to have been as devastating as the Black Death of A.D. 1349[cal] (Davies 1993:69; Morris 1996:223). It is likely that scholars have down played the causal effects of the plagues in this era because they have almost entirely ignored the related climatic disturbances along with resulting amplifying effects of famine and social disturbance.

There is general agreement that the plagues struck the Britons hard without affecting the Anglo-Saxon populations to any great extent (Davis 1982:10; Morris 1996:223; Wacher 1995:412-413; Whittock 1986:205). This is plausible for several reasons:

1) As the archaeological evidence shows, the Britons were trading directly with the eastern Mediterranean—the source of the plagues, while the Anglo-Saxons traded exclusively with the North Sea regions during the first plague episode, and were not trading at all in the second. When they do start trading again after mid-sixth century, it is with the area of northwestern Gaul. Gregory of Tours recorded the progress of the Justinian plagues—which he himself narrowly escaped as a child—as they traveled along the trade routes through central and southern Gaul to the upper and middle Rhine area in A.D. 544 and 545[cal]. He does not note the occurrence of plagues in the lower Rhine area or in northwestern Gaul (Morris 1996:222).

2) Epidemics are largely urban phenomena. Many Britons were still participating in town life during the first plague event of the mid-fifth century, and some continued to inhabit the former towns in the first half of the sixth century (Snyder 1996; Whittock 1986). The Anglo-Saxons, by contrast, were living entirely in a settlement pattern of dispersed family farms.

3) There was virtually no interaction (not even fighting) between the British and Anglo-Saxon populations at the time of the plagues, thus the epidemic could not have easily spread from one to the other. Gildas mentions the 'unhappy partition' of Britain that prevented British pilgrims from visiting Christian shrines located in Anglo-Saxon held areas (Morris 1996:223).

4) The Anglo-Saxon populations, even if decimated to some degree by famine and plague, were replenished by immigrants from the Continent (Wacher 1995:412) due to the severe climatic conditions discussed above, while the British population was further depleted by the emigration to Brittany.

A theory of selective catastrophic epidemic arising from drastic climate change would explain away the apparent contradictions in the historical and archaeological records outlined in the annihilation vs. assimilation argument. A British population seriously decimated due to disease helps to explain the lack of linguistic and cultural borrowing, and the dearth of British material in the archaeological record, while at the same time explaining the abandonment of towns and villas with no evidence of destruction.

Archaeological evidence for plague exists from this period at Cirencester in the form of unburied bodies lying in the streets of an abandoned town (Snyder 1996:39; Wacher 1995:413, 417). Again at Cirencester, it may be the case that the town was reestablished within the Roman amphitheater outside the city by survivors of the earlier plagues who were fleeing the disease (Snyder 1996:39-40).

Other evidence pertaining to bubonic plague are the remains that have been found in the archaeological record of this period of the particular species of rat (black rat) and rat flea (*xenopsylla cheopis* and *nosopsyllus [cerator phyllus] fasciatus*) that are carriers of the bubonic plague (Wacher 1995:421; Whittock 1986:203-204). (Until recent archaeological studies, it was thought that the black rat had not reached Britain until the eleventh century—Whittock 1986:204.) These rats were probably transported to Britain via ships from the Mediterranean (Morris 1996:223).

The selective epidemic theory also explains why the Anglo-Saxons did not settle in the towns they conquered. It was formerly thought that the Anglo-Saxons were superstitious of the towns and avoided them as the home of ghosts (Hills 1980:86). They called towns 'tombs surrounded by nets,' which makes sense if there were plague victims left lying in the streets unburied by the British survivors who abandoned the town when they fled the epidemic (Wacher 1995:414). An early Anglo-Saxon poem, *The Ruin*, describes an abandoned British town using these words,

days of pestilence came; death swept away all the bravery of men; their fortresses became waste places; the city fell to ruin. The multitudes who might have built it anew lay dead on the earth. (Hills 1980:86-7)

This easily serves as an apt description of a town deserted on account of plague.

Epidemics explain both the timing and the ease of the Anglo-Saxon successes against a larger, better-organized British population living in impregnable fortified settlements. Sources record that one of the victims of the plague was Maelgwn, King of Gwynedd (died A.D. 549^{cal}), the most powerful British king at the time (Ellis 1994:71). Deaths of leaders like these would have seriously crippled the British ability to organize a concerted defense.

The plague theory similarly helps to explain the timing and massive scale of the migrations to Wales and Brittany as groups fled plague-ridden areas (Wacher 1995:415). Brittany, along with the rest of northwestern France did not suffer greatly from the plagues, and stories have been handed down of British refugees who came to Brittany expressly at that time to escape contagion (Morris 1996:222).

The theory further provides plausible factors relating to the end of British trade with the eastern Mediterranean. In addition to the breakdown of economic organization and production following the plague episode, it could very well be that after receiving plague-bearing ships, ships from the eastern Mediterranean were no longer welcome.

Climate as a Funnel of Process

[Britain; climate funnel metaphor; spheres of social interaction, ethnic difference, aggregated urban/town settlement, dispersed rural settlement; A.D. 536]

In proposing the selective epidemic hypothesis, I want to avoid speaking of climate and the resulting epidemics in a deterministic way. An "event," such as the climatic disruption of A.D. 536^{cal}, acts as a funnel through which on-going processes are channeled, and are in some sense transformed. The event, along with its resulting famine and disease, sped up the existing trends toward dispersed settlement patterns and the reduction of town life in which the Britons were already heading, due to other factors such as reduced trade and warfare. Disease was only one factor in the depopulation of the countryside, which amplified the effects of genocide, slavery, and emigration.

Importantly, we want to notice that the event affected two groups of people living side by side, in very different ways due to their differing settlement patterns and differing cultural relationships. The event of A.D. 536^{cal} was a disaster from which the British never recovered. After the event, there was no more unified British resistance, no more life in towns, and thus no more Romano-Mediterranean culture that the town life had supported. The Britons divided into small, independent, warring kingdoms after the mid-sixth century, soon becoming isolated by the Anglo-Saxons to the point of developing separate languages (Welsh, Cornish, Breton) (Figure 2.2C). This is another possible influence on the ending of the Mediterranean trade. Things "Roman" would have lost their cachet and symbolic meaning, as the common—not only Roman, but—British culture disappeared in a new, decentralized and dispersed context.

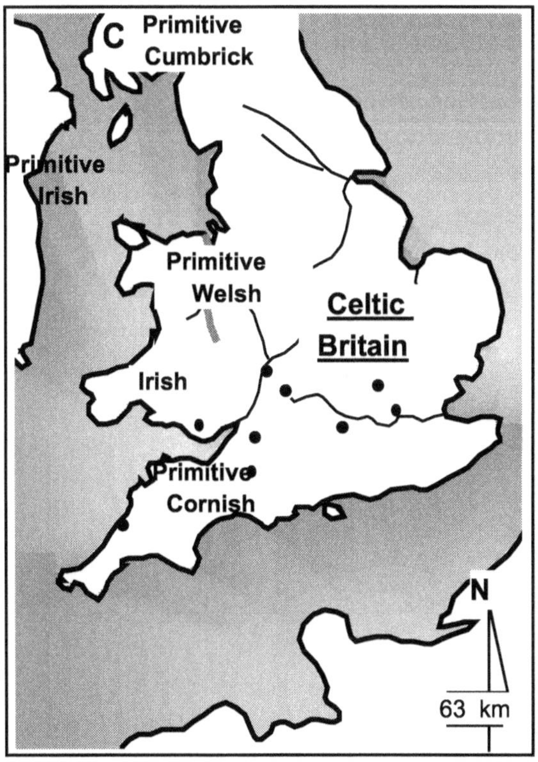

Figue 2.2. Settlement and Artifacts Distributions in Dark-Age Britain. (C) Distribution of languages in Celtic areas of Britain ca. A.D. 700 (Adapted from Wendy Davies 1984:66).

Yet the same event was a boon for the Anglo-Saxons who found a new homeland. By A.D. 600^{cal} they had conquered nearly all of the present territory of England, and the royal lineages were established that developed into the early Anglo-Saxon kingdoms (Campbell 1982:27; Davis 1982:10-11; Hills 1980:87) (Figure 2.2D).

Cultural Memory, A.D. 536, and King Arthur

[Britain, Gaul; cultural memory, folk memory, history, literature, archaeology; Celtic kingship, King Arthur legend, death of King Arthur, end of Roman Britain; Early Middle Ages, A.D. 470-537^{cal}]

Having examined how the A.D. 536^{cal} event might have appeared archaeologically, I will now turn to how it could have been recorded in cultural or folk memory. This brings us to the second of the highly-contested questions of Dark-Age Britain—the problem of the historical King Arthur. The historical documentary evidence for Arthur is scant, to say the least, but does point to the existence of an actual person. The earliest mention, a Welsh poem around A.D. 600^{cal}, mentions the prowess of another warrior, but then says that he was no Arthur and was therefore killed in battle (Campbell 1982:27). The ninth-century *Historia Brittonum,* attributed to Nennius, speaks of Arthur as the "*Dux Bellorum*" (war commander) and lists 12 battles he is supposed to have fought ending with the battle of Badon Hill (Ashe 1995:2). The third reference, the twelfth-century Welsh *Annales Cambriae,* mentions the battle of Badon Hill and Arthur's death along with Medraut in the Battle of Camlann in A.D. 537^{cal} (Campbell 1982:27).

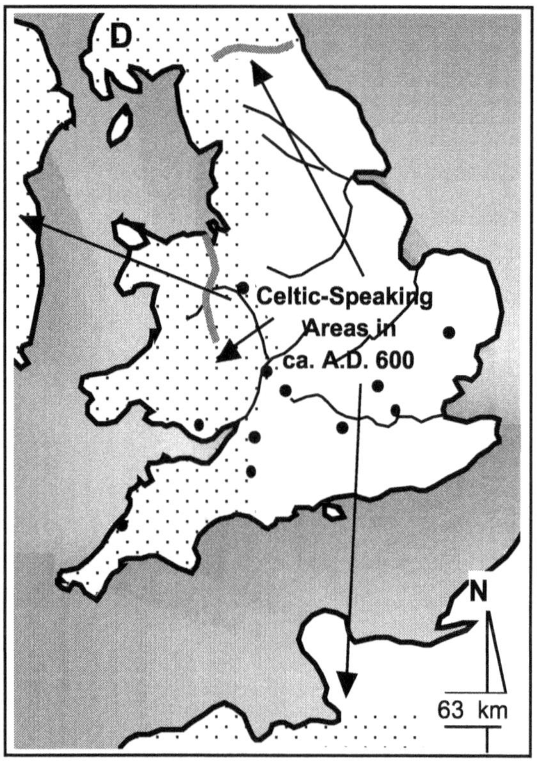

Figue 2.2. Settlement and Artifacts Distributions in Dark-Age Britain. (D) Celtic-speaking areas of Britain around A.D. 600^{cal} (Adapted from James 1993:167).

The great bulk of information about Arthur that figures in later legend comes not from these limited historical sources, but from Geoffrey of Monmouth's twelfth century literary history of the kings of Britain. This work has been regarded as largely fictional, but Geoffrey Ashe (1995:9), combining archaeology with detailed literary and historical criticism, has shown that although Geoffrey of Monmouth exaggerates, he does not make episodes up entirely. The core of what he says always has at least some basis in folk memory.

Much of what Geoffrey of Monmouth reports about Arthur rings true and corresponds to the King of the Britons who is recorded in Continental sources as Riothamus. This king at the request of the Anthemius the western colleague of Emperor Leo I, brought an army across to France to help save Gaul from the Visigoths. Riothamus was betrayed by the imperial prefect "Morvandus," and after losing to the Visigoths disappeared in friendly Burgundian territory headed toward Avallon (Ashe 1995:13). Ashe identifies Riothamus as Arthur. I will not go into all Ashe's interesting arguments for his identification of Arthur as Riothamus here, other than to say that "Riothamus" is not a name, but a title, like Vortigern, meaning "High King" (Ashe 1995:15; Campbell 1982:37). Ashe places his disappearance at around A.D. 470-471^{cal} (1995:15), which would have given him enough time to begin the British resurgence around A.D. 460^{cal} in Britain before his expedition to France. The very fact that a king of the Britons could feel secure enough to take an army out of Britain shows that the Anglo-Saxon threat had been curtailed to a substantial degree (Davies 1993:57).

Most of Arthur's activities make sense in the context of the fifth century (Ashe 1995:17-19). Additionally, the places associated with Arthur—Tintagel, Glastonbury, and South Cadbury/Camelot have been shown archaeologically to have been occupied at just the right time (Ashe 1995:14). Cadbury-Camelot, especially, was refortified in mid-fifth century on scale unknown for any other site of the period—it is estimated that it would have taken a force of about 870 men to defend the ramparts alone (Snyder 1996:27). This is an organization of men and materials on a scale requiring the resources of a king (Ashe 1995:4; Alcock 1987:153-4).

The problem is that the dates given for Arthur's death in the sources at A.D. 537^{cal} in the *Annales Cambriae* or A.D. 542^{cal} by Geoffrey of Monmouth, come far too late, resulting in a nearly 100 year-old king who dies fighting on a battlefield (Ashe 1995:3). Ashe is able to explain Arthur being credited with the victory of Badon Hill around A.D. 500^{cal} by the fact than in late Roman times military units were often named after individuals, and continued to function after the individual's death. Welsh poems speak of battles being won by "Arthur's men" who become in legend the Knights of the Round Table, and it is possible that it was "Arthur's men," not Arthur, who led the way at the battle of Badon Hill (Ashe 1995:21). The list of battles preceding Badon Hill and attributed to Arthur by Nennius are located all over England, against all of the "emerging English kingdoms" (Davies 1993:59), and may well have been the work of several different leaders among "Arthur's Men." Plausible as it may sound, this scenario does not solve the problem of the even later dates given for Arthur's death, coming two generations after Badon Hill. Ashe gives the unsatisfying explanation that the late death dates must be due to a multiplying of scribal error (Ashe 1995:11, 14-15).

Other scholars, on the basis of the late death date, have reconstructed a northern sixth-century Arthur in Scotland (e.g. Goodrich 1986). But this Arthur could not have been responsible for the mid-fifth-century battles. Nobody's reconstruction of Arthur could have been responsible for all the dates that are attributed to him.

It is also curious that Gildas, writing at the time of Arthur's supposed sixth-century death does not mention his name at all, even though he does cover the British resurgence and the Battle of Badon Hill. If Arthur was active instead, around the mid-fifth century, this omission would not be so strange (Ashe 1995:8).

I myself doubt that the A.D. 537-542^{cal} date range is correct for the historical Arthur's actual demise, but I think that in consideration of the climatic event of A.D. 536^{cal}, there are very good *cultural* reasons for his death to be attached, centuries later in oral and literary traditions, to the period of the climatic disturbance. Remember that the circumstances of his death or disappearance were mysterious, free-floating, and available to be attached to a significant, meaningful event. Some said he never died, while centuries later in Wales some were still waiting for his return from France (Barber 1993:30; Taylor 1996:381-382).

There are actually three reasons that the A.D. 536 event particularly suits Arthur. The first is that the death of Celtic kings was traditionally tied to the appearance of comets and great storms. For example, Plutarch "After mentioning the visit of Demetrius to the sacred islands of the British coast, ...adds:"

> So after his arrival a great disturbance of the atmosphere took place, accompanied by many portents, by the winds bursting forth into hurricanes, and by fiery bolts falling. When it was over, the islanders said that some one of the mighty had passed away (Loomis 1927:212).

A climatic event on the scale of the one at A.D. 536^{cal} could only be explained in a Celtic context by the death of someone of Arthur's stature.

The second reason for attaching Arthur's death to the A.D. 536^{cal} event is that Celtic kings, as in many ancient cultures, were tied and ritually "married" to the land in such a way that the body and attributes of the king were reflected in the condition of the land itself (Loomis 1927:185; Rees and Rees 1961:74). For this reason kings in Ireland and Wales had to be physically perfect (Hutton 1991:172). Thus a man could eliminate a rival for the kingship by simply cutting of a finger or biting off a nose. As late at the tenth century in Wales, rivals for the throne were being blinded or castrated (Walker 1990:6). As an extrapolation of this idea, the fifth-century St. Patrick is recorded as demanding physical perfection as a necessary quality in a candidate for bishop (Charles-Edwards 1993:106-107). A healthy, just, and victorious king would be reflected in a fertile land and healthy subjects—or as the Irish said in describing the qualities of a rightful king, for him "the trees bend low with the weight of their fruit, the rivers and the sea team with abundance of fish, and the earth brings forth rich harvests" (MacCana 1983:119). An injured or maimed king in body or character would be reflected in a blighted, barren landscape (Loomis 1927:185). The classic example of this in Arthurian tradition is the maimed Fisher King whose realm will remain a waste land until he can be healed. Arthur himself is identified with the Fisher King as he lays wounded in Avalon awaiting healing (Loomis 1927:193-4). Geoffrey of Monmouth has Merlin explaining the wounding of the king and predicting his eventual return. So long as the king remains wounded, Merlin says that

> the stars would turn away from them...and face away from them and alter their wonted courses across the ecliptic. In their wrath the very stars would wither the standing fields of grain and forbid the rain to fall from heaven (Goodrich 1986:332).

This would not be a bad description of the climatic event and dry fog of A.D. 536^{cal}.

The third reason the A.D. 536^{cal} event was appropriate for the death of Arthur is that the event and its repercussions signaled the death of Roman Britain—the end of Camelot. Arthur was the foremost symbol of Roman Britain. Even his name, Artorius, was Roman. His reign was remembered as the golden age of the Britons—the British period of recovery between the revolt of the Saxons and the renewal of aggression after around A.D. 550^{cal}. After the mid-sixth century there were no more "kings of the Britons." By the seventh century, there were no more Britons—they were replaced by the Welsh, the Cornish, and the Bretons. Thus it was apt that his death be remembered as related to the catastrophes following A.D. 536^{cal}.

Integration, Linking and Complex Local Responses

[Britain, globe; interdisciplinary integration, multisourcing, analogies; secondary climatic effects, settlement patterns, spheres of social interaction; A.D. 536^{cal}]

A focus on the climatic event of A.D. 536^{cal} reveals the interrelationships among various aspects of Dark-age Britain that are often studied in isolation, and remain mysterious in that context. Logically, we know that past phenomena are interrelated, but it is difficult to find a perspective from which to grasp the complex whole of a past situation in a way that the links may be seen. The dramatic disruption caused by the event of A.D. 536^{cal} lays bare many of these interconnections and exposes links between folk legends, historical traditions, the archaeological record, settlement patterns, trade relationships, linguistic changes, and climate. The event and its aftermath provide plausible explanations for the disappearance of British culture in Anglo-Saxon areas, and for the contradiction between the dating of King Arthur's activities and his death. Furthermore, the event shows how the two problems are related to each other.

Additionally, this particular historical example seems to show that one of the most important causal links between dramatic climate changes and human society may be through the secondary effects of disease, and that the causal links are not simple or direct. As shown by the case of the Britons and Anglo-Saxons, groups living in the same environmental zone may be affected differentially depending on such things as their settlement patterns and spheres of social interaction. This is an important point to bear in mind when examining other historical contexts (e.g. see Young this volume for the effects of these same plagues on yet a third population, that of sixth-century Gaul), as well as in the current discussions of global warming.

ACKNOWLEDGMENTS: I would like to thank Joel Gunn for introducing me to the A.D. 536 "event" and for encouraging me to write a paper for A.D. 536 and Its Aftermath: The Years Without Summer symposium at the 1996 Southeastern Archaeological Conference.

References

Alcock, Leslie (1987) *Economy, Society and Warfare among the Britons and Saxons*. University of Wales Press, Cardiff.

Ashe, Geoffrey (1995) The Origins of the Arthurian Legend. *Arthuriana* 53:24.

Baillie, M.G.L. (1994) Dendrochronology Raises Questions about the Nature of the AD 536 Dustveil Event. *The Holocene* 4:212-217.

Baillie, M.G.L. (1995) Patrick, Comets and Christianity. *Emania* 13:69-78.

Baillie, Mike (1999) Exodus to Arthur: Catastrophic Encounters with Comets. B. T. Batsford, London.

Barber, Richard, editor and trans. (1993) *The Arthurian Legends*. Barnes and Noble, New York.

Campbell, James, Edited by (1982) *The Anglo-Saxons*. Phaidon, Oxford.

Campbell, Ewan, and Alan Lane (1993) Excavations at Longbury Bank, Dyfed, an Early Medieval Settlement in South Wales. *Medieval Archaeology* 37:15-77.

Charles Edwards, T.M. (1993) *Early Irish and Welsh Kinship*. Clarendon Press, Oxford.

Davies, John (1993) *A History of Wales*. Allen Land, The Penguin Press, London.

Davies, Wendy (1984) Picts, Scots and Britons. In *The Making of Britain: The Dark Ages*, edited by Lesley M. Smith, pp. 63-76. Macmillan, London.

Davis, K. Rutherford (1982) *Britons and Saxons: The Chiltern Region 400-700*. Phillimore, Chichester.

Ellis, Peter Berresford (1994) *Celt and Saxon: The Struggle for Britain AD 410-937*. Constable, London.

Falkus, Malcolm, and John Gillingham, eds. (1987) *Historical Atlas of Britain*. Crescent Books, New York.

Fulford, Michael G. (1989) Byzantium and Britain: A Mediterranean Perspective on Post-Roman Mediterranean Imports in Western Britain and Ireland. *Medieval Archaeology* 33:16.

Gilbert, Martin (1968) *Atlas of British History* Cartography by Arthur Banks. Dorset Press, London.

Gilchrist, Roberta (1988) A Reappraisal of Dinas Powys: Local Exchange and Specialized Livestock Production in 5th-7th-century Wales. *Medieval Archaeology* 32:50-62.

Goodrich, Norma Lorre (1986) *King Arthur*. Franklin Watts, New York.

Graham-Campbell, James (1980) The Celtic Contribution: Picts, Scots, Irish and Welsh. In *The Northern World: The History and Heritage of Northern Europe*, edited by David M. Wilson, pp. 95- 126. Harry N. Abrams, Inc., New York.

Hills, Catherine (1980) The Anglo-Saxon Settlement of England. In *The Northern World: The History and Heritage of Northern Europe*, edited by David M. Wilson, pp. 71-94. Harry N. Abrams, Inc., New York.

Hutton, Ronald (1991) The Pagan Religions of the Ancient British Isles: Their Nature and Legacy. Blackwell, Oxford.

James, Simon (1993) *The World of the Celts*. Thames and Hudson, London.

Keys, David (1999) Catastrophe: A Quest for the Origins of the Modern World. Ballantine Books, New York.

Linden, Eugene (1996) Global Fever. *Time* July 8:56-57.

Loomis, Roger Sherman (1927) *Celtic Myth and Arthurian Romance*. Columbia University Press, New York.

MacCana, Proinsias (1983) *Celtic Mythology*. Peter Bedrick Books, New York.

Morris, John (1996) *The Age of Arthur: A History of the British Isles from 350 to 650*. First published 1973 by Weidenfeld and Nicolson. Barnes & Noble Books, New York.

Pearce, Susan (1978) *The Kingdom of Dumnonia*. Lodenk Press, Padstow, Cornwall.

Rees, Alwyn, and Brinley Rees (1961) *Celtic Heritage: Ancient Tradition in Ireland and Wales*. Reprinted 1990. Thames and Hudson, London.

Snyder, Christopher A. (1996) *Sub-Roman Britain (AD 400-600)A Gazetteer of Sites*. BAR British Series 247. Tempus Reparatum, Archaeological and Historical Associates, Ltd., Oxford.

Taylor, Beverly (1996) Legends of Arthur's Return. In *The New Arthurian Encyclopedia*, edited by Norris J. Lacy, pp. 381-382. Revised edition. Garland Publishing, Inc., New York & London.

Thomas, Charles (1993) *Book of Tintagel: Arthur and Archaeology*. English Heritage Series. B.T. Batsford, Ltd., London

Thomas, Charles (1986) *Celtic Britain*. Thames and Hudson, London.

Todd, Malcolm (1984) After the Romans. In *The Making of Britain: The Dark Ages*, edited by Lesley M. Smith, pp. 21-33. Macmillan, London.

Wacher, John (1995) *The Towns of Roman Britain*. 2nd revised edition. B.T. Batsford, Ltd., London.

Walker, David (1990) *Medieval Wales*. Cambridge Medieval Textbooks Series. Cambridge University Press, Cambridge.

White, Roger (1991) Scrap or Substitutes: Roman Material in Anglo-Saxon Graves. In *Anglo- Saxon Cemeteries: A Reappraisal*, edited by Edmund Southworth, pp. 125-152. Proceedings of a conference held at Liverpool Museum 1986. Alan Sutton Publishing, Wolfeboro Falls, NH.

Whittock, Martyn J. (1986) *The Origins of England 410-600*. Croom Helm, London.

3. CLIMATE AND CRISIS IN SIXTH-CENTURY ITALY AND GAUL

Bailey K. Young

Department of History, Eastern Illinois University, Charleston, Illinois 61920

The Mediterranean Basin in the Sixth Century

[Mediterranean, Rome, Constantinople, Carthage; literature, historiography; politics, famine, population decline, dry fog, invasion, cultural change; sixth century A.D.]

A.D. 536[cal] was an ominous year, a year of decisive change for Rome and the Ostrogothic Kingdom of Italy to which it belonged. It began with the capture of Syracuse, chief city of Sicily, by the formidable Belisarius, the Byzantine general; he had recently destroyed the Vandal kingdom of Africa and reunited its provinces with the Eastern Roman Empire of Constantinople. Constantinople's young Emperor Justinian seemed bent on reviving the Roman Empire in the West. In July Belisarius captured Naples in southern Italy and pressed north. The Goths, who had ruled Italy for over a generation with at least the tacit agreement of Constantinople and with the active co-operation of the Roman aristocracy and their venerable institution the Senate, reacted by deposing King Theodohad (Wolfram 1988:337-342). His treacherous murder of his own wife Amalasuintha had provided the apparent justification for the Byzantine invasion. Although the new King Witiges promptly executed his predecessor, the war continued. Belisarius's army, which had easily occupied the Eternal City by December, found itself in turn besieged there in February A.D. 537[cal] by the Goth (Figure 3.1; Bury 1958 2:151-198).

Around that time the highest-ranking civilian administrator in Italy, the Praetorian Prefect Magnus Aurelius Cassiodorus Senator, sent one of his subordinates, Ambrosius, a letter that relates disorders in nature; as if Belisarius's were not enough to contend with, strange and disruptive phenomena were appearing in the environment. Cassiodorus Senator reminded Ambrosius that these and other changes in the political order come ultimately under God's providence.

> Those who survey the changed order of things are often troubled men, since the clearly unusual is frequently a portent. For nothing is done without a reason, nor is the world involved in fortuitous happenings; but all that we see brought to a conclusion must be the plan of God. When kings have changed their decrees, men are in suspense, lest things should go in a guise other than use has accustomed them to. But who will not be disturbed, and deeply curious about such events, if something mysterious and unusual seems to be coming on us from the stars? For, as there is a certain security in watching the seasons run on in their succession, so we are filled with deep curiosity, when we see that such things are changing.

Following this preamble, Cassiodorus Senator details current conditions that prompted these political and religious admonitions.

> How strange it is, I ask you, to see the principal star (the sun), and not its usual brightness; to gaze on the moon, glory of the night, at its full, but shorn of its natural splendor? All of us are still observing, as it were, a blue-coloured sun; we marvel at bodies which cast no mid-day shadow, and at that strength of intensest heat reaching extreme and dull tepidity. And this has not happened in the momentary loss of an eclipse, but has assuredly been going on equally through almost the whole year. How fearful it is, then, to endure for so long what will terrify a people, even when it passes quickly! So, we have had a winter without storms, spring without mildness, summer without heat. Whence can we now hope for mild weather, when the months that once ripened the crops have been deadly sick under the northern blasts? For what will give fertility, if the soil does not grow warm in summer? What will open the bud, if the parent tree does not absorb the rain? Out of all the elements, we find these two opposed to us: perpetual frost and unnatural drought. The seasons have changed by failing to change; and what used to be achieved by mingled rains cannot be gained from dryness only.
>
> And therefore, from the crops of the past, your prudence is to defeat the future dearth; for such was last year's fortunate abundance that provisions will also suffice for the coming months. Everything that is sought for as food must be put in store. The private person will easily find what he needs when the public supply system has filled itself.
>
> But, lest the present situation should be tormenting you with deep doubts, return to pondering the order of nature: what seems mysterious to the marveling crowd should be reasonable to you. For it has certainly been arranged by divine ordinance, the stars of this year have so met in their houses by joint operations, that winter is rendered drier and colder than its wont. Hence, the air, condensed from snow by excessive cold, is not thinned by the sun's fire; but it endures in the density that it has

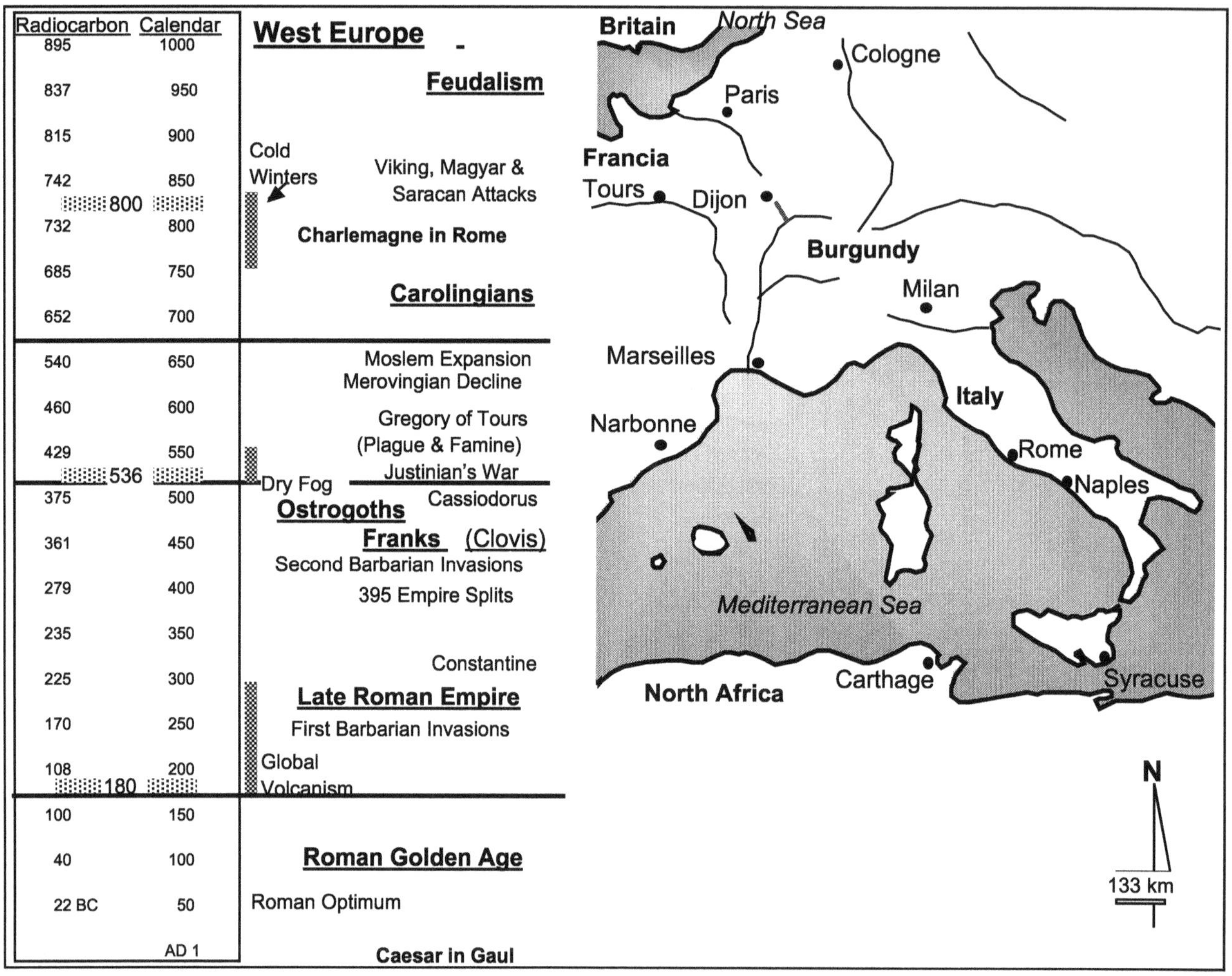

Figure 3.1. Map and Chronology of Italy and Gaul.

acquired, obstructs the heat of the sun, and cheats the gaze of human frailty. For things in mid space dominate our sight, and we can see through them only what the rarity of their substance allows. For this vast inane, which is spread between earth and heaven as the most tenuous element, allows us to see clearly as long as it is pure, and splashed with the sun's light. But, if it is condensed by some sort of mixture, then, as if with a kind of tautened skin, it permits neither the natural colours, nor the heat of the heavenly bodies to penetrate. In other ages, too, this has often temporarily happened with a cloudy sky. Hence it is that, for so long, the rays of the stars have been darkened with an unusual colour; that the harvester dreads the novel cold; that the fruits have hardened with the passage of time; that the grapes are bitter in their old age.

But, if this is to be ascribed to divine providence, we should not be troubled, since, by God's own command, we are forbidden to look for a sign (Matthew, xvi. 1-4). However, we understand that all this is plainly harming the fruits of the earth, when we fail to see our own customary foods nourished by their own natural law. Therefore, your care must see to it that one year's dearth does not throw us into confusion, since the first administrator of our office (Joseph) took care that past plenty should suffice to mitigate succeeding scarcity." (Book XII, Letter XXV, pp. 179-81, Barnish translation) (Barnish 1992)

The modern reader, accustomed to accounts of events that seek to keep "objectively observed facts" distinct from opinion and speculation, is likely to be puzzled by the elliptical and allusive nature of this document, and wonder if it has serious claims as real history or refers to something that really happened. Yet despite the highly "literary" nature of this text, it is in fact a bureaucratic directive from a top-ranking Cabinet officer of Rome to a major department head. Its author, Cassiodorus Senator, was a man as distinguished in his day for intellectual achievements as for his political career (O'Donnell 1939:28-52). He is recognized in our times as a major source of history and is here clearly discussing natural events both real and worrisome. Unusually cold weather and drought, continuing from winter through the growing season, have ruined crops and created famine. Furthermore, there has been persistent dry fog thick enough to eliminate shadows in daytime, and giving the full moon a blue appearance. These conditions have lasted almost an entire year. His letter shows Cassiodorus reacting as an administrator, ordering his deputy to alleviate the famine by distributing food kept in reserve under his authority. In accord with the universal assumptions and the literary conventions of the times, the administrator also writes as a Christian who sees supernatural causes in the natural events, and who seeks to reassure his subordinate, and through him the local elites under his influence to whom the Prefect's letter will be read, that these disturbing events in the natural realm are all part of God's mysterious plans; as well as, he holds, those in the political realm at which he can only hint.

A striking corroboration of Cassiodorus Senator's observations comes from the pen of the Byzantine official on the other side of the battle line, as it were: the historian Procopius of Caesarea, one of the principal sources for the history of the earlier sixth century (Bury 1958 2:417-430). Highly educated in the classical tradition, Procopius was serving on General Belisarius's staff, and in A.D. 536^{cal} was probably based in Carthage, the city recently recovered from the Vandals, as noted above. He describes what he witnessed in these terms:

> And it came about during this year that a most dread portent took place. For the sun gave forth its light without brightness, like the moon, during the whole year, and it seemed exceedingly like the sun in eclipse, for the beams it shed were not clear nor such as it is accustomed to shed. And from the time when this thing happened men were free neither from war nor pestilence nor any other thing leading to death. And it was the time when Justinian was in the tenth year of his reign. (Procopius: *History of the Wars* IV, xiv.36, 39-42)

Another source on A.D. 536^{cal} is Bishop John of Ephesus, who seems to be ascribing freakish cold and drought in Mesopotamia, and also testifies to a long year without normal sunlight: "The sun was dark and its darkness lasted for eighteen months; each day it shone for about four hours; and still this light was only a feeble shadow...the fruits did not ripen and the wine tasted like sour grapes." (Gill 1994:312)

Two sources are found from the city of Constantinople. One, identified as Lydus, holds: "The sun became dim in the course of the recently passed fourteenth indictim, for nearly a whole year, when Belisarius had the highest honor, so that the fruits were killed at an unreasonable time." The other ("an anonymous chronicler, usually associated by tradition with the name of Zacharias of Mytilene") says: "in the year fourteen...the Sun began to be darkened by day and the Moon by night...from the 24th of March in this year until the 24th of June in the following year." (Gill 1994:312-313)

Although there appear to be some chronological problems in reconciling these five sources, they substantially agree with one another on the conjoined phenomena of a darkened sun for well over a year, drought and famine. (Procopius refers to Justinian's tenth year of reign, the anonymous source to the year fourteen; Procopius's "whole year" becomes eighteen months in John of Ephesus.) A much later thirteenth century Jewish source, Bar Hebraeus reports: "And in the year eight hundred and forty-eight (A.D. 537^{cal}) there was a sign in the sun the like of which had never before appeared. The sun became dark and this darkness lasted for eighteen months. Each day the middle of the heaven shone with a shadowy light, and every man decided that (the sun) would never recover its full light. That year the fruits did not ripen and the wine tasted like urine." Bar Hebraeus's six hundred year perspective on A.D. 536-537^{cal} suggests that it was a rare or unique condition (Gill 1994:313).

These sources are cited by investigators of atmospheric phenomena from a number of disciplines. They include Stothers (1984), an atmospheric scientist, in his search for the source of the so-called "dry fog" phenomenon and its magnitude, Baillie (1994), a tree ring chronologist, who observed narrow growth rings in the years following A.D. 536^{cal} and the 15 years following, and Gill (1994), a student of Maya archaeology. Gill was searching for evidence to corroborate his thesis that a major volcanic eruption in A.D. 536^{cal} not only triggered droughts in Central America but had a major impact on global climate. It is of interest that these authors from various disciplines should draw attention to sixth-century written sources from the Mediterranean not hitherto interpreted in this sense by scholars of late antiquity. It will be the purpose of the rest of this chapter to provide more detailed context for understanding the crisis that was transforming the late Roman world, particularly Italy and Gaul, in the years surrounding and succeeding this possible "global climactic event" and so to consider what might be its implications for cultural change, the end of the classical world and the beginning of the Middle Ages.

Historical Perspectives on the "Decline and Fall of the Roman Empire"

[Western Mediterranean; historiography; Ancient-Medieval Transition, Gothic Wars; sixth century, late Roman empire, early Middle Ages]

The notion that the sixth century was a time of ominous change in the Roman west, particularly in its most central and best-known regions of Italy and Gaul, has been a truism among historians for at least two hundred years, ever since Edward Gibbon published his celebrated study *The Decline and Fall of the Roman Empire* (Gibbon 1909-1914, 1963). True, Gibbon was looking at a rather long perspective on a vast scale: he opens his story around A.D. 100^{cal} when the Roman Empire was at its height, and carries it forward as far as 1453^{cal}, when the city of Constantinople, the last remnant of the Byzantine Empire, fell to the Turk. Even if we narrow the problem to the collapse of the *Western Roman Empire*, serious problems develop as early as the third century with barbarian invasions compounded by the ravages of chronic political instability, economic crisis and civil war. A period of recovery followed in the fourth century; and finally, a new wave of invasions in the fifth century submerged the Imperial government in the west (A. H. M. Jones 1966).

Gibbon indeed declared the official "end" of the Roman empire in A.D. 476^{cal} when the last Emperor was deposed by Odoacar, the barbarian Master of Soldiers (Gibbon 1963:608-611). But, as the great Belgian historian Henri Pirenne argued (Pirenne 1955), this event was hardly a hallmark change; by then the Western emperor had long been a mere puppet, with real military and administrative authority shifted into the hands of Odoacar (in Italy) and other barbarian military chieftains in the other western provinces: Africa, Spain and Gaul (Musset 1975). Pirenne further argued that we must not be misled by mere political power shifts; instead, one should seek when and how the changes occurred that effectively brought an end to the ancient way of life along with its deeper social, cultural and economic institutions (Havinghurst 1958:18-31). From all appearances, the Roman empire weathered the barbarian invasions pretty well, Pirenne suggested, for the so-called "barbarians" were eager to become part of it and contribute their warlike prowess to keeping it intact.

An outstanding example of barbarian kings maintaining Roman institutions occurred during the period with which we are concerned; soon after the deposition of the last Western emperor in A.D. 476^{cal}, Theodoric, King of the Ostrogoths, took over Italy (Bullough 1965:157-164; Wolfram 1988:278-290). He claimed to be the legal and official agent of the eastern Emperor in Constantinople; he showed great respect for the old Roman Senate, which continued to meet and vote laws as it had for a thousand years; he enforced Roman laws and governed with the help of highly qualified Roman officials (Wolfram 1988:286-327; Hodgkin 1923). These included Cassiodorus Senator's father, who served as Theodoric's Praetorian Prefect from ca. A.D. 503-507^{cal}, and Cassiodorus himself, who moved up the career ladder from Quaestor to Master of Offices under the old king. Theodoric's administrative and diplomatic correspondence, much of it drafted and edited by Cassiodorus, shows the Ostrogothic regime maintaining the roads, the water system, and other public infrastructure, and asserting leadership amongst the various post-Roman regimes of the West through the traditional diplomatic techniques of marriage, alliance, and gift-exchange (Barnish 1992; L. W. Jones 1966). Theodoric also built for himself in his capital of Ravenna a splendid mausoleum in the style and manner of the highest Roman aristocracy; it still stands today. Recent research on other barbarian rulers, such as the Burgundian Gunthchramn (Geary 1988:121-122) in Gaul and, most strikingly, the Frankish Clovis, Theodoric's greatest rival in the west (Rouche 1996), confirms Pirenne's basic contention that these men saw themselves as part of an ongoing Roman Empire whose amenities and institutions they did their best to preserve, and they did so successfully during the first quarter of the sixth century.

Recent researchers also tend to agree that, these pro-Roman sentiments and policies notwithstanding, in the sixth century, things went very wrong for Roman amenities and institutions in the West, so wrong that a very different type of world came to replace it. The term "feudal" is generally applied to this world where life tended to be nasty, brutish, and rural in character. Effective political power devolved upon a noble and warrior class that was predominantly local, or at best regional, in its power base and outlook.

The reasons for the marked change in the sixth century, and even its chronology, remain swbjects of dispute. Pirenne himself thought the major reason for the change was the expansion of Islam (Pirenne 1955:140-150). Islam was then a new prophetic religion that seized control of the Mediterranean Sea during the eighth century. This ended most long-distance trade and effectively isolated the old Roman west from Constantinople and the other civilized cities of the east. For Pirenne, then, the ultimate causes of change were religious, political and economic. However, this part of his thesis has been widely questioned by historians. By the eighth century, most early medieval scholars would agree, the ancient world, in most if not all of its essential aspects, had already ended. A few years ago I took part in a symposium that examined the seventh century as the critical time of rupture with Antiquity. Among the signs of this rupture were the decline of effective monarchical power and the rise of vigorous and contentious local and regional aristocracies; the decay of Roman towns and the creation of vast rural monasteries; the breakdown of Mediterranean trade and the development of new trading networks around the North sea. It is clear that this seventh-century world, in Gaul and Italy at any rate, was not just a weakened version of the Late Antique civilization in which Romanized barbarians like Theodoric and Clovis had come to power in the late fifth century: European society was already fundamentally different by the seventh century (Brown 1996).

With Antique ways clearly surviving in the early fifth century, and already gone by the seventh century, the sixth century is clearly marked for closer scrutiny. Can we find evidence therein of changes that are violent and far-reaching enough to have fundamentally disrupted the structures of Late Antique civilization? The key question in regard to the issues of this book is: were late Antique social, political and economic structures still fundamentally intact on the eve of A.D. 536^{cal}?

There is no lack of evidence for growing violence. Warfare was at its most brutal and devastating. As the opening of this chapter stresses, the year A.D. 536^{cal} was indeed pivotal for Italy. Belisarius's successful capture of Syracuse, Naples and Rome in that year launched a full-scale assault on the Ostrogothic successors of Theodoric. The resulting so-called "Gothic Wars" went on for almost 20 years and turned much of Italy—which hitherto had been little affected by the civil wars, power struggles and invasions of the past two centuries—into ruins and wasteland (Bury 1996:180-270). The city of Rome was besieged and changed hands four times. The resulting devastation was so great that the Roman Senate, after a millennium, was finally dissolved; its members abandoned the city for safer rural estates as far away as Sicily. By mid-century the former metropolis, which had boasted a population of a million in its Imperial heyday, was mostly ruins interspersed with some fields and gardens; they supported the remaining 10,000 or so inhabitants (Krautheimer 1980; Llewellyn 1971). Most of these were connected to the churches that were becoming the main reason for any urban survival at all in the west. This was truly a change from times when ships regularly docked at Ostia bringing loads of grain from the distant breadbaskets of Egypt, the littoral of North Africa and Sicily to feed Rome's populations.

Historical sources, which it must be admitted are both vague and scanty, suggest that not much of the administrative and technical infrastructure of Rome and the other towns of Italy survived into the second half of the sixth century. What they do show is that Bishops, hitherto figures concerned mostly with religion and religious property, were stepping into a power void. Bishops assumed such responsibilities as poor relief, local defense, or negotiating with Imperial officials when they did happen to show up—in other words, they were exercising effective local power. In the case of Rome, this state of affairs was rendered all the more necessary by the ultimate failure of Emperor Justinian's grand strategy, which had been to re-establish direct Imperial political control over Italy and the West. Justinian's generals did succeed in eliminating the Ostrogoths by A.D. 550^{cal}. This done, they turned to see a new group of barbarians, the Lombards, invade from over the Alps in A.D. 568^{cal} (Bullough 1965:165-174). The Lombards were Germans and far less Romanized than the Ostrogoths had been. They were mostly pagans, and they followed warband chiefs instead of a king whose authority based on that of a Roman general. They did extensive damage to whatever urban infrastructure and rural landscapes had been spared in the Gothic wars. The Byzantines succeeded in holding the seacoasts and some major towns, but the Lombards seized and held much of the north and the mountainous spine of Italy. War was henceforth endemic. Thus was created the state of political and cultural division, animated by frequent warfare, that became the Italian way of life for the succeeding thousand years (Hodgkin 1896).

Gaul in the Sixth Century

[France, Burgundy, Brittany, Gaul; historiography; Gregory of Tours (A.D. 539-594^{cal}), Clovis, Franks, prodigies, cold, ice, plagues, famine; Merovingians]

North of the Alps in Gaul, the year A.D. 536^{cal} also witnessed the beginning of sweeping political changes --but less apparently catastrophic than in the Mediterranean. Between A.D. 481^{cal} and 511^{cal} the Merovingian Clovis had forged the fractious Frankish war bands into an effective monarchy; he gained the support of the Gallo-Roman aristocracy when he converted to their Catholic religion instead of the Arian heresy of the Goths (Rouche 1996). This alliance helped him consolidate his power over three quarters of modern France when he defeated the Visigoths and expelled them from Aquitaine in A.D. 507^{cal} (Geary 1988:82-88; Wood 1994). However, the Ostrogothic forces of Theodoric checked any further expansion southward at that time. The Byzantine attack on the Ostrogoths was thus a golden opportunity for Clovis's grandsons: in A.D. 536^{cal} they successfully attacked the Burgundian kingdom (an Ostrogothic protectorate) and seized Provence, effectively bringing all of Gaul under Merovingian Frankish rule. What is more, as archaeological studies now make clear, they proceeded to extend Frankish hegemony over the Germanic peoples east of the Rhine River (Thuringians, Alemani) and north of the Danube River (Bavarians) (Geary 1988:117-119; James 1982, 1988). Certainly this Frankish success, playing against the growing paralysis of Italy, was at the origin of the long-term shift in political, economic and cultural power from the Mediterranean basin of Roman days to the northwest Atlantic seaboard and Rhineland of the Middle Ages. Could the impact of cataclysmic environmental events felt in the aftermath of A.D. 536^{cal} have played a role in hastening the decline of surviving late Antique cultural institutions in Gaul? Did they contribute to the growth of new—more rural, local, and "feudal"—institutions to replace them?

Were it not for the writings of Gregory of Tours, there would be little an historian could say to such a question. There are some written sources for the sixth century and they do suggest that great changes were taking place. However, these documents are much concerned with wars, murders and power struggles, with the lives of saints and their miracles, with decisions made by churchmen when they met in synods to discuss church policy and, to a lesser extent, with property rights, particularly when these were transferred to churches that kept written records; some of these records managed to survive the long centuries ahead. Early medieval written sources for western Europe are as a rule extremely selective in regard to the kind of information they preserve. They recall A.D. 536^{cal} as a year when political events of obvious consequence, immediate and long-lasting, occurred. They don't mention cataclysmic natural events. Does this mean that none occurred?

Who was this Gregory of Tours? He was born probably in A.D. 539^{cal} in the town today called Clermont-Ferrand in the mountainous center of France (Gregory of Tours 1974:7-16). He came from a distinguished aristocratic Gallo-Roman family that claimed an ancestor among the earliest converts to Christianity. One of these was a man who had indeed been thrown to the lions in the days of persecution. In recent times the family counted several high-ranking church men and women among its members. Gregory was educated to join them, and in A.D. 573^{cal} he succeeded his uncle as Bishop of Tours, a town on the Loire River. Tours's greatest claim to fame was the grave of an earlier Bishop, Martin (died A.D. 397^{cal}), the missionary saint who had led the conversion of many people in Gaul to Christianity two centuries earlier. Miracles often occurred at Martin's tomb, attracting many pilgrims from near and far, from peasants to poets to Frankish royalty. The great Clovis had honored the saint's tomb after his victory over the Visigoths, and his widow had spent her years of retirement near it. As Bishop, Gregory was an avid promoter of the cult of Saint Martin. His activities are documented because he was a prolific writer, composing books that told the story of saints and their miracles, and books of history that have become our most detailed and complex source material for Gaul during the sixth century. Although Gregory's ten books of history came to be known as the *History of the Franks* as early as the tenth century, his own conception—as historians are only just now realizing—was that of a Christian moralist, relating events to help explain God's providence and human behavior, or more often, misbehavior (Goffart 1988:112-234). Although some of the events he relates took place in the distant past, most of them occurred during his lifetime, or not long before. Gregory probably died in A.D. 594^{cal}. He narrates many stories in which he was directly involved or of which he had direct testimony. Among the kinds of events he mentions are natural disasters, or prodigies (Gregory of Tours 1974:50-52). None of these appear to related directly to the year A.D. 536^{cal}, so all must be considered a part of the over half century following. Some may be related to A.D. 536^{cal} or its ramifications. For example, Gregory relates that the first ten years of his tenure as Bishop were consumed by the rebuilding of burned churches (Gregory of Tours 1974:601-602). One has to ask if this was the result of ongoing conditions of the time, the early churches being made of wood, or if it was a result of a condition, perhaps unrest following on the heels of A.D. 536^{cal}. Future research should be directed at resolving this question.

Gregory records many prodigies and their dates; much of his work is almost conversational in nature and natural phenomena fit well with his penchant for moralizing. For example, one night in Paris in A.D. 578^{cal} he saw some twenty rays of light in the night sky, starting in the east and moving around to the west. "One was longer than the others and shone high above them: it reached right up into the sky and then disappeared, and the others faded away too. In my opinion they were a presage of Merovech's death..." (Gregory of Tours 1974:282). Here Gregory is alluding to the outcome of one of the civil wars within the Merovingian royal family, and echoing the assumption noted above in Cassiodorus's letter: that unusual events in the natural world reflect something, usually disastrous, happening in human history.

Many of Gregory's observations were astronomical in nature. Eight years later (A.D. 586^{cal}) during a trip to the Ardennes forest in eastern France, Gregory was awestruck by "rays of light toward the north, making the clouds blood red on both sides"(Gregory of Tours 1974:449). He watches in wonder as the sky fills up with them in all four quarters, making him think of the striped pavilions of a military tent. Later he hears about two islands in the sea that were set alight by fire from heaven around that time, a fire said to have burned for seven days, destroying everything, and he wonders if what he had seen in the sky had not been reflections of this conflagration. Elsewhere he reports fiery globes crossing the night sky. "So bright a light illumined a wide spread of lands in the middle of the night"—he writes in A.D. 590^{cal} —"that you would have thought it was high noon." (Gregory of Tours 1974:581) For Gregory such astronomical events are portents, signs from Heaven of some momentous event taking place or about to occur. This might be a natural disaster: for example, in Auvergne in A.D. 571^{cal} there was a three quarters eclipse of the sun after which "a star, which some call a comet, appeared over the region for a whole year, with a tail like a sword..." and at this time the region was ravaged by a deadly plague (Gregory of Tours 1974:226). Often the portended event is political: in A.D. 561^{cal}, for example, King Lothar's death was announced by lightening flashes (Gregory of Tours 1974:247). When in A.D. 585^{cal} Gregory saw "a great beacon traversing the heavens before dawn and in the north a column of fire hung in the sky for two hours, with a great star perched atop it", he took this for signs announcing the defeat and death of the pretender Gundovald, who had been trying to set himself up as a king in Aquitaine, far to the south. There was an earthquake too, for good measure (Gregory of Tours 1974:395). Portents aside, such phenomena could be examined by astronomers for relations to the A.D. 536^{cal} aftermath; Baillie reports as numbers of questions about the timing of the A.D. 536^{cal} and proposes that there may have been several events. If they continued through the century, Gregory may have been observer to them.

Since Gregory was born after A.D. 536^{cal} and was a mature, writing adult after the immediate consequences of A.D. 536^{cal} had faded, the information he supplies pertains to the first century of aftermath. A number of his observation suggest out-of-the-ordinary conditions. Besides the earthquake of 585^{cal} (and another one five years later) he mentions droughts, thunderstorms, rains so excessive as to provoke flooding, and excessive cold. In A.D. 548^{cal}, for example, "the mountain torrents were frozen solid and people walked across them as if they were on dry ground. (Gregory would have been nine years old.) The snow lay deep and the birds were numbed with cold and famished with hunger, so that they could be taken by hand without any need of snares." (Gregory of Tours 1974:193) In A.D. 582^{cal} wolves entered the town of Bordeaux, showing no fear of humans, and ate the dogs (Gregory of Tours 1974:193). There were plagues of locusts that destroy harvests. There were elder-trees producing grapes instead of berries, roses blooming in January, fruit trees blossoming with a new, second crop late in the fall (Gregory of Tours 1974:51-52).

Before modern transportation systems became universal in Europe, famine was a common occurrence, particularly in conjunction with war. Gregory's enumerations of troubled climate need to be weighted against those of other centuries to determine if their frequency of occurrence was unusual. Elizabeth Jones (see Jones chapter) is beginning a dissertation on demography in Burgundy since the sixteenth century. Since this is one of the most difficult periods of the Little Ice Ages Minima, the results of her studies could be compared with the sixth century.

Of particular interest for understanding the aftermath, I think, is the evidence of hunger and disease. In the year A.D. 585^{cal}, Gregory notes: "almost the whole of Gaul suffered from famine. Many people made bread out of grape pips or hazel catkins, while others dried the roots of ferns, ground them to powder and added a little flour...Many others, who had no flour at all, gathered grasses and ate them, with the result that they swelled up and died...The poor sold themselves into slavery to obtain something to eat." (Gregory of Tours 1974:427) I alluded earlier to the great plague that ravaged Auvergne in A.D. 571^{cal}. Bodies were so numerous as to be beyond counting, Gregory says, but tells us that some 300 were counted on one Sunday in St. Peter's church, and there was a shortage of tombstones and coffins. In A.D. 582^{cal} an epidemic ravaged Narbonne, near the Mediterranean coast; after three years it came back, and this time spread north to the Albi region and beyond (Gregory of Tours 1974:346). Indeed it seems to have become endemic, for in A.D. 590^{cal} Viviers and Avignon, in the Rhone Valley to the east, were attacked, and the very last entry in Gregory's chronicle, dated April A.D. 591^{cal}, shows the plague killing people in his own town of Tours in the Loire Valley, and striking Nantes down river to the west as well. This entry concludes on a generally somber note: "the epidemic killed wild animals as well as domestic cattle. Throughout the forest glades a great number of stags and other beasts were found lying dead in places difficult of access. The hay was destroyed by incessant rains...Acorns grew, but never ripened..." (Gregory of Tours 1974:593)

There were also some decidedly unnatural events reported—suitable for Halloween. In A.D. 582^{cal} in the Paris region "real blood rained down from a cloud, falling on the clothes of quite a number of people and so staining them with gore that they stripped them off in horror..." (Gregory of Tours

1974:346) In A.D. 586^{cal}, on an island off the coast of Brittany, the water in a fishpool turned to blood a yard deep, so that "day after day vast packs of dogs and flocks of birds assembled there to drink, returning home at nightfall completely sated." (Gregory of Tours 1974:455) In A.D. 587^{cal} Gregory reports "Some said they had seen snakes drop from the clouds. Others maintained that an entire village had vanished into thin air, taking the houses and the men who lived in them...." (Gregory of Tours 1974:483) These reports need to be examined for credibility, and if they are, understanding made of their unusual nature, perhaps having to do with the A.D. 536^{cal} aftermath.

Gregory of Tours paints the picture of a world under attack, a world in decline. How reliable is this picture? For a long time he was regarded as an unsophisticated, even naive writer, himself the product of the decline of Roman culture, expressing himself in a crude and direct Latin style because he didn't know how to write with the refinement of a Cassiodorus—as though he were some Gallo-Roman Sergeant Joe Friday earnestly conveying "just the facts" as he happened to see or hear them. Recent re-evaluations depict Gregory as a much more complex and subtle author, who chooses his material carefully to illuminate a moralizing and theological vision of a world that is a battleground between Good and Evil, a place where Evil—thanks to Original Sin—often wins (Goffart 1988:112-234). Such a view of Gregory need not impeach, I would argue, the credibility of his observations of the world around him. There is no reason not to believe him when he describes natural events or astronomical phenomena he himself witnessed (when he reports what others have said, it is another matter). His powers of precise and vivid description make it a great shame, from the viewpoint of this book that he was born shortly *after* A.D. 536^{cal}, and that most of his book deals with events in the A.D. 570s and 580s^{cal}, when he was the mature and active Bishop of Tours. However, he was writing of the years subsequent to the A.D. 536 event. Presumably these aftermath years, judging by Gregory's account witnessed a continuing unfolding of negative environmental problems.

Cassiodorus, Gregory, and the End of Antiquity

[Italy, Gaul; historiography, archaeology; Late Roman Empire-Early Middle Ages Transition, Late Roman construction, Late Roman trade, macrocosm; A.D. 536^{cal}, sixth century]

We have thus come, at the very end of our reflections, to the question of the historical sciences and the contribution they might make to the problem defined for this book: the possible impact, short and longer term, on human societies of cataclysmic interludes, such as the postulated A.D. 536 "climatic event." Thus far, there has been no contribution, for the problem itself had not been identified. Let me sum up the situation as it looks, I would argue, from the viewpoint of the literary sources and the historiographical traditions since the time of Edward Gibbon. Until the A.D. 530s^{cal}, the structures of late Roman civilization in Italy and much of Gaul seem to have remained relatively vigorous, despite more than a century of periodic instability and frequent warfare. Literary sources show so-called "barbarian" kings like Theodoric the Ostrogoth, Gunthram the Burgundian, or Clovis the Frank operating more like late Roman generals and political leaders than like barbarian invaders. They did so with the help highly cultured Roman aristocrats like Cassiodorus Senator.

Extensive urban excavations and related research over the past thirty years in Gallic towns like Geneva, Lyons, Arles, Autun, Rouen, and Grenoble indicate that ambitious, complex and highly decorated new religious buildings were renewing and transforming these urban environments from the time of the first Christian emperors in the fourth century through the days of these "post-Roman" monarchs in the fifth century (Young forthcoming; Loseby 1996). Artifacts from excavations also show that the port of Marseille was a flourishing center for the import and diffusion of products from North Africa and the eastern Mediterranean during this period (Loseby 1992). There is further evidence that some of this activity continued in the days of Gregory of Tours in spite of the mid-sixth century conflagration in Italy. Gregory himself, for example, was a great embellisher of the religious monuments he inherited in the city of Tours. Most notable was the great pilgrimage basilica of Saint Martin, which was a century old then, and he built a number of churches in the smaller towns of his diocese (Gregory of Tours 1974:601-602). He also confirms long-distance commercial activity in the port of Marseille, and refers to merchant communities in such other towns as Bordeaux, Paris and Orleans.

But what are we to make of Gregory's overwhelmingly negative view of his own age? Some of the evidence I have discussed earlier, that of the famines and the epidemics in particular, suggests that his insistence that he is living in a world in decline is not just the *parti pris* of a moralizing religious pessimist. One can of course blame wars for this sad state of affairs—and Gregory does so, castigating the Merovingian rulers of his day for their incessant attacks on one another, and detailing the bloody acts of violence committed by their agents and by the land owning magnates on their own (Gregory of Tours 1974:253-254, 403). But these wars in Gaul hardly seem as endemic and as destructive as the Gothic Wars and the succeeding Lombard-Byzantine standoff, which so clearly ended ancient civilization in Italy at that time. Perhaps in drawing attention to destructive weather, hunger, earthquakes and alarming heavenly portents Gregory is trying to put his finger on visible signs of macrocosmic changes whose natural causes no man in his day had the means to understand, though they felt their effects.

It is not surprising that Gregory would attribute the ultimate causes to the moral and theological realm where, in the traditions he and Cassiodorus shared with all other educated men of the day, ultimate reality was lodged. Later students of history who did not share those assumptions, from Gibbon's day to our own, have proposed explanations for the those vast transformations of European culture labeled "the decline and fall of the Roman Empire" drawing on political events, economic developments, social conflicts, and cultural transformations (Hodges and Whitehouse 1983; White 1966; Brown 1996). The evidence for a "global climactic event" in A.D. 536^{cal} adds a macrocosmic perspective that seems to integrate well with the problem as defined from these other traditions. Elizabeth Jones' (see Jones chapter) interesting suggestion that pestilence may have played a significant part in weakening the hitherto strong resistance of Roman Britons associated with the Arthur tradition to Anglo-Saxons, might fit well, for example, with evidence of the spread of plague in Justinian's empire in the 540s^{cal} and in the Gaul of Gregory of Tours, and a link between a macrocosmic event and plague might be demonstrable. Clearly, students of late Antique and early Medieval European history need to consider these possibilities seriously.

References

Baillie, M. G. L. (1994) Dendrochronology Raises Question about the Nature of the AD 536 Dust-Veil Event. *The Holocene* 3:212-217.

Barnish, S.J.B. (1992) *The Variae of Magnus Aurelius Cassiodorus Senator.* Translated Texts for Historians 12, Liverpool University Press, Liverpool.

Brown, Peter (1996) *The Rise of Western Christendom. Triumph and Diversity* AD 200-1000, Blackwell Press, Oxford.

Bullough, Donald (1965) Germanic Italy: the Ostrogothic and Lombard Kingdoms. In *The Dawn of European Civilization*, edited by David Talbot Rice, pp 157-74, McGraw-Hill, New York.

Bury, J. B. (1958) History of the Later Roman Empire, from the Death of Theodosis I to the Death of Justinian. 2 vols, Dover, New York.

Geary, Patrick (1988) Before France and Germany. The Creation and Transformation of the Merovingian World. Oxford University Press, Oxford.

Gibbon, Edward (1909-1914) *The Decline and Fall of the Roman Empire.* Edition of J.B. Bury, Methuen, London. Also, abridged version by Frank C. Bourne (1963) Dell Paperback, New York.

Gill, R.B. (1994) *The Great Maya Droughts.* Unpublished Ph.D. dissertation, the University of Texas at Austin.

Goffart, Walter (1988) *The Narrators of Barbarian History (AD 550-800).* Princeton University Press, Princeton.

Gregory of Tours (1974), *The History of the Franks.* Translated with an Introduction by Lewis Thorpe, Penguin, Baltimore.

Havinghurst, A.F. (1958) *The Pirenne Thesis.* Holt, Rinehart &Winston, Boston.

Hodges, Richard (1982) Dark Age Economics: the Origins of Towns and Trade A.D. 600-1000. St. Martin's Press, New York.

Hodges, Richard and David Whitehouse (1983) *Mohammad, Charlemagne and the Origins of Europe.* Cornell University Press, Ithica.

Hodgkin, T. (1896) *Italy and Her Invaders.* Second edition, Oxford University Press, Oxford.

Hodgkin, T. (1923) *Theodoric the Goth.* Second edition, MacMillan, London.

James, Edward (1982) The Origins of France. From Clovis to Charlemagne. MacMillan, London.

James, Edward (1988) *The Franks.* Oxford University Press, Oxford.

Jones, A.H.M. (1964) *The Late Roman Empire, 284-602*, 4 vols. Oxford University Press, Oxford.

Jones, A.H.M. (1966) *The Decline of the Ancient World.* Longman Press, London.

Jones, Leslie W. (1966) An Introduction to the Divine and Human Readings by Cassiodorus Senator. Octogon Books, New York.

Krautheimer, Richard (1980) *Rome: Profile of a City, 312-1308.* Princeton University Press, Princeton.

Llewellyn, Peter (1971) *Rome in the Dark Ages.* Praeger, New York.

Loseby, Simon (1992) Marseille: a Late Antique Success Story? *Journal of Roman Studies* 82:165-185.

Loseby, Simon (1996) Arles in Late Antiquity: Gallula Roma Arelas and Urbs Genesii. In Town in Trasition. Urban Evolution in Late Antiquity and the Early Middle Ages, pp. 45-70, Scholar Press, Aldershot, England.

Lyon, Bryce (1974) *Henri Pirenne—a Biographical and Intellectual Study.* E. Story Scientia, Ghent, Belgium.

Musset, Lucien (1975) *The Germanic Invasions.* Pennsylvania State University Press, College Park, Pennsylvania.

O'Donnell, J. J. (1979) *Cassiodorus.* University of California Press, Berkeley.

Pirenne, Henri (1955) *Mohammed and Charlemagne.* Barnes and Noble, New York.

Procopius of Caesarea (1971) *History of the Wars, Secret History and Buildings.* 6 vols, London, Heineman.

Rouche, Michel (1996) *Clovis.* Fayard, Paris.

Stothers, R. B. (1984) Mystery cloud of Ad 536. *Nature* 307(5949):344-345.

White, Lynn Jr. (1966) *The Transformation of the Roman World* University of California Press, Berkeley and Los Angeles.

Wolfram, Herwig (1988) *History of the Goths.* Translated by T. Dunlap, University of California at Berkeley.

Wood, I. N. (1994) *The Merovingian Kingdoms.* Longmans, New York.

Young, B.K. (forthcoming) Medieval France, in *The Garland Encyclopedia of Medieval Archaeology.* Garland Press, New York.

II. CHANGES IN POWER: ECONOMIC AND MILITARY BALANCES IN THE MAYA LOWLANDS

The Maya lowlands have recently emerged as historic cultures because the code of Maya writing has finally been cracked. However, direct evidence of unusual climate circumstances such as the dry fog accounts of Afro-Eurasia remain to be identified. Nevertheless, the era following A.D. 536^{cal} is accompanied by general unrest and notable changes in political balances among regional states. The chapters in this concept set explore the character and global-regional linkages made evident by newly found history and a long tradition of archaeological exploration in the Maya lowlands. The examination of the A.D. 536 question is made from the perspective of two prominent cities, Tikal and Caracol, whose fortunes reversed over the course of the aftermath, more than once. The underlying issues appear to be similar to those of shift in power loci in Europe discussed in the previous chapters.

Tikal is on the easiest portage across the Yucatan Peninsula between the Usumacinta and Hondo river systems, and therefore its reason d'être and the reason why other cities were willing to challenge Tikal. Hugh Robichaux's chapter stresses the global conditions and broader Mesoamerican issues, primarily from a Tikal perspective. In keeping with the structure of the book, I thought it would be helpful to have a chapter from the Caracol perspective. The Tikal perspective represents that of a regional trading city across whose history fell the heavy hand of Teotihuacán, possibly the largest middle first millennium city in the world, and the usual accouterments of trade and elite wealth, which include hyperhierarchical social organization. In the apparent aftermath of the A.D. 536 event, Caracol, in league with another prominent competitor of Tikal, Calakmul, surged to prominence. Arlen and Diane Chase find through their years of work at Caracol a more distributed economic and social order.

The rise of Caracol to prominence poses questions of why, during the aftermath episode, are apparently variant social and economic forms advantaged, forms that contrast with those prevalent during the previous centuries of Tikal prominence? Was a distributed cultural system of greater use in a world of less stable environment? Diminished climatic stability seems to be one of the characteristics of a cooled world, the thermodynamic consequences of less atmospheric moisture. The historic accounts from Europe could be interpreted as such. In western Europe, the transition from antiquity to the modern era has been argued about since the 1930s. Rather by a process of elimination (see Young's chapter), the sixth century has fallen out as the most likely pivotal moment. The train of events in China could be interpreted as representing similar implications (see Houston's chapter). In the Maya lowland, the situation is in many regards the opposite one. Rather than a vast transitional moment, A.D. 536^{cal} seems to appear when those cultures were on a growth trajectory. However, there are distinctive differences between the Early and Late Classic periods that are clearly observable both at Tikal and Caracol, as is thoroughly document in the following chapters. A.D. 536 appears more as a pause than a full stop.

Hugh Robichaux is interested in the fact that the year A.D. 536^{cal} initiates the so-called Early-Late Classic Hiatus in the Maya Lowlands. Hugh was a co-author with me on an article in *Geoarchaeology* (Jan95) on Maya Lowland climate change. The A.D. 536 story broke just in time to prevent us from making some unwarranted statements about the Hiatus. Robichaux's chapter makes convincing arguments for the interactions of environmental degradation, warfare, and climate. I especially like the two-gap hypothesis he poses as characterizing life in Tikal during the Hiatus. Also at issue is Tikal's relationship with Teotihuacán. Those working at Teotihuacán report that there are data being analyzed that indicate the center of the city burned around A.D. 550^{c14}. If this is true, it strengthens Robichaux's case by pointing to the conditions under which Teotihuacán's influence at Tikal waned.

The revelation of Caracol as an economic giant through the work of Arlen and Diane Chase casts the southern Maya lowlands into a quite interesting pattern relative to A.D. 536^{cal} and its 300-year aftermath. A.D. 536^{cal} was one of the occasional years-without-summer that has global scale ramifications for both environmental and socio-cultural conditions worldwide. The Caracol discussion involves presenting the alternative conditions that appear to have followed the year A.D. 536^{cal}, especially distributed social and economic forms. It consists of a discussion of supporting information, including the internal causeway system within the city of Caracol that suggests the bones of the city's dispersed economic structure. This is contrasted with an alternative structure in Tikal. During the aftermath, Tikal's preeminence in Maya lowland power structure falls to that of Caracol.

Robichaux offers a hypothesis for the Early–Late Classic Hiatus, first noticed as a cessation of stela erection at Tikal. The phenomenon, however, appears to be much more broadly observable than simply at Tikal. Robichaux incorporates interaction between Tikal and Caracol, a neighboring powerful regional state, with modeled and observed effects of the A.D. 536 event, to arrive at a more comprehensive explanation of the Hiatus than has heretofore been possible. The modeling is accomplished by analysis of modern river discharge data. Drought detection through the use of lake cores is also used to control environmental factors in the Maya Lowland milieu. Robichaux concludes that the A.D. 536 event could have been a significant factor in disturbing the ongoing balance between Caracol and Tikal.

Robichaux's double hiatus hypothesis, like Jones' selective epidemics hypothesis, lays claim to validity through parsimonious explanation of events that adds serendipitous understanding to current wisdom. By incorporating global climate effects, he demonstrates that a previously recognized but little understood double gap in stela erection at Tikal could be very significant in illuminating the evolution of state formation and political dominance in the Maya Lowlands.

The Maya lowland chapters carry the weight of discussion about social order and global disturbances, in these cases, a contrast between distributed and hierarchical societies. Crumley (1993) has examined this question in Europe for some time by comparing Roman and Celtic economies. The Mesoamerica case studies push the argument further in a culturally independent context from the Eurasian examples, both in Crumley's work and in the Young and Jones chapters in this book. The reader can reflect on the implications of the European data through the catalyst of the Tikal-Caracol contrast. Because of a system of causeways and their ritual and political ramifications, Mesoamerica bears the heterarchy-hierarchy argument more efficiently than Eurasia because of the highly structured and iconic nature of the Maya culture. The causeways become a landscape hieroglyph that is lacking in Europe after the collapse of the Roman road system.

Global Change Assessment. There is a rich variety of lessons to be learned from these two intensive studies of Maya Lowland interactions with the A.D. 536 event. Perhaps the most striking one is that a variety of cultural systems help to insure that ongoing human life in a region. The conditions of A.D. 536^{cal} have not escaped comparison with the more evident tragedy of the grand Maya lowland collapse of the ninth century and do not in the volume (see Chase's chapter). Could it be that one of the contributing factors that made recovery from an atmospheric disturbance in the sixth century, as opposed to the cataclysm of the ninth, was a difference of cultural heterogeneity?

Looking more closely at the consequences of the A.D. 536 event, suggests that societies with more distributed social systems and economies function more efficiently during times of environmental instability. Caracol has signatures of both heterarchy, a broadly enjoyed economy, and hierarchy, a solar causeway pattern. Perhaps it was in this ambiguity that propelled Caracol during the so-called Maya Hiatus.

Crumley, Carole L.
1993 Analyzing Historic Ecotonal Shifts. *Ecological Applications* 3:377-384.

4. THE MAYA HIATUS AND THE A.D. 536 ATMOSPHERIC EVENT

Hubert R. Robichaux

The University of the Incarnate Word, San Antonio, Texas

Tikal Regional State and the A.D. 536 Hypothesis

[Petén; cultural history; social destabilization, complexity; Maya Early–Late Classic Transition, Hiatus]

The ancient lowland Maya of southeastern Mexico and northern Central America (Figure 4.1) achieved a high level of cultural complexity, and the status of "civilization", during their Classic period, which extended from about A.D. 250 to 900^cal^. The general pattern of growth and fluorescence that prevailed during the Classic period was interrupted, at least in a portion of the Maya lowlands, during the period known as the "Hiatus," or "Early-Late Classic Hiatus." The Hiatus has traditionally been dated to A.D. 534-593^cal^.

This chapter defines the Hiatus and examines it from the perspective that it began at about the same time as a major atmospheric disturbance of uncertain origin. It caused adverse climatological effects across the eastern hemisphere, and provoked chaos and instability for many cultures extending from Europe to China (see Jones, Young, Schmidt, Houston chapters). A hypothesis, building on the earlier suggestion of Gill (1994), is developed that accounts for the Hiatus by a climate-provoked destabilization of the Tikal regional state. The Tikal state was apparently the most expansive regional-scale polity in the Maya lowlands at the on-set of the Hiatus. The gaps in monument erection at Tikal associated with the Hiatus are integrated with the A.D. 536^cal^ atmospheric disturbance, paleoclimate reconstructions, and culture-historical occurrences in the Maya lowlands.

History of the Hiatus

[Petén; epigraphy; stela, Hiatus; Maya Early–Late Classic Transition, Hiatus]

The intrepid American Maya archaeologist, Sylvanus G. Morley spent a large part of his career crisscrossing the dense tropical jungles of the Maya lowlands in search of carved stone monuments (stelae and altars) bearing inscriptions written in the hieroglyphic script of the ancient Maya. He was the first archaeologist to note the occurrence of the Maya "Hiatus." In 1938, Morley observed that beginning after A.D. 534^cal^ (the Maya year 9.5.0.0.0), there was a decline in the number of stone monuments being dedicated at ancient Maya sites (Morley 1938: Vol 4, 333). These monuments were erected by the Maya ruling elite. The subject matter of the texts on them usually commemorated various aspects of rulers' lives, such as their position in dynastic lineages, their parentage, inauguration, claims of military victories, and other important milestones. Morley perceived the Hiatus to have ended about A.D. 613 (9.9.0.0.0), after which there was again an increase in the number of stela erected.

Tatiana Proskouriakoff (1950:111-112), in a study of Maya sculpture, described the Hiatus as the period from A.D. 534 (9.5.0.0.0) to 593^cal^ (9.8.0.0.0), which she considered to be a "dark period in the history of Maya sculpture." She noted that when monuments began to be regularly erected again after the Hiatus, a new style of depicting the rulers on the monuments appeared. She argued that the decline in monument dedications during the Hiatus reflected "some (unknown) momentous historical event."

Twenty-five years later, Gordon Willey (1974) reexamined the Hiatus. He accepted the approximate dating suggested by Proskouriakoff, and also offered a hypothetical explanation for the Hiatus. Willey suggested that the withdrawal of Teotihuacan influence (a great city in the highland Basin of Mexico, see below) from the Maya lowlands, which is understood to have occurred around the time of the Hiatus, could have caused political disruptions of sufficient magnitude to disturb the normal pattern of periodically erecting carved monuments (Willey 1974).

Peter Mathews's recent study of Early Classic and Hiatus period monuments confirms that the Hiatus actually occurred, and is reflected in the archaeological record by diminished numbers of carved stone monuments dedicated during the period A.D. 534 (9.5.0.0.0) to about 593^cal^ (9.8.0.0.0) (Mathews 1985).

Both Mathews and Linda Schele see the Hiatus as being a downturn focused within the Tikal polity (Culbert 1991b:316). Others, including Willey and Culbert (Culbert 1991b:316-317) maintain that the Hiatus affected a larger area, including the central and western lowlands.

There has been, in the last few decades, considerable advancement in understanding the Maya hieroglyphic script. Many previously unreadable texts are now partially or substantially readable. Also, a few new monuments have been found. Altogether, these developments have provided new insights and details on the dynastic histories of ancient Maya regional- and city-states, and on the Hiatus.

One of the newly discovered monuments, a ball court marker at the large site of Caracol in west-central Belize, has a text that has been interpreted as providing another possible explanation for the Hiatus. Before examining this Caracol monument, however, it is appropriate to discuss the important site of Tikal in Guatemala during the Early Classic period and before the Hiatus.

Tikal During the Early Classic Period

[Tikal, Petén; ; Tikal region state, trade route, Teotihuacan, stela, core area; Early Classic]

The Early Classic roughly fills the first half of the Classic period. The Hiatus time period has been viewed as an interlude between the Early Classic and the subsequent Late Classic period. As we shall see, there is now reason to limit this perspective.

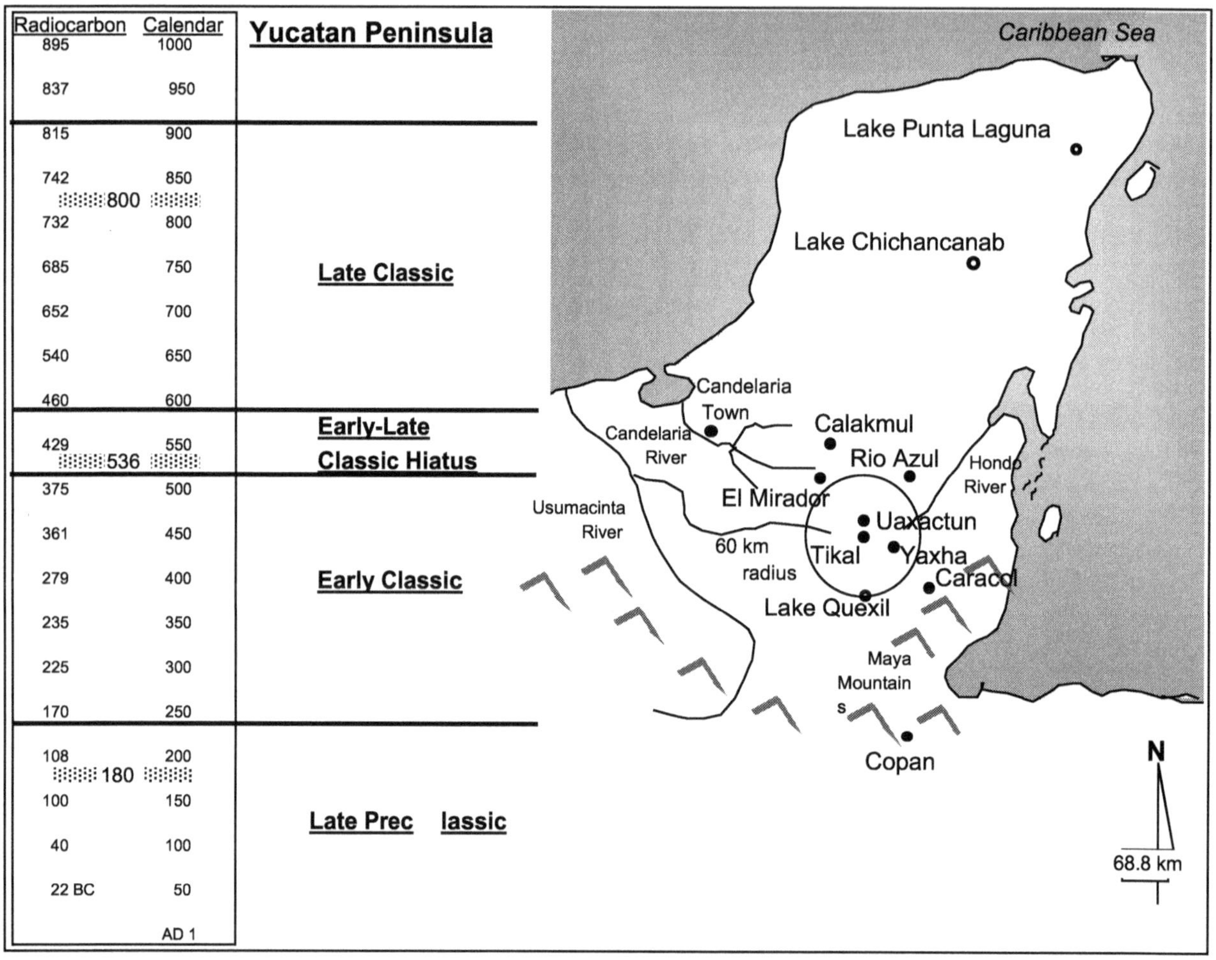

Figure 4.1. Map and Chronology of the Maya Lowlands and Yucatan Peninsula.

It is widely accepted that the city of Tikal in the northeastern Petén Department of Guatemala was the preeminent community of the lowland Maya world during the Early Classic period, until the start of the Hiatus (Culbert 1991a:128-134; Schele and Freidel 1990:159; Matthews 1985:31; Sharer 1994:207; Grube 1994:106). While Tikal appears to have experienced strong political competition during the preceding Late Preclassic period from the nearby city of El Mirador, El Mirador faded suddenly around A.D. 150-200[cal] (Matheny 1987:321; Gunn, Folan, and Robichaux 1991, 1994, 1995; Gill 1994) leaving Tikal with apparent unchallenged hegemony over an area that was regional in scale. It seems likely that the rulers of Tikal at this time oversaw a hierarchy of communities of various sizes extending out for a moderate distance of at least 60 km (see Figure 4.1).

That regional-sized states existed within the Maya lowlands has become increasingly apparent in recent years (Adams and Jones 1981; Marcus 1992; Culbert 1991a:140; Martin and Grube 1994, 1995). The preeminence of Tikal during the Early Classic is supported by its large size during the Early Classic, abundant public architecture, and prolific Early Classic dynastic monuments. Adams and Jones (1981), based upon nearest-neighbor and rank-size analysis, suggest that the Tikal Regional State during the *Late Classic* period, covered an area of 21,095 square kilometers, and included a multitude of large-, medium-, and small-sized communities as components of its regional state. At this later time, it coexisted with, and competed with, several other comparable regional states. During the Early Classic, however, Tikal appears to have been unique in size and likely political domain.

In mapping the locations of sites actively erecting carved monuments at different times during the Early Classic, Mathews also found further evidence supporting the view that Tikal had hegemony over a domain of regional scale (Mathews 1985:27-28). At the start of the Early Classic period around A.D. 250[cal], and for nearly two centuries thereafter, most of the cities erecting carved monuments were located within 60 kilometers of Tikal. This probably reflects the precociousness of cultural development in the region, which is considered to be the "core" area of lowland Maya civilization. It is likely that, in this incipient period of the Maya stela "cult," most of these cities were independent and pursued different political trajectories.

During the succeeding 100 years of the Early Classic, monuments began to be raised in other regions of the Maya lowlands, such as in the vicinity of Copan in Honduras, and along the Usumacinta River to the west of Tikal (see Figure 4.1). This indicates that cultural traits that developed in the core area were gradually being adopted in the broader Maya area.

However, of the sites within 60 kilometers of Tikal that were previously erecting monuments, all had ceased to do so except for Uaxactun and Yaxha, suggesting that most of these sites could have lost their independence and become subordinated to Tikal. Tikal's monuments proliferate at this time. While Uaxactun continued to erect monuments during the latter part of the Early Classic, texts at both Uaxactun and Tikal indicate that Uaxactun was subjugated by Tikal in about A.D. 378^{cal}, and it became part of the Tikal state. While there is no direct evidence that Yaxha became part of the Tikal regional state at this time, it has been noted that iconography at Yaxha involving Teotihuacan symbolism closely resembles that at Tikal, hinting at some form of relationship between the two political centers. Also, Tikal and Yaxha are the only known sites that constructed a particular form of architectural complex known as the "Twin Pyramid complex." Twin Pyramid complexes are highly specific in their configuration, and their use only at Tikal and smaller Yaxha strongly suggests a Tikal-Yaxha linkage.

Evidence also exists for Tikal's influence extending beyond its 60 kilometer ring of control. At Rio Azul some 70 kilometers northeast of Tikal, both Stela 1 and the Tomb 1 burial texts and iconography link Rio Azul to Tikal; Rio Azul could have been apart of the Tikal political system as early A.D. 392-417^{cal} (Adams and Robichaux 1992:417,426). Additionally, recent interpretation of texts found at Caracol, some 80 kilometers southeast of Tikal, suggests that Tikal also had significant political influence over Caracol during the later Early Classic (Grube 1994:106).

Overall, the evidence is clear that during the course of the Early Classic, Tikal grew in power and influence, and absorbed many of its neighbors into its political domain. This attainment represented a political system of a size previously unknown in the Maya world. The precocious importance of Tikal could have derived in large measure from economic activities involving external trade, principally with the huge city of Teotihuacan in the distant highland Basin of Mexico, and internal trade within the Maya region (Hoopes 1985; Sharer 1994). Tikal is strategically located in the vicinity of rivers that facilitate east-west portage across the southern Yucatan peninsula, and in a central position between major population and resource centers to the north and south.

There is abundant archaeological evidence indicating a strong Tikal-Teotihuacan political and economic connection during the Early Classic (Coggins 1975,1979; Culbert 1991a:131-133; Millon 1993:28; Sharer: 1994:187-191). The Teotihuacan presence at Tikal started about A.D. 250 and lasted until about 550^{cal}. The evidence for Teotihuacan involvement at Tikal includes the presence at Tikal of Teotihuacan style architecture, a Teotihuacan style monument, depictions of elite individuals, including at least one Tikal ruler, in Teotihuacan attire, Teotihuacan iconography, Teotihuacan ceramics, and green obsidian originating from near Teotihuacan. Later, various elements of Teotihuacan attire and symbolism seem to have diffused from Tikal to other Maya sites. The Teotihuacan presence at Tikal is commonly interpreted to have served as a catalyst for the rapid ascendancy of Tikal during the Early Classic. From this background springs Willey's reasonable suggestion that Teotihuacan's withdrawal from the Maya lowlands around the time of the Hiatus could have caused the Hiatus.

The diachronic pattern of monument erections at Tikal through the Early Classic and Hiatus is as follows. Monuments (stela and altars) were erected with increasing frequency from A.D. 292 to 527^{cal} (Mathews 1985:22-24). (Note that the stela and altar numbers reflect the sequence in which they were found by archaeologists, and *not* their chronological sequence of erection.) Stela 10 and 12, which were both erected in A.D. 527^{cal}, are the last at the end of the Early Classic, until Stela 17, which was dedicated in A.D. 557^{cal}. Following Stela 17, there are no known monuments at Tikal until A.D. 692^{cal} (Culbert 1991b:316-317). While the

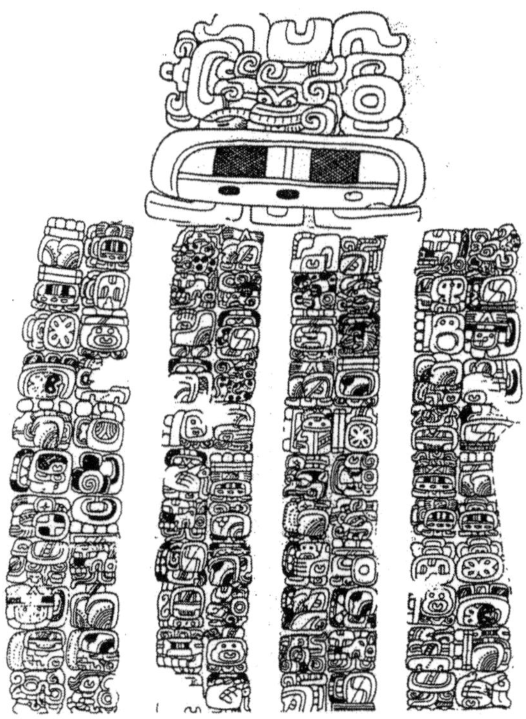

Figure 4.2. Part of text on Tikal Stela 31 dating to ca. A.D. 435 (after Jones and Satterthwaite 1982).

Hiatus has been traditionally dated to A.D. 534-593^{cal} based on lowland-wide data (Willey 1974), some have interpreted the Hiatus at Tikal as extending from A.D. 557^{cal}, the date of Stela 17, until A.D. 692^{cal}, the date of Steel 30 (Schele and Freidel 1990:172). However, as Culbert (1991b:316-317) has pointed out, there are actually two gaps in the monument record at Tikal, one between A.D. 527 and 557^{cal}, and the other between A.D. 557 and 692^{cal}. After A.D. 692^{cal}, there is a full restoration of the monument erection practice at Tikal.

Having set the scene for the Hiatus, it is now appropriate to return to a consideration of Altar 21 at Caracol.

The Caracol-Tikal Connection as an Explanation for the Hiatus

[Caracol, Maya Mountains; Epigraphy; Gap 1, Gap 2, Altar 21, War; Early Classic, Hiatus]

During the Hiatus, both the number of monuments erected, and the number of sites raising monuments, decreased (Mathews 1985). The notable exception in this trend was the city of Caracol in west, central Belize (see Figure 4.1). Caracol first erected monuments during the latter part of the Early Classic shortly before the Hiatus (Figure 4.2). A break in recorded monument dates at Caracol appears between A.D. 534 and about 553^{cal} (Houston 1987:87, Figure 68; Grube 1994:111, Table 9.1). This period approximates the Gap 1 break in monument erections at Tikal. Neither Tikal nor Caracol had monuments erected at this time. Following A.D. 553^{cal}, and continuing until about A.D. 650^{cal}, a multitude of dates appeared in the Caracol textual records, and many monuments were erected there. Thus, while Tikal was experiencing Gap 2 in monument erections (A.D. 557-692^{cal}), Caracol engaged in an unprecedented period of erection signaling a vigorous dynastic rulership (Houston 1987:97; Chase and Chase 1987).

Possibly the most important of the Caracol monuments with regard to understanding the Hiatus is Altar 21, which was discovered in 1986. The text on this monument contains at least 128 glyphs. Of these, a third have been destroyed by erosion. Still other glyphs have been partially obliterated. The segments of text that remain have been interpreted to indicate that Caracol twice victoriously engaged in hostilities with Tikal, first in A.D. 557^{cal} (9.6.2.1.11) in a small scale conflict, and in A.D. 562^{cal} (9.6.8.4.2) in a major encounter (Houston 1987:93; Schele and Freidel 1990:173). A variety of archaeological evidence at Tikal seems compatible with Caracol having had some degree of influence at Tikal for a while after this date. This evidence includes the presence of ceramics, monuments, and burials at Tikal that are more consistent with material culture at Caracol than with earlier patterns at Tikal (Coggins 1975, 1979; Chase and Chase 1987; Montgomery 1988). In a recent reexamination of the text on Altar 21, Grube (1994:106) interpreted the text to indicate that Caracol was joined by another Maya center, probably Calakmul (Site Q, see Figure 4.1), in defeating Tikal in the major conflict of A.D. 562^{cal}. Importantly, as noted above, Grube sees texts at Caracol as implying that Caracol was under Tikal political influence at some time *prior* to the Hiatus.

The apparent military defeat of Tikal in A.D. 562^{cal} by Caracol and its likely ally, Calakmul, has been used to explain Gap 2 (A.D. 557-692^{cal}) in monument erections at Tikal (Chase and Chase 1987; Culbert 1991b:317; Schele and Freidel 1990:173). It has been suggested that upon the defeat of Tikal, the elite people there were subjugated to some degree by their conquerors. This situation would have precluded the erection of any new monuments relating to the previous Tikal ruling dynasty. The disruption caused by this conquest is considered to have lingered until about A.D. 682-692^{cal}, by which time a full restoration of local rule had occurred at Tikal.

The only hypothesis that has been offered for Gap 1 in monuments at Tikal between A.D. 527 and 557^{cal} is that by Christopher Jones who proposed that any stela erected at Tikal during this period could have been deliberately destroyed by the conquerors (C. Jones, as cited in Culbert 1991b:317).

The A.D. 536 Atmospheric Event

[Globe, Eastern Hemisphere; Ice layers, Alternative Explanations, Parsimony, Serendipity; Meteors, Volcanoes; A.D. 536]

The discussion thus far has detailed the preeminence of Tikal during the Early Classic, and also outlined the two, heretofore most viable explanations for the Hiatus, i.e., the withdrawal of Teotihuacan influence from the Maya lowlands, or the defeat of Tikal by Caracol in a military conflict. Recent research, however, has disclosed another potential component of the Hiatus. This new component could have precipitated simultaneously both the Teotihuacan withdrawal, and provided the opportunity for Caracol to challenge the previously dominant and expansive Tikal regional state.

Since approximately 1984, there has been an increasing realization that a major, not well-defined, atmospheric event occurred about A.D. 536^{cal}. This event had a debilitating effect on many societies across the eastern hemisphere (Stothers 1984; Baillie 1994, 1995; Gill 1994). Historic accounts from Europe, the Near East, North Africa and China described its adverse environmental and cultural consequences (see Young, Jones, Schmidt and Houston chapters in this volume). The most common effect noted was an extreme dimming of the sun's light. Accounts from Italy and China indicate the phenomenon lasted for approximately 18 months. The darkening of the sky by a "dust-veil" (Baillie 1995:72) was accompanied by colder temperatures and crop failures, which in some places caused massive starvation. For example, frosts and snows were recorded in China during July and August of A.D. 536^{cal}. Crops were killed provoking widespread starvation (Gill 1994:310-314; Baillie 1995:70-74). Northern China experienced a political collapse during the period A.D. 535-545^{cal}, and southern China suffered social and economic calamity during the A.D. 540s (Houston, this volume). These developments in China upset 200 years of cultural progress (Keys 1995, as reported in Baillie 1995:74).

A major, unknown volcanic eruption has been assumed as the origin for the dust cloud (Stothers 1984:344; Gill 1994:318; Baillie 1995:72). Baillie (1995:72) notes, however, that the study of ice layers from the Greenland glacier cores does not provide support for the occurrence of a volcanic eruption. Baillie hypothesizes that the explosion of cometary debris in the Earth's atmosphere, or the impact of a large meteorite in an ocean could account for the A.D. 536 event.

Paleoclimate reconstructions based upon tree rings from both Europe and North America indicate a cold spell following the 18 month long A.D. 536 event. It persisted for approximately 10 years, or through about A.D. 545[cal] (Baillie 1995:70-71; Scuderi 1993). The years A.D. 535, 536, and 541[cal] were the second, third, and fourth coldest years in North America during the last two thousand years based upon analysis of Sierra Nevada tree ring data (Scuderi 1993:1435). Thus, while the dust-veil reported in eastern hemisphere accounts was discernible after the 18-month-long A.D. 536 event, the cold temperatures associated with it lingered on for about eight years in Europe and North America, and possibly in other areas of the northern hemisphere. Based upon this data, a more appropriate label for the "atmospheric event" would be the "A.D. 536-545 event," rather than the "A.D. 536 event."

What evidence, we can ask, is there from the Maya lowlands to suggest that the A.D. 536-545 event had impact there? We consider this question next.

Attempted Reconstructions of Maya Lowland Paleoclimates

[Globe, Maya Lowlands; Candelaria model, A.D. 536 event, Lake cores; Paleoclimate, Resolution/detection threshold; Hiatus, Classic, Terminal Classic, Postclassic]

As early as 1917 Huntington suggested that climate change had affected the cultural trajectory of the ancient Maya. A gradual escalation of efforts to define paleoclimates affecting the ancient Maya has occurred since. Some inferences have been made based upon archaeological data, such as the presence of structures in areas presently subject to seasonal inundation (Dahlin 1983:251), but increasingly there is a reliance upon non-archaeological physical data that serves as a surrogate for past climatic conditions. This section reviews paleoclimate reconstructions that illuminate the Hiatus period. In some cases these data are very spatially localized in their implications, while in others inferences for broader landscapes are possible.

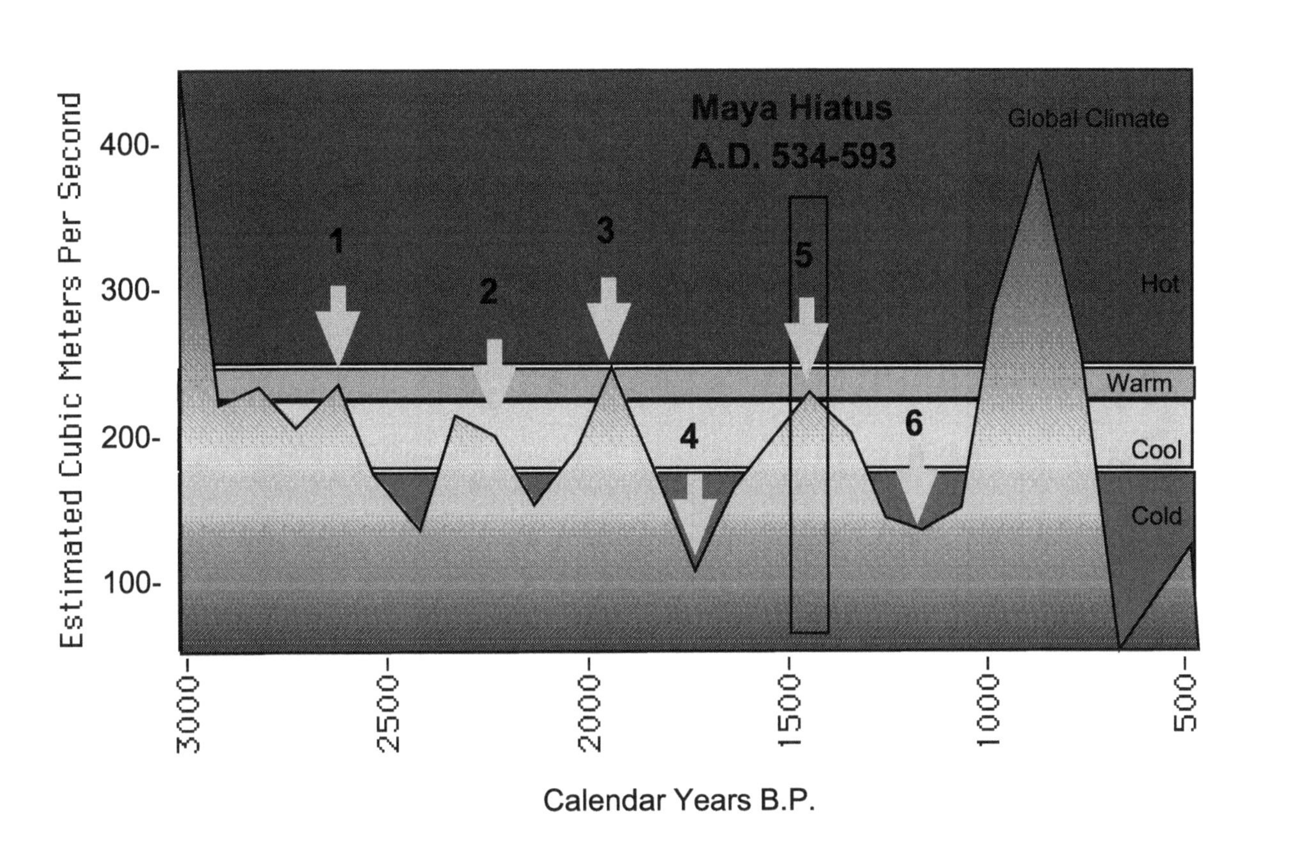

Figure 4.3. Retrodicted discharge of the Candelaria River and major cultural events. 1. Occupation of Guiro, Nakbe & Tintal. 2. Rise of Calakmul and El Mirador, 3. El Mirador Maximum development. 4. Decline of El Mirador. 5. First Calakmul Stela. 6. Last Calakmul Stela, Maya Collapse.

Gunn et al. (1991, 1994, and 1995) used a model that related the average temperature of the northern hemisphere troposphere to available moisture in the Candelaria River watershed in the southwestern corner of the Yucatan peninsula, as measured by monthly river flow (Figure 4.3). River flow data taken at the town of Candelaria was available on a monthly basis for the period 1958 through 1990. The model utilized estimates of atmospheric temperature for the Late Holocene period that were based on solar emissions, "El Nino" oscillations, volcanism, and atmospheric chemistry, to retrodict hydrology within the Candelaria River for the period 1050 B.C. to A.D. 1550^{cal}. The Candelaria River has a catchment basin area of 10,840 square kilometers (see Figure 4.1). Two large ancient Maya communities, El Mirador and Calakmul, are located along the periphery of the watershed, and Tikal is about 60 miles southeast of it. This model indicates the existence of a general climatological trend in which "hot" global temperatures generate increased amounts of moisture in the Candelaria River catchment basin, while "cold" climates, on average, produce the least amount of moisture. "Warm" and "cool" global temperatures are associated with moderate amounts of moisture (see Figure 4.3).

Figure 4.3 shows inferred moisture levels during the Preclassic, Classic, and Postclassic periods based upon the model. For the period including the year A.D. 536^{cal}, the model indicates that a major change in the trajectory of moisture levels occurred near this time. The trend starting shortly after A.D. 536^{cal} is of cooling temperatures and increasing aridity. The model indicates that separate, extended drier periods accompanied Maya cultural downturns during the Late Preclassic, the Terminal Classic (the Maya "Collapse" at the end of the Late Classic), and the Postclassic. For the Hiatus period of A.D. 534-593^{cal}, and the early part of the Late Classic, however, the Candelaria River model suggests that moisture levels were in a moderate range, which has been interpreted as compatible with positive cultural development (Gunn et al. 1995:36). With regard to this model, it should be pointed out that if, indeed, Baillie's hypothesis that the A.D. 536 event was caused by a meteorite is correct, that event would not be reflected in the Candelaria model retrodiction because such a parameter was not included among the model's influencing variables.

To my knowledge, Richardson B. Gill was the first Mayanist to become aware of the eastern hemisphere A.D. 536 event, with its adverse effects, and recognize that it was synchronous with the Maya Hiatus, and further, that it presented another possible explanatory framework for the Hiatus. In a 1994 dissertation, Gill marshaled evidence from a wide variety of sources, including the above described model, to build a case that very cold temperatures in Europe and North America were accompanied by drought conditions in the Maya lowlands. From this premise, Gill reasoned that if the A.D. 536 event provoked unusually cold temperatures in the eastern hemisphere and very cold temperatures were also experienced in North America (Scuderi 1993), then a severe drought in the Maya lowlands could have accompanied the Hiatus. Gill's overall analysis suggests, in general consonance with the Candelaria River model, that major dry periods occurred in the Lowlands concurrently with significant ancient Maya cultural downturns. The two reconstructions differ only for the Hiatus, where the Candelaria River model indicates mild temperatures and moderate-moisture conditions prevailed.

Paleoclimates for the northern and central Yucatan peninsula were recently estimated based on Oxygen 18/Oxygen 16 isotope ratio analysis of sediment cores obtained from two lakes (see Figure 4.1, Figure 4.4). Hodell et al. (1995) used a 4.9 meter long lake bed core from Lake Chichancanab in the central Yucatan peninsula to estimate relative moisture availability during the period starting about 8,000^{c14} years ago until recent times. Subsequently, Curtis et al. (1996) obtained a 6.3 meter core sample of lake bed sediments from Lake Punta Laguna in northeastern Yucatan peninsula whose earliest component dated to about 1,700^{c14} B.C.

The two lake studies produced generally similar results (see Figure 4.4). They indicate that the period A.D. 165-1020^{c14} was, overall, drier than times before or after it within the study period. Particularly dry spells were noted for A.D. 862, 986, and 1051^{c14} (Curtis et al. 1996:37). These latter dates match the dry period retrodicted by the Candelaria River model, and by Gill for the Maya Collapse time about A.D. 900^{cal}. For the year A.D. 536^{cal} and a few years after, however, the data from both lakes indicate moisture was at a *moderate* level in the vicinity of the lakes. This is the same result indicated by the Candelaria River model. Gill's reconstruction, as noted above, suggests that conditions could have been colder and drier than the other models indicate.

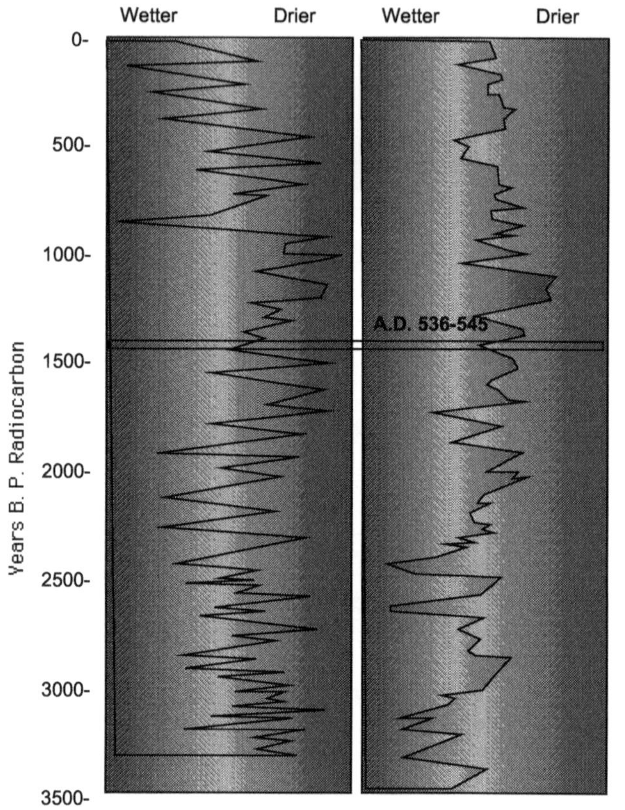

Figure 4.4. Oxygen isotope determinations from lakes Punta Laguna and Chichancanab (after Curtis et al. 1996:43).

Overall, there is a high degree of congruence between the Candelaria River model's and Gill's reconstructions for the timing of major arid periods in the Maya lowlands. Only for the time around the Hiatus are they at variance. The oxygen isotope analysis for the two Yucatan lakes is also in clear agreement concerning the occurrence of a major dry period at the time of the Maya Collapse, and indicates variable, but mostly dry conditions extending back to about. A.D. 165^{c14}.

The variation in results among the models for the Hiatus period cao probably be attributed to limitations in temporal resolution for the Candelaria River and lacustrine reconstructions. Of the three, the Lake Punta Laguna construction has the greatest claimed temporal resolution, decadal. The Lake Chichancanab reconstruction has about 20-year resolution, and the Candelaria River model retrodiction was based upon 100 year estimates of atmospheric temperature. The very cold conditions associated with the A.D. 536 event appear to have lasted for, at most, 10 years, with the earliest 18 months being the coldest. This event would, if detected by the Candelaria River and lacustrine paleoclimate reconstructions, appear as a nearly vertical downward spike, indicating a very rapid descent to colder temperatures and drier conditions, and an equally rapid ascent back to milder temperatures and moderate rainfall. This spike would be, in the general climatological trends of centuries and millennia, an aberration, a short-lived, yet powerful phenomenon provoked by a quite rare event. It seems likely the event escaped direct detection in the Candelaria and lacustrine reconstructions simply because it was so short lived.

The Hiatus Period Maya and the A.D. 536-545 Atmospheric Event

[Petén; Lake cores/phosphorus, ; Hiatus detection, Population, Abandonment Rio Azul, Tikal regional state/Vulnerability, Leadership replacement, Trade sphere disruption, Caracol Gap 1 disruption, ; Maya Hiatus]

The Early-Late Classic Maya Hiatus has primarily been recognized as a hiatus in the erection of carved monuments within the Maya lowlands. Here and there, however, there are signs of a broader decline. For example, phosphorus loading studies at Lake Quexil, some 60 kilometers south of Tikal, indicate that population levels around the lake were lower during the Hiatus than in either the preceding Early, or subsequent Late Classic periods (Binford 1983:201; Brenner 1983:206; Gill 1994). Rio Azul, a sizable community that was associated with Tikal, appears to have been nearly abandoned around the Hiatus time (Adams 1992:28). These locations provide some support for the hypothesis that the political disarray associated with the Maya Hiatus stemmed from a more fundamental underlying condition, perhaps a significant agricultural deficit caused by the A.D. 536-545 event.

If the 18 month long obscuration of the sun's light noted in the eastern hemisphere was indeed experienced in the Maya lowlands, a severe deterioration in agricultural production would have occurred over approximately a two year period. This would have undoubtedly provoked some degree of food deprivation for the Maya, and could have depleted seed reserves for planting in subsequent years. Seven or eight additional years of colder than average temperatures in the region would have slowed a recovery from the two-year food production deficit because of drought.

The scene in the southern lowlands prior to the onset of the Hiatus, as we have seen, was one in which the Tikal regional state had steadily grown, and outstripped all other polities in the Maya region. Tikal's political influence apparently extended outward to Rio Azul and Caracol prior to the Hiatus (Adams and Robichaux 1992; Grube 1994). Much of its prominence derived from its strategic importance as a major trade hub servicing both the Maya lowlands, and external areas such as central Mexico and the Maya highlands.

The Tikal regional state, due to its large size and structural complexity, and the inherent limitations of a pedestrian command and communication system, would have been the most vulnerable political entity in the Maya world to a major environmental crisis. A food shortage calamity could have destabilized the Tikal polity and effectively distracted the rulership from less vital issues such as monument erections until a full recovery was made. This would account for Gap 1, A.D. 527-557^{cal}, in the Tikal monument erection pattern. It is interesting to note that the text on Tikal Steel 17, which was dedicated in A.D. 557^{cal}, comments retrospectively on the inauguration of a new Tikal ruler named "Double Bird" in December of A.D. 537^{cal} (9.5.3.9.15), approximately the time that the 18 month dust-veil ended in the eastern hemisphere. Could the assumption of power by a new ruler at this time have been a response to difficult challenges facing the Tikal polity?

Any chaos associated with the A.D. 536-545 event would also have undermined routine trade in elite and utilitarian goods in the lowlands. This would have adversely affected smaller Maya polities within the Tikal trade sphere, adding, for them, to the negative impact of the environmental downturn. This could have provoked the general reduction in monument erections that occurred across the lowlands. Teotihuacan's apparent withdrawal from the Maya region around this time could also have derived from economic and social stress related to the event. Their retreat would have further exacerbated the Maya destabilization.

These calamitous events would, very plausibly, also have diverted Tikal's attention away from concerns about political control in the more distant reaches of its domain. The community of Caracol, situated at the outer fringe of Tikal influence, could have used the disorder accompanying the A.D. 536-545^{cal} event as an opportunity to disassociate itself from Tikal. Caracol, as noted earlier, erected no monuments during the period A.D. 534-553^{cal}, suggesting that it too suffered the immediate impacts. But Caracol seems to have recovered quickly and vigorously. The second gap in monument erections at Tikal (Gap 2: A.D. 557-692^{cal}, see Figure 4.2) can reasonably be attributed to Caracol (Culbert 1991; Schele and Freidel 1990; Grube 1994:106), which seems to have attained full autonomy after the A.D. 536-545 event. Subsequently, Caracol grew, formed alliances, and strengthened sufficiently to successfully challenge, and defeat Tikal militarily in A.D. 562^{cal}. Chase and Chase (1987) suggest that Caracol could have had a more heterarchical social system, and thus it could have been better able to adjust to the post-A.D. 536 environmental instability. It could have also been sheltered by its mountainous habitat.

Conclusion

[Maya Lowlands; Parsimony, Serendipity, Future research/literature reexamination; A.D. 536 event, Political disorder, Famine, Cold, Teotihuacan influence, Caracol war, China analogy; Maya Hiatus, Maya Classic]

This chapter has reviewed the Maya Hiatus within the context of the major atmospheric event that began in A.D. 536cal. The event was consistently described in historical accounts from Europe, North Africa and Asia as a dust-veil phenomenon that obscured the sun's light for periods as long as 18 months. Across the eastern hemisphere, the event provoked cold temperatures, food shortages, and varying degrees of political disorder. The nature of the Early-Late Classical Maya Hiatus, a political downturn detected principally through a diminution in the number of carved monuments erected during the Hiatus period, was described. The Hiatus was examined in relation to the Tikal regional state, the largest and most powerful polity in the Maya lowlands prior to the onset of the Hiatus. The specific temporal gaps in the erection of carved monuments at Tikal were considered in relation to the timing of the atmospheric event, and culture-historical occurrences such as the withdrawal of Teotihuacan influence from the Maya lowlands, and the A.D. 562cal military victory of Caracol over Tikal. The timing of all of these phenomena match in a coherent sequence supporting an interpretation that the A.D. 536-545 event did actually impact the Maya region. The evidence suggests that a scenario very similar to that which was historically documeoted for China (see Houston's chapter) could apply to the Maya. The failure, thus far, to find direct, non-archaeological physical evidence confirming the event's occurrence in the Maya region seems likely to be due to the event's brief time duration.

Finally, it seems likely that the cultural consequences of the event's occurrence in the eastern hemisphere would not yet be fully recognized, except for the fact that people who lived through it wrote descriptions that have survived to tell us of it. It seems quite possible that, as in the eastern hemisphere, the literate Maya could also have made some written comments relating to that remarkable 18 month long dimming of the sun's light, and the adverse environmental and cultural developments that accompanied it. Given the present knowledge that this dramatic event could indeed have been experienced by the Maya, a reexamination of surviving Maya texts dating to the Hiatus period and later times could yet reveal such references.

ACKNOWLEDGMENTS: I thank Richard E.W. Adams who in 1994 provided me a copy of a newspaper article referring to the A.D. 536 event, eastern hemisphere accounts of it, and Baillie's hypothesis concerning it. I thank Joel Gunn to whom I sent the newspaper article. He took that "seed" and nurtured it into the symposium upon which this volume is based. Finally, I would like to express my gratitude to the University of the Incarnate Word for providing financial support. In particular I would like to thank Dr. Roger Barnes, Dr. Gilberto Hinojosa, and Dr. Eduardo S. Paderon for their support.

References

Adams, R.E.W. (1992) Ixcanrio Regional Archaeological Project: 1991 Interim Report. Unpublished manuscript. University of Texas at Austin.

Adams, R.E.W., and R. Jones (1981) Spatial Patterns and Regional Growth among Maya Cities. *American Antiquity* 46:301-322.

Adams, R.E.W., and H. Robichaux (1992) Tombs of Rio Azul, Guatemala. *National Geographic Research and Exploration* 8:412-427.

Baillie, M. G. (1995) Patrick, Comets and Christianity. *Emania: Bulletin of the Navan Research Group* No. 13.

Binford, M. (1983) Paleolimnology of the Petén Lake District, Guatemala. Erosion and Deposition of Inorganic Sediment as Inferred from granulometry. *Hydrobiologia* 103:199-203.

Brenner, M. (1983) Paleolimnology of the Petén Lake District, Guatemala. Mayan Population Density and Sediment and Nutrient Loading of Lake Quexil. *Hydrobiologia* 103:205-210.

Chase, A., and D. Chase (1987) *Investigations at the Classic Maya City of Caracol, Belize: 1985-1987.* Pre-Columbian Art Research Institute, Monograph 3, San Francisco: Pre-Columbian Art Research Institute.

Coggins, C. (1975) Paintings and Drawing Styles at Tikal: An Historical and Iconographic Reconstruction. Ph.D. Dissertation, Harvard University, Cambridge MA.

Coggins, C. (1979) A New Order and the Role of the Calendar: Some Characteristics of the Middle Classic period at Tikal. In *Maya Archaeology and Ethnohistory*, edited by N. Hammond and G. Willey, 38-50, University of Texas Press, Austin.

Culbert, T.P. (1991a) Polities in the northeast Petén, Guatemala. *In Classic Maya Political History*, edited by T.P Culbert, pp. 128-146, School of American Research, Santa Fe.

Culbert, T.P. (1991b) Maya Political History and Elite Interaction: A Summary View. *In Classic Maya Political History*, edited by T.P. Culbert, 311-346, School of American Research, Santa Fe.

Curtis, J.H., D. Hodell, and M. Brenner (1996) Climate Variability on the Yucatan Peninsula (Mexico) during the Past 3500 Years, and Implications for Maya Cultural Evolution. *Quaternary Research* 46:27-47.

Dahlin, B. (1983) Climate and Prehistory on the Yucatan Peninsula. *Climatic Change* 5:245-263.

Gill, R.B. (1994) *The Great Maya Droughts.* Unpublished Dissertation, The University of Texas at Austin.

Grube, N. (1994) Epigraphic Research at Caracol, Belize. *In Studies in the Archaeology of Caracol, Belize*, edited by D. Chase and A. Chase, 83-122, Pre-Columbian Art Research Institute Monograph 7, Pre-Columbian Art Research Institute, San Francisco.

Gunn, J., W. Folan, and H. Robichaux (1991) An Analisis of Discharge Data from the Candelaria River System in Mexico: Insights into Paleoclimates affecting the Ancient Maya Sites of Calakmul and El Mirador. Unpublished Paper, Garrow and Associates, 1921 Billabong Lane, Chapel Hill NC 27516.

Gunn, J., W. Folan, and H. Robichaux (1994) Un Analisis Informativo sobre La Descarga del Sistema del Rio Candelaria en Campeche, Mexico: Reflexiones acerca de Los Paleoclimas que Afectaron a Los Antiguos Sistemas Mayas en Los Sitios de Calakmul y El Mirador. In *Campeche Maya Colonial*, edited by W. Folan, pp. 174-197, Universidad Autonoma de Campeche, Campeche Mexico.

Gunn, J., W. Folan, and H. Robichaux (1995) A Landscape Analysis of the Candelaria Watershed in Mexico: Insights into Paleoclimates Affecting Upland Horticulture in the Southern Yucatan Peninsula Semi-Karst. *Geoarchaeology: An International Journal* 10:3-42.

Hodell, D, J. Curtis, and M. Brenner (1995) Possible Role of Climate in the Collapse of Classic Maya Civilization. *Nature* 375:391-394.

Hoopes, J. (1985) Trade and Exchange. In *A Consideration of the Early Classic*, edited by G. Willey and P. Mathews, pp. 145-160, SUNY Albany.

Houston, S. (1987) Notes on Caracol Epigraphy. *In Investigations at the Classic Maya City of Caracol, Belize*, edited by A. Chase and D. Chase, 85-100, Pre-Columbian Art Research Institute, Monograph 3, San Francisco: Pre-Columbian Art Research Institute.

Huntington, E. (1917) *Maya Civilization and Climatic Change.* XIXth International Congress of Americanists, Washington DC.

Jones, C., and L. Satterthwaite (1982*) The Carved Monuments and Inscriptions of Tikal: The Carved Monuments.* Tikal Report 33A. The University of Pennsylvania, The University Museum, Philadelphia.

Marcus, J. (1992) Dynamic Cycles of Mesoamerican States. *Research and Exploration* 8:392-411.

Matheny, R. (1987) El Mirador: An Early Maya Metropolis Uncovered. *National Geographic* 172:317-339.

Martin, S., and N. Grube (1994) Classic Maya Politics within a Mesoamerican Tradition: An Epigraphic Model of "Hegemonic" Political Organization. Unpublished Paper presented at the Primer Seminario de las Mesas Redondas de Palenque, Mexico.

Martin, S., and N. Grube (1995) Maya Superstates. *Archaeology*, November/December pp. 41-46.

Mathews, P. (1985) Maya Early Classic Monuments and Inscriptions. *In A Consideration of the Early Classic Period in the Maya Lowlands*, edited by G. Willey and P Mathews, 5-54, Institute of Mesoamerican Studies, SUNY Albany.

Millon, R. (1993) The Place Where Time Began: An Archaeologist's Interpretation of What Happened in Teotihuacan History. *In Teotihuacan: Art from the City of the Gods*, edited by K. Berrin and Esther Pasztory, pp. 16-43, Thames and Hudson, New York, NY, and the Fine Arts Museums of San Francisco.

Montgomery, J. (1988) An Investigation into Middle and Late Classic Caracol Influence at Tikal, Guatemala. An unpublished manuscript in the possession of the author.

Morley, S. (1938) *Inscriptions of the Petén*, 5 Volumes. Publication 437, Carnegie Institute of Washington, Washington DC.

Proskouriakoff, T. (1950) *A Study of Classic Maya Sculpture*. Publication 593, Carnegie Institute of Washington, Washington DC.

Schele, L., and D. Freidel (1990) *A Forest of Kings*. W. Morrow and Co., New York.

Scuderi, L.A. (1993) A 2000-Year Tree-Ring Record of Annual Temperatures in the Sierra Nevada Mountains. *Science* 259:1433-36.

Sharer, R. (1994) *The Ancient Maya*. Stanford University Press, Stanford.

Stothers, R.B. (1984) Mystery Cloud of A.D. 536. *Nature* 307:344-45.

Willey, G. (1974) The Classic Maya Hiatus: A Rehearsal for the Collapse? *In Mesoamerican Archaeology: New Approaches*, edited by N. Hammond, pp. 417-444, University of Texas Press, Austin.

5. SIXTH AND SEVENTH CENTURY VARIABILITY IN THE SOUTHERN MAYA LOWLANDS: CENTRALIZATION AND INTEGRATION AT CARACOL, BELIZE

Arlen F. Chase and Diane Z. Chase

University of Central Florida

While differences in environment, resources, and cultural development have been extensively contrasted for the highlands and lowlands of Mesoamerica, archaeological data have more recently been used to demonstrate that substantial cultural and environmental variability also existed throughout the lowland areas of prehistoric Mesoamerica. This environmental and cultural diversity could have played a role in the differential successes of individual polities during times of social and/or environmental stress. The archaeological records of Caracol, Belize and Tikal, Guatemala (both sites were the focus of extended archaeological projects) can be compared and contrasted to explore socio-political variations that could help explain their different developmental trajectories. The A.D. 536 event is viewed as one potential catalyst in the rise of Caracol, Belize between A.D. 562^{cal} and A.D. 680^{cal}.

Could global climatic events have impacted the course of Maya civilization? The answer must be "certainly." But world-wide or even hemispheric concerns are not a standard issue of concern in Maya studies. Considerations of change in Maya civilization are either painted in the broadest of strokes (Messenger 1990) or, more usually, in the smallest of lines. With few exceptions (Sharer 1991; Webster 1988), such expositions also do not usually emphasize contemporary variation in cultural trajectories. Environmental change has been considered a factor primarily in explanations for the eighth and ninth century onset of the "Classic Maya Collapse" (Adams 1973:23). Excessive ground clearing for farming has been viewed as causing soil exhaustion/erosion or leading to the dominance of savanna over forest (Cook 1921, Cooke 1931; Abrams and Rue 1988) and climate-triggered drought has been suggested as playing a role in the Maya collapse (Gunn and Adams 1981; Dahlin 1983; Hodell et al. 1995). With the exception of the "collapse," however, there has been little consideration of the differential impact of global phenomena on the diverse environmental and cultural landscape of the Maya lowlands.

Yet, advances have been made toward correcting this outlook. Americanists have long been concerned with identifying relationships among various prehistoric Mesoamerican cultures and their neighbors. The lowland Maya of all time periods have been viewed as receptors for outside cultural influences and impacts, which were sometimes thought to have been expressed locally as horizon styles (cf. Stone-Miller 1993). Proposed interactions have had a major impact on the interpretation of Maya prehistory, although many of them have been called into question by more recent data (see, for example: the Maya Protoclassic development out of the "Q Complex" [Vaillant 1935]; Early Classic Teotihuacan, Mexico influence at Tikal, Guatemala [Sanders and Price 1968]; Terminal Classic Tula, Mexico relationships with Chichen-Itza, Mexico [Tozzer 1930, 1957]; and the extent and impact of the Postclassic "Mixteca-Puebla" style [Nicholson 1960]). Rathje (1971), in contrast, proposed a model that combined ideology and economy to explain the largely internal development of lowland Maya civilization. Wallerstein's (1974) world systems model introduced an interactive and more multi-directional approach that has been used to consider the interconnectedness of Mesoamerica as a whole. Following this perspective, a change in one part of the Mesoamerican system (whether highland Mexico or the Maya lowlands) could have consequences for, or be related to, events and processes elsewhere in the system. Such an inter-related approach has been applied to Classic warfare arenas (Chase and Chase 1992), Postclassic economics (Kepecs et al. 1994), and overall Mesoamerican development (Blanton et al. 1996). However, even these applications are predominantly within the realm of human endeavor rather than global environmental events. They are generally not concerned with worldwide patterns or phenomena. Yet, they could be.

The Maya thought of time and experience as being embedded in ever larger cycles - flowing from one into another - all working together; cyclical time influenced and conditioned Maya perceptions and actions (Puleston 1979; Chase 1991). Archaeological considerations of causality can be regarded as somewhat analogous to Maya views of cyclical time. Immediate causes of events are often superficially identified, but these events could be embedded in other, broader contexts that are not always immediately accessible or apparent. They are, therefore, not given due consideration. And, past perceptions and conclusions about causality cycle back and condition modern interpretations, assumptions, and models. Even though Maya archaeologists have made impressive strides in understanding the problems of using analogies and models for interpreting the past (Marcus 1995 and Issac 1996), once an immediate field - and possibly a regional arena - has been recognized, we may never look for or consider broader contexts. The A.D. 536

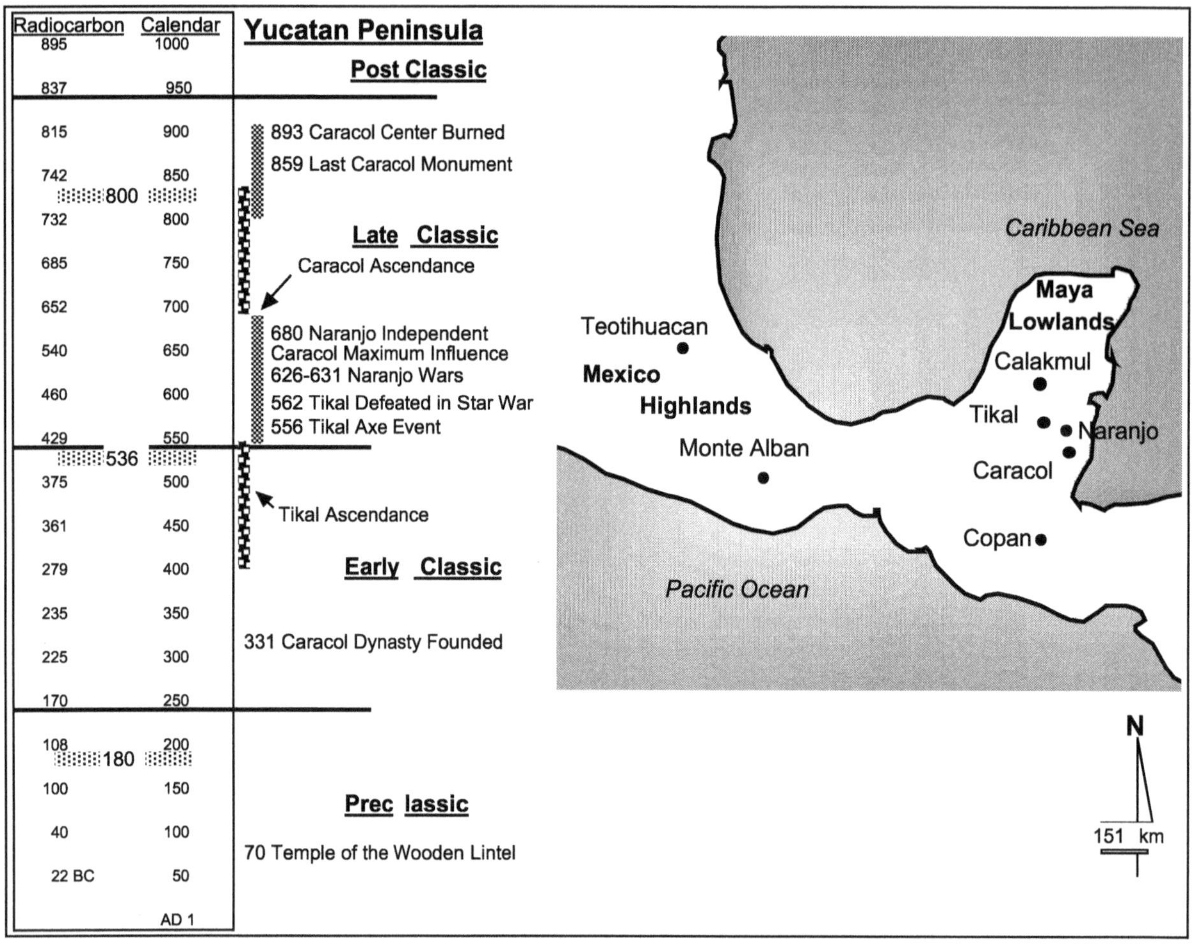

Figure 5.1. Map and Chronology of the Maya Lowlands and Caracol.

event (Baille 1994; Gunn, this volume) may well be one of these broader contexts.

Evidence of Trouble in the Sixth Century Maya Lowlands

Tree-ring and historic writings reflect a climatic change from Ireland through Mediterranean Europe and into mainland China for the decade following the recognized, but still unexplained, A.D. 536 event. The entire northern hemisphere was seemingly affected. The areal extent and length of the associated climatic problems and other catastrophes that can be seen in the Eastern Hemisphere of A.D. 540^{cal} would indicate that the Western Hemisphere may well have experienced similar ill effects. And, indeed, the Maya area the mid-sixth century is one that was characterized by turmoil - an era that was once described by Gordon Willey (1974) as being a practice run for the eventual Maya collapse (Figure 5.1). This "hiatus" was initially viewed as a Maya "dark ages" and was identified by a decline in carved stone monument erection at Maya sites between the years A.D. 534^{cal} and A.D. 593^{cal} (Proskouriakoff 1950:111); this absence of monuments was especially evident and prolonged at Tikal, Guatemala, where a lack of carved stelae and altars spans from approximately A.D. 558^{cal} through A.D. 692^{cal}.

One current explanation ascribes this troubled era to an archaeologically documented shift in political fortunes between Tikal, Guatemala and Caracol, Belize (Chase 1991; Chase and Chase 1987:59-60; see Figure 5.1). A previous perspective held that the troubled times in the Maya lowlands resulted from the economic and political collapse of Teotihuacan in central Mexico (Rathje 1977; Willey 1982), but more recent (and later) dating of Teotihuacan "collapse" (Millon 1988; Diehl and Berlo 1989), as well as concerns over the priority of Teotihuacan versus Tikal developments (Demarest and Foias 1993; Laporte and Fialko 1990:58-59, 1995:66; Valdes 1995:85), have to a large extent disadvantaged this earlier model. An even broader context in which to interpret the political events that unfolded in the sixth century Maya lowlands could be in terms of differential responses to the disastrous world-wide climatic event of A.D. 536.

Many early models of Maya development were predicated on a largely uniform lowland people existing in a redundant environment (Meggers 1954; Sanders 1973); however, we now recognize a great amount of environmental diversity within the Maya lowlands (Sanders 1977; Rice 1993). It is evident as well that the lowland Maya area was characterized by substantial linguistic and cultural diversity (Chase et al. 1991; Schele and Freidel 1990:51; Sharer 1991; Webster 1988). And while this diversity has implications for differential responses and development by the lowland Maya, these potential differences have been neither completely explicated nor explored, partially because of the difficulty in gathering appropriate data and formulating archaeological interpretations.

Likely a direct result of early Carnegie research strategies geared towards investigation of one key site for each time horizon (cf. Chase 1990:149), reconstructions of Maya history have often sequentially focused on a series of primate centers. In past paradigms this primate center was often perceived as being Tikal, Guatemala (Figure 5.2) - the locus of long-term investigations by U.S. and Guatemalan researchers (Coe 1967; Coe and Haviland 1982; Laporte and Fialko 1990, 1995). By partial analogy to the Mexican examples, it was once assumed that Tikal's excavation would provide a framework for understanding the totality of development in the Southern Maya lowlands. We now know that this is not the case. Rather than having a single mega-site, the Maya area can claim at least half a dozen - and all provide amazingly different and diverse adaptations and developmental trajectories. Caracol, Belize (Figure 5.3) was one of these other Maya mega-sites.

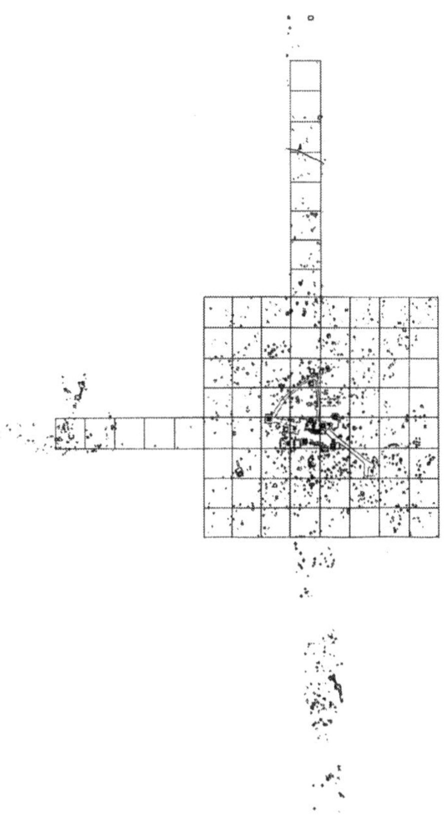

Figure 5.2. Map of Tikal, Guatemala, showing locations of architecture and causeways. This map combines information from Carr and Hazard (1961) and Puleston (1983) and is designed to be easily compared to the Caracol map. Each grid square represents 500 square meters. Grid is oriented to magnetic north, which is at the top of the page.

Caracol, Belize

The site of Caracol is a massive settlement covering approximately 177 square kilometers and located in the western foothills of the Maya Mountains of Belize. While within the Southern Maya lowlands, Caracol is at an elevation of approximately 500 meters above sea level. In its proximity to the Maya Mountains, Caracol had easy access to many resources not found elsewhere in the Southern lowlands (Graham 1987). Caracol also receives heavier rainfall amounts than nearby sites to its west and north (such as Tikal - Rice 1993: Fig. 4), Thus, micro-environmental variations may have been a factor in the ultimate development of Caracol's distinctive history - particularly with regard to the impact of the A.D. 536 event.

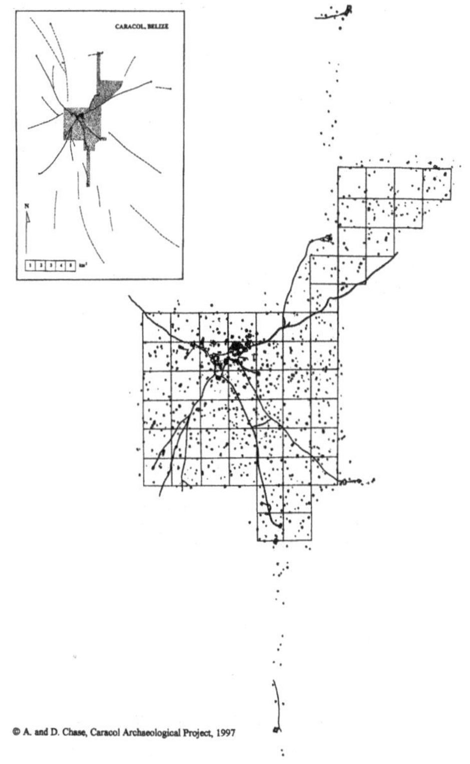

Figure 5.3. Map of Caracol, Belize, showing locations of architecture and causeways. Each grid square represents 500 square meters. Grid is oriented to magnetic north, which is at the top of the page. Inset shows the wider Caracol causeway system. Dashed lines indicate causeways visible on landsat imagery that have not yet been ground-checked.

The University of Central Florida Caracol Archaeological Project began investigations at the site in 1983 and 1984. Formal field seasons started in 1985 and have continued on an annual basis. As a result of these research efforts, a substantial body of data has been collected that document the development of one of ancient Mesoamerica's largest urban centers (A. Chase and D. Chase 1987, in press; D. Chase and

A. Chase 1994). Of the 177 square kilometers that the city of Caracol occupied, approximately 17 square kilometers have been mapped by transit (see Figure 5.3). The central 16 square kilometers of this settlement manifest over 4,400 structures in 903 discrete groups. Extensive excavations have been undertaken in the site epicenter and in the surrounding core and indicate that the site's settlement history begins between B. C. 900 and 600^{cal} and ends by A.D. 1200^{cal}. During 14 seasons of work, 107 plaza groups (12 percent of the mapped sample) have been investigated - some in great detail. Of the 107 groups examined, 102 (95 percent) of them provided evidence of having been occupied at the height of Caracol's prominence in the Late Classic Period between A.D. 550^{cal} and 800^{cal}; the remainder of the investigated groups (n=5) represent contemporary (Late Classic Period) special function architectural complexes. At its peak in A.D. 650^{cal} the city itself consisted of over 36,000 structures in an estimated 9,000 plaza groups housing some 150,000 people. At this time Caracol was the primate center of a small empire that spread over approximately 12,000 square kilometers of the Southern Maya lowlands and exerted some type of control over a much larger population - presumably in the neighborhood of 1,000,000 or more people.

Methodological Concerns in Interpreting Caracol's History

The consideration of Caracol's past requires the conjunction of several bodies of data: calendric, hieroglyphic, and archaeological. Within Maya studies, problems in interpretation exist for each of these data classes.

One of the most critical factors in the interpretation of Maya history is the correlation of the Maya and contemporary calendars. The Maya maintained their own calendar which recorded the passage of time in a cyclical fashion in a base-20 system from an initial starting date centuries before the time of Christ. The correlation of Maya and Christian calendars used by most Mayanists and in this paper (for convenience) is the 11.16.0.0.0 Goodman-Thompson-Martinez correlation (Thompson 1950). Many other potential correlations exist (Kelley 1983; Chase 1986). An 11.3.0.0.0 or 12.9.0.0.0 correlation (both of which have occasional adherents) would change the dates used here by 256 years in either direction. Under a 12.9.0.0.0 correlation, the A.D. 536 event would conceivably be linked with the onset of the Maya collapse at 9.18.2.0.0. Under an 11.3.0.0.0 correlation, the A.D. 536 event would be linked with the onset of the Early Classic Period in 8.12.2.0.0. Under the 11.16.0.0.0 correlation that is generally used by Mayanists, the A.D. 536 event occurred in 9.5.2.0.0, which correlates with the very beginning of the Maya Late Classic era (conventionally dated as starting about A.D. 550^{cal}).

Setting aside the correlation question by assuming it to be 11.16.0.0.0, information concerning Caracol's past history comes from both Maya hieroglyphic texts and archaeological investigations. Yet, the inter-relationships between the hieroglyphic and archaeological records also are very complicated. Because they provide tangible information about royal people and events, the glyphic records often constitute the frame for our perceptions of Maya history. Yet, these records are a form of political history that may be contradicted by other epigraphic interpretation or by archaeological data. For instance, Schele and Freidel (1990:130-164) use the hieroglyphic record to argue for the conquest of Uaxactun by Tikal in A.D. 378^{cal}. Laporte and Fialko (1995:58) question that such an event ever took place; they suggest that extant archaeological data from the well-excavated sites of Uaxactun and Tikal show that it did not. Thus, the hieroglyphic records and the archaeological records must be carefully related to each other in order to understand past Maya history.

The value of conjoining archaeological and epigraphic data are amply demonstrated by examining the Tikal star-war recorded on Caracol Altar 21. The interpretation of hieroglyphic texts are not always straight forward and small changes in readings can have substantial impact on overall interpretation. Grube (1994:106) has recently reinterpreted the warfare recorded on Caracol Altar 21 - in which Tikal was defeated - as not being directly carried out by Caracol (cf. Houston 1987, 1991), but rather by Calakmul. However, there is additional epigraphic and archaeological information pertaining to these interpretations. First, the hieroglyphs and emblem on Altar 21 following the agency expression for this event cannot be assumed to be Calakmul, as Grube (1994:106) implies; the glyphs are too eroded to ascertain this. Grube's (1994:107) dynastic rgconstruction identifies Batz Ek, Caracol ruler K'an II's mother, as coming from Calakmul, but the key texts on Caracol Stela 3 relating to this interpretation do not actually comment on her place of origin. And the use of the "k'ul yax ahau" title by Batz Ek at Caracol, seen as supporting evidence by Grube (1994:107) for a Calakmul association, substantially predates any usage at Calakmul. These same texts also note that K'an II and the Site Q rulers were "brothers" (Grube 1994:108), rather than in a hierarchical relationship as has been implied (Martin and Grube 1995); David Stuart (personal communication, 1997) has re-interpreted this passage to simply note that the Site Q ruler witnessed K'an II's actions. Also problematic for any direct Calakmul connection is the fact that another location for the epigraphically known Site Q has recently been identified at a site in northwest Guatemala, quite distant from Calakmul (Graham 1997; Schuster 1997; Stuart, personal communication, 1997). Furthermore, the geographic distance between Calakmul and Caracol may also be used to imply that the two sites were not in a direct hierarchical relationship (Chase and Chase 1998a). Finally, the archaeological record indisputably shows that Caracol directly benefited from the Tikal star-war with its population growing from 19,000 people to over 120,000 people within one hundred years of this A.D. 562 event (Chase and Chase 1989, 1996a:68; 1997). The only record of this star-war is at Caracol and Caracol's dramatic growth and expansion followed this event. Such a linkage suggests that Caracol was both instrumental in and directly involved in the Tikal defeat, as was originally suggested (Chase 1991; Chase and Chase 1987:33,60; Houston 1987:93, 1991). Thus, a conjoined analysis of archaeological and epigraphic data are necessary to view relationships among sites and the potential impact of any global environmental phenomena.

Site History: Caracol's Rise to Dominance between A.D. 562 to A.D. 680

The Maya of Caracol carved stone monuments and left us an epigraphic political history that extends from A.D. 331cal to A.D. 859cal in an 11.16.0.0.0 correlation. For Caracol most dates of historical significance are derived from Maya texts on over 60 carved stone monuments, on a handful of stuccoed building facades, and on the walls or roofs of a few painted tombs. Individual death dates in tombs at Caracol are recorded for unnamed individuals in A.D. 537^{cal}, A.D. 577^{cal}, and A.D. 634^{cal}; the dedication of tomb chambers is noted in capstone texts for A.D. 582^{cal}, A.D. 614^{cal}, and A.D. 696^{cal}. Unfortunately, none of these dates currently can be correlated either with recorded dates on Caracol's monuments or with the deaths of any of Caracol's dynastic kings. However, these dated tombs can be used to anchor other, less precise, dates that are obtained by traditional archaeological means (e.g., stratigraphy, seriation, C^{14}). Assuming that methodological difficulties have been overcome, such a conjunctive approach permits fine-tuned dating of ancient activities. The archaeological excavations carried out by the UCF Caracol Project confirm some aspects of the site's epigraphic history as well as considerably expand and amplify our understanding of the site's past. Significantly, investigations suggest that times with limited hieroglyphic history are not always times of decline, but rather sometimes correlate with substantial prosperity and growth. The combination of Caracol's inscriptions and archaeology provides us with a detailed view of the development of this distinctive Southern lowland polity.

Archaeology informs us that Caracol had an extensive early occupation. Preclassic Period (ca. 900 B.C. - A.D. 250^{cal}) remains dating to at least 600^{cal} B.C. have been recovered in the outlying settlement. Epicentral remains in several locations date to before the time of Christ. By A.D. 150^{cor} an early version of Caracol's tallest complex - Caana - rose to over 38 meters in height and the Temple of the Wooden Lintel had already been built. The substantial prosperity enjoyed by epi-Preclassic Caracol is revealed in the simple burial of a woman that dates to this era: it was accompanied by at least 32 pottery vessels and over 7,000 jadeite and shell beads. Ritual deposits dating to A.D. 70cor (Chase and Chase 1995) contain cached items that are not seen at Tikal for another 200 years (Krejci and Culbert 1995:113). Thus, Preclassic Caracol was an important site, although not the demographic giant that it became in the Late Classic era.

While the archaeology of Early Classic Caracol (A.D. 250-500^{cal}) suggests that the site prospered, few epicentral texts are known from this era. A Late Classic monument dating to A.D. 798^{cal} records the founding of the Caracol dynasty in A.D. 331^{cal}, but the detailed histories of the site's early rulers have yet to be recovered. Elaborate Early Classic tombs found in the epicenter and its immediate settlement area presumably include members of the royal family, but are not associated with hieroglyphic texts. Contemporary ritual offerings suggest Mesoamerican-wide patterns and trade connections.

Towards the end of the Early Classic era - definitely by A.D. 495^{cal} - Caracol began a tradition of erecting monuments every 20 years (a Maya "katun"). This carved record was particularly focused on "Giant Ahau" altars (a Maya day sign that corresponded to the end of a katun - Beetz and Satterthwaite 1981) and the tradition of erecting these katun markers at Caracol continued uninterrupted from A.D. 495 through 652^{cal}. Stelae, carved upright rectangular stones that often accompanied these katun ceremonies, provide most of what we know about Caracol's political history. The births and enthronements of various Caracol rulers are found in these texts, as are records of their successful wars.

Scattered dynastic interpretations have been made concerning Caracol's early history (Houston 1987:87-88; Grube 1994:105-106). Events dated to A.D. 484^{cal} and A.D. 487^{cal} can be linked to an earlier king or kings, whose name or names are not known. A Giant Ahau altar was set in A.D. 495^{cal} and Stela 13 was erected in A.D. 514^{cal}. Epigraphically, we know that lord K'an I acceded to the throne in A.D. 531^{cal} and erected two katun monuments in A.D. 534^{cal}. His son, Lord Water (or, more correctly, Yahaw Te K'inich), acceded to the throne in A.D. 553^{cal}, possibly under the aegis of a Tikal king (Grube 1994:106). Yahaw Te K'inich enjoyed a relatively long reign and erected katun monuments in A.D. 554, 573, and 593^{cal}. The first non-katun event that can be associated with Lord Water is an A.D. 556^{cal} "axe event" involving Tikal and Caracol. The next event we know about is the Tikal star-war in A.D. 562^{cal}, which effectively proclaimed Caracol's independence.

Following the Tikal star-war, Lord Water remained in power for at least 35 more years. One of his sons, Smoke Ahau, acceeded to Caracol's throne in A.D. 599^{cal}. Little is known, however, about his reign because of the eroded conditions of his monuments. A second son of Lord Water, K'an II, acceeded to the throne in A.D. 618^{cal}. K'an II extended Caracol's military success by incorporating the Guatemala site of Naranjo into the Caracol polity between A.D. 626 and 631^{cal} (with a final, possibly symbolic, event in A.D. 636^{cal}). The archaeological record of Caracol for this era confirms that this was a time of unprecedented population growth at Caracol. Based on a stucco hieroglyphic text on a buried building, K'an II died in A.D. 658^{cal}; 29 days before his death he was succeed by Smoke Skull, whose relationship to K'an II is unknown.

Even before the death of K'an II in A.D. 658^{cal}, Caracol's formal inscriptions began to fall silent. We suspect that the Caracol bureaucracy usurped much of the dynastic power of the kingship following the death of K'an II and that archaeologically visible manifestations of kingship, such as stelae, were largely suppressed. We know, however, from a stucco text on a buried building that Naranjo regained its independence from Caracol in A.D. 680^{cal}. Based on the inscriptions (and in spite of the later lack of them), it is archaeologically clear that Caracol reached its greatest extent and influence between A.D. 631 and 680^{cal}.

Even though only a single carved stone monument (Stela 21 dating to A.D. 702^{cal}) is known from Caracol between A.D. 652^{cal} and 798^{cal}, other epigraphic dates relating to this era occur in a tomb (A.D. 696^{cal}), on stucco building texts (A.D. 658, 662, 680, and about 750^{cal}), and on carved monuments at the Guatemalan site of La Rejolla (A.D. 662 and 672^{cal}) that record Caracol dynastic information. In A.D. 692^{cal} Tikal, Guatemala erected its first monument in well over a century, signaling a new upsurge in that center; interestingly, the Tikal altar of A.D. 692^{cal} is in Caracol style and is the only Giant Ahau altar to occur at that site. The archaeological record indicates that Caracol prospered through the remainder of the Late Classic era, in spite of a dearth of inscriptions to affirm this. There is continued economic activity with the bulk of Caracol's population having access to most goods and services; Caracol simultaneously exhibits a managerial style that does not overly focus on dynastic control.

The final sequence of monuments at Caracol begins in A.D. 798^{cal}, when H'ok Kauil is named as the 27th king in the Caracol royal dynasty, and extends through A.D. 859^{cal}. The epigraphy of this period demonstrates a resurgence of dynastic power, perhaps gained through the manipulation of ideology. This is most likely seen in the iconography of Caracol Stela 18 which aggrandizes the royal dynasty through covering the face of the stela with a huge vision serpent, a motif related to dynastic ritual. Whether manipulated or not, the epigraphic record ceases in A.D. 859^{cal}. The archaeological record indicates that during this late part of Caracol's history, more plaza groups made use of carved and uncarved monuments, but that, in general, goods were not evenly distributed throughout Caracol's society, as they had been for most of the Late Classic Period. The royal elite seemingly had reclaimed control of Caracol's economic system and did not share certain goods with the general populace. The epicenter of the city, occupied by the royal elite, prospered unabated for almost 40 years beyond the last dated monument. At this point, however, the epicenter was burnt in a sudden conflagration that left many artifacts and a dead child on the floors of Caracol's central palaces. Yet, non-epicentral settlement apparently continued for another two centuries in Caracol's rich agricultural fields.

In summary, then, while Caracol was an important site throughout its history, it clearly blossomed after the A.D. 562^{cal} defeat of Tikal and came to dominate the Southern Maya lowlands through approximately A.D. 680^{cal}. Caracol continued to prosper for minimally 200 years beyond this point. Caracol's distinctive cultural and political systems, which functioned until the Terminal Classic era (ca. A.D. 790^{cal}) and which are likely directly responsible for its long-term success, are perhaps best embodied in the site's unusual layout (described below).

Site Layout: Caracol's Distributed Economic System

The site of Caracol is organized in a solar fashion with a series of causeways radiating out from a central concentration of plazas and architecture (see Figure 5.3 inset). When combined with archaeological data, this spatial pattern has implications for social and economic policies at Caracol that can be contrasted with other city plans in the Maya lowlands.

The city of Caracol occupied a huge metropolitan area. Its largest plaza and groups occur in the epicenter of the city (see Chase and Chase 1987:51-54 for definitions of "epicenter," "core," and "mantle" relative to Caracol). Caracol's core settlement is continuous for a distance of 8 to 10 kilometers in any direction from this epicenter. Five main roadways lead out from the epicenter. Most of these split into multiple roads within 1.5 kilometers from the site center; there is some suggestion of further bi- and tri-furcations in these road systems at a distance of 3 kilometers from the epicenter. The resulting causeways bind the core of Caracol together, connecting both high-status groups directly to the epicenter as well as large formal plaza complexes (Chase and Chase 1996b). Two rings of causeway termini can be identified. The first occurs at a distance of approximately 3 kilometers and consists of special function architecture and plazas that appear to be purposefully placed in their loci during the earlier part of the Late Classic era (post A.D. 560^{cal}). The second ring of termini exists at a distance of 5 to 9 kilometers from the epicenter and appears to represent the incorporation of pre-existing centers, and in a few cases elite residential units, into the Caracol urban system. Again, this was accomplished largely during the early part of the Late Classic era - the period noted above as being one of explosive growth. It is important to note that the ring of termini located 3 kilometers distant from the epicenter do not serve as connecting nodes for termini located 7 to 10 kilometers away from the Caracol site center. Instead, each terminus is directly connected to the epicenter by its own causeway branch; none connect with other termini. Thus, the layout is not so much "dendritic" as it is "solar." This layout is also extremely centralized; the site epicenter is interconnected with each terminus independently without an intermediate connector.

Site Variability: Contrasting Caracol's Layout with Other Maya Centers

Caracol's layout is one feature that sets it apart from other Southern lowland Maya centers. The centrally-focused, solar causeway layout contrasts with that at Tikal where causeways connect major epicentral architecture. However, it would be a mistake to conclude that all of Caracol's large architecture is epicentrally bound, for, unlike Tikal, Caracolenos purposefully placed large plaza areas and structural groups throughout its core area, joining these architectural complexes to the epicenter by means of long causeways.

Just as the Caracol causeway system is well-defined and seemingly unique in the Southern Maya lowlands, other aspects of the Caracol settlement and archaeological record are also relatively unusual compared to what we have come to expect as being "typical" of the Classic era Maya. Most notably, this variability stands in strong contrast to what we have come to expect of the Southern Maya from a Tikal perspective. Differences include features that are visible in the layout of the site and other items that are manifest in the archaeological record.

At Caracol, residential plaza groups do not cluster together and are rarely situated immediately adjacent to each other; rather, they are rather fairly regularly spaced over the landscape, especially given the karst backdrop. Because this distribution contradicts an inherent tendency of kin to live near kin, which is observed in the city plans of both Tikal (Haviland 1972) and Copan (Henden 1991), we interpret this feature as indicative of zoning control at Caracol with regard to group location. It, thus, reflects a level of control of the landscape by Caracol's central bureaucracy that is not generally seen at other Classic Maya sites.

The extensive terracing that characterizes most parts of the site and exists between and about almost all plaza groups is also seemingly unique to Caracol and its vicinity. These terraced fields abut the epicenter on all sides and are ubiquitous both in low broad valleys and on steep hillsides throughout the site (Chase and Chase 1998b). Caracol terracing is also associated with features that served to control the flow of water. The intensive and all-encompassing nature of these terrace systems has been interpreted as an attempt by the Caracol Maya at agricultural self-sufficiency, consistent with the site's role as an expansionist imperial capital (Chase and Chase 1996b).

The archaeological remains are also indicative of a variant bureaucratic strategy for controlling the Caracol population. Caracol integrated its society through fostering a sense of nationalism and ethnic identity by permitting widespread access to both goods and rituals that were generally restricted at other Classic Maya sites (Chase and Chase 1996a; see also A. Chase 1992). This interpretation is derived from the extensive settlement work that has been undertaken by the Caracol Project. The settlement mapping and excavation analysis has demonstrated that most Maya households at Caracol had access to tombs and to buildings that functioned as mausoleums within living groups. These same outlying households contain specially-made ritual goods in the form of cache vessels and incense burners that are generally thought to be socially restricted in use at other excavated Maya sites. Polychrome ceramics, many being trade items, appear in simple burials throughout Caracol's outlying Late Classic households. Obsidian, another trade item, occurs within all household groups. And, most household groups also once contained individuals who had their teeth inlaid with jadeite or hematite, a trait once thought to be associated with higher status (Krejci and Culbert 1995:105). Thus, the ancient Caracolenos had access to ritual items and social markers that were highly restricted at other Maya cities. In fact, we believe that the widespread distribution of archaeological tradewares at Caracol is the result of a highly administered economy in which the Caracol epicenter utilized its termini as market areas for the distribution of these items in a further effort to better integrate its population and "nation-build."

Site Summary: Nation Building and its Consequence

As reflected archaeologically, Caracol's development between A.D. 562 and 800^{cal} indicates a highly organized and centralized administration that permitted access to wealth and ritual for a wide proportion of its population. While the components for this organization (and its associated practices) had to be in place prior to A.D. 562^{cal}, they were successfully employed by a series of strong charismatic rulers who used successful warfare and its spoils, as well as a shared internal identity, to integrate a large population. While consolidating its nation-building program, there was also a concerted effort to gain the self-sufficiency desired by an imperial capital through the careful control of both agriculture and settlement in Caracol's extended urban system. Sometime after A.D. 658^{cal}, dynastic power may have been subverted by mid-level bureaucrats who managed the huge metropolis and its economy. The resurgence of Caracol's dynasty at the end of the Classic Period - at least as expressed in carved stone monuments - is correlated to a large extent archaeologically with the cessation of social and economic policies that had formed the bulwark for Caracol's original rise to political power in the Southern lowlands some 250 years earlier.

Archaeology, Epigraphy, and the "Hiatus" at Tikal and Caracol

The need for viewing contemporary variation among sites is nowhere clearer than in the spot-light of the Maya sixth-century "hiatus." Evidence from Caracol and other sites have dispelled the notion that this era was a Maya "dark ages." The archaeological data from Caracol, in fact, document a distinctive and independent Maya cultural adaptation in the Southern lowlands that can be compared and contrasted with contemporary Tikal. When combined with a consideration of the global A.D. 536cal phenomenon, a more complete and interesting interpretation of Maya prehistory can be devised.

What was the situation in the Maya lowlands before, at, and after the A.D. 536 event? In order to understand this era, it is necessary to critically examine both the hieroglyphic records and the archaeological contexts at Tikal and Caracol.

At both Tikal and Caracol, the hieroglyphic records for this period are much disturbed. At Caracol, the critical monuments (Stelae 13,14,15, and 16) were all assembled in front of Structure A4 and some were buried beneath a later altar dating to A.D. 652^{cal} (Beetz and Satterthwaite 1981:52). Other early monuments were also interred beneath altars elsewhere at the site (Grube 1994:91), suggesting that intentional burying of early stelae was a common practice at Caracol. At Tikal the site's earlier monuments were also greatly disturbed (Satterthwaite 1958). They were moved about, reset upside down, as well as placed in the interior of purposefully buried buildings (Jones 1991; Laporte and Fialko 1990). Ancient monument disturbance at Caracol occurred at various times and this activity cannot be ascribed to any single event; however, it has been suggested that extensive movement and disturbance at Tikal was a direct outcome of the Tikal-Caracol conflict (Chase 1991:35-36; Jones 1991:117). Thus, for whatever reasons, a number of key epigraphic records that bracket the global A.D. 536 event at Caracol and Tikal were disturbed in antiquity.

In spite of the ancient Maya disruption of archaeological contexts, there is substantial information on this time. It is evident that the epigraphic records of both sites are continuous before and following A.D. 536^{cal}. Nothing indicative of the A.D. 536 event appears in the texts of either site. Relations between Caracol and Tikal were initially amicable (Grube 1994:106). At Caracol, Lord Water's father, K'an I, came to power in A.D. 531^{cal} and was succeeded 22 years later in A.D. 553^{cal} by Lord Water. Caracol's even earlier dynastic history is presently unclear, but monuments exist at the site that were erected during this era (cf. Stelae 2, 20, and 23) and carved altars were erected on schedule (each katun or 20-year period) at A.D. 534^{cal} and

A.D. 554^{cal}. Stability, however, is evident in the relatively long reigns of Caracol's mid-sixth century royalty (K'an I [acceded A.D. 531cal] - 22 years; Lord Water [acceded A.D. 553^{cal}] - 46 years).

At Tikal, Double Bird came to power in A.D. 537^{cal} and celebrated 20 years of rule in A.D. 557^{cal} on Tikal Stela 17. Six other rulers are thought to have reigned between Jaguar Paw Skull, the 14th ruler, who is noted as being in power in A.D. 488^{cal} on Tikal Stela 7, and Double Bird, the 21st ruler, whose accession to power in A.D. 537^{cal} is recorded on Tikal Stela 17. The names and accession dates of the six rulers during this 49-year timespan are unknown, and the total length of their reigns is a matter of debate (Schele and Freidel 1990:167; Jones 1991:115); Jones (1991:115) suggests that it is possible that all six of these rulers could actually have been sandwiched into a single decade between A.D. 527^{cal} and A.D. 537^{cal}. Even though this era at Tikal is characterized by a great many stelae, these many monuments are largely eroded, broken, and out of context - presumably as a result of the A.D. 562cal star-war. New monuments relating to this era and even earlier periods are still being located at Tikal (Valdes, Fahsen, and Munoz C. 1997), and it can be expected that more dynastic details covering this turbulent time at Tikal will eventually be revealed.

Thus, Tikal's dynastic history was continuous, but volatile for either one or four decades before the A.D. 536 event. Six rulers occupied a span ranging from 10 to 49 years (Jones 1991:115). This turbulent era was followed by the stability of Double Bird's minimally 25-year rule. It is unclear as to how long Caracol's early rulers were in office; while some presumably have been found archaeologically (Chase and Chase 1996c), they are not associated with any hieroglyphic texts. K'an I, who came to power in A.D. 531cal, presumably ruled for 22 years.

The combined evidence from Tikal and Caracol suggests that if the A.D. 536 event had an impact on the Southern Maya lowlands, it could have been to encourage dynastic stability. If the climate worsened and made agricultural production more difficult, then stable dynastic rule could have solidified these polities in the face of environmental problems. Looked at from this standpoint, however, it would appear that Caracol's adaptation - which stressed participatory nation-building through co-opting much of populace into both a highly centralized administered economy and state-sanctioned ritual at the household level - was more successful and "forward-looking" than Tikal's adaptation - which seemingly continued with older, less participatory, social strategies. The success of Caracol's adaptation to environmental stress is suggested in at least three specific lines of evidence: by Caracol's public works (e.g. causeways and agricultural terracing), by Caracol's defeat of Tikal, and by Tikal's political absence from the realm of formal inscriptions for almost 130 years.

In sum, then, the Maya "hiatus" is a misnomer. Rather than suffering a decline - almost on the order of the Maya collapse (Willey 1974) - Maya society was vibrant, changing, interacting, and continuous. Dynastic political fortunes waxed and waned. Even when no monuments are known, such as at Tikal, large constructions continued to be built and elaborate burials continued to be made (cf. Jones 1991:114-119). More recent archaeological data and epigraphic readings suggest that the previous interpretation of a "Maya hiatus" resulted from an incomplete record and from historical accident.

Conclusion

To return to, and to paraphrase, the question posed at the beginning of this discourse, did the A.D. 536 event have an impact on Caracol? Possibly. Did it lead to calamity at Caracol? Probably not. For, there is no evidence for any disaster at Caracol immediately following A.D. 536^{cal} and there is no monument gap. As noted above, during the fifth and early sixth centuries, Caracol monuments were erected once every katun (20 years) and Caracol continued in this process, as expected, during the time of the A.D. 536 event by erecting monuments in A.D. 534^{cal} and in A.D. 554^{cal}. It could be that Caracol's already somewhat distinctive features - particularly its agricultural terracing and potentially higher rainfall inherent in its mountain siting - had buffered it from atmospheric disturbances and given it an advantage that Tikal and other neighboring centers lacked.

Could the A.D. 536 event account for Caracol's success? Indirectly, yes. By causing environmental turmoil in the Southern Maya lowlands, the A.D. 536 event could have contributed to, or even caused, a volatile situation to develop. The differences in cultural backgrounds and in the variable environments of diverse Maya cities could have spawned differing adaptations to a perceived calamity. The higher elevation and higher yearly rainfall of Caracol could have permitted it to follow a different path than that taken by Tikal, ultimately resulting in the A.D. 562^{cal} confrontation. "Just-so stories" of a quarter-century agricultural famine at Tikal, combined with the forced, long-term tribute of Caracol's agricultural products could have resulted in the final conflict. But we, as archaeologists, cannot tangibly deal with these mythical scenarios beyond using them as hypotheses against which to contemplate data. Rather, we can only order known events and then attempt to imply causal relationships. But, this requires studies to establish linkages between local cultural and climatic processes and global changes - work that still remains to be done for the western flanks of the Maya Mountains.

The value of viewing the lowland Maya relative to global phenomena such as the A.D. 536 event go far beyond the development of another "just so story" for a troubled time in Maya history in that it forces the consideration of micro-variations in Maya cultural development. The core-periphery model developed by Rathje (1971) to explain the rise and development of Maya civilization was based on the assumption that all lowland Maya lived in an ecologically redundant lowland environment with little opportunity for agricultural intensification; in that model, the lowland Maya traded ritual thoughts, practices, and specific symbols (i.e., ideology) for necessary household commodities (i.e., material items from the highlands). There was no consideration of internal variation in the Maya lowlands. While few would subscribe to a strict interpretation of Rathje's (1971) original model, the "core"-"periphery" juxtapositioo that is commonly found in broader applications of Wallerstein's world-systems models (cf. Kepecs et al. 1994) implies a simplification of variability that we now know not to be the case. Caracol and Tikal, with epicenters located only 76 kilometers distant from each other, are merely two examples of the cultural and political diversity that once existed within the Sowthern Maya lowlands and of the potential impact that these differences could have had on long- and short-term cultural survival.

Acknowledgements: The University of Central Florida Caracol Project has been assisted by many individuals, institutions, and foundations during its existence. The University of Central Florida and the Government of Belize have been particularly instrumental in ensuring the success of the Project. Over the years major funding for the project has been obtained from numerous sources: private donations to the University of Central Florida (annually); the Harry Frank Guggenheim Foundation (1988, 1989); U.S.A.I.D. and the Government of Belize (1989-1992); the Government of Belize (1993); the National Science Foundation (1988, 1994-1996); the Dart Foundation (1996); the Foundation for the Advancement of Mesoamerican Studies, Inc. (1997); the Stans Foundation (1997, 1998); the J.I. Kislak Foundation (1998); and the Ahau Foundation (1998). Joel Gunn kindly provided editorial comments and inspiration for the final version of this manuscript.

References

Abrams, Elliot and David Rue (1988) The Causes and Consequences of Deforestation Among the Prehistoric Maya. *Human Ecology* 16:377-395.

Adams, Richard E.W. (1973) The Collapse of Maya Civilization: A Review of Previous Theories. *The Classic Maya Collapse*, edited by T.P. Culbert, pp. 21-34, University of New Mexico Press, Albuquerque.

Baille, M.G.L. (1994) Dendrochronology Raises Questions about the Nature of the A.D. 536 Dust-veil Event. *The Holocene* 4(2):212-217.

Beetz, Carl P. and Linton Satterthwaite (1981) *The Monuments and Inscriptions of Caracol, Belize*, University Museum Monograph 45, The University Museum, Philadelphia.

Blanton, Richard E., Feinman, Gary M., Kowalewski, Stephen A., and Peter N. Peregrine (1996) A Dual-Processual Theory for the Evolution of Mesoamerican Civilization. *Current Anthropology* 37(1):1-14.

Carr, Robert F. and James E. Hazard (1961) *Tikal Reports Number 11: Map of the Ruins of Tikal, El Peten, Guatemala*, The University Museum, Philadelphia.

Chase, Arlen F. (1986) Time Depth and Vacuum: The 11.3.0.0.0. Correlation and the Lowland Maya Postclassic. in J.A. Sabloff and E.W. Andrews, V., Eds., *Late Lowland Maya Civilization: Classic to Postclassic*, pp. 99-140, University of New Mexico Press, Albuquerque.

Chase, Arlen F. (1990) Maya Archaeology and Population Estimates in the Tayasal-Paxcaman Zone, Peten, Guatemala. in T.P. Culbert and D.S. Rice, *Precolumbian Population History in the Maya Lowlands*, pp. 149-165, University of New Mexico Press, Albuquerque.

Chase, Arlen F. (1991) Cycles of Time: Caracol in the Maya Realm. in M.G. Robertson and V.M. Fields, Eds., *Sixth Palenque Round Table, 1986, Vol. VII*, pp. 32-42, University of Oklahoma Press, Norman.

Chase, Arlen F. (1992) Elites and the Changing Organization of Classic Maya Society. in D.Z. and A.F. Chase, Eds., *Mesoamerican Elites*, pp. 30-49, University of Oklahoma Press, Norman.

Chase, Arlen F. and Diane Z. Chase (1987) *Investigations at the Classic Maya City of Caracol, Belize: 1985-1987*, Monograph 3, Pre-Columbian Art Research Institute, San Francisco.

Chase, Arlen F. and Diane Z. Chase (1989) The Investigation of Classic Period Maya Warfare at Caracol, Belize. *Mayab* 5:5-18.

Chase, Arlen F. and Diane Z. Chase (1992) El Norte y el Sur: Politica, Dominios, y Evolucion Cultural Maya. *Mayab* 8:134-149.

Chase, Arlen F. and Diane Z. Chase (1995) External Impetus, Internal Synthesis, and Standardization: E Group Assemblages and the Crystalization of Classic Maya Society in the Southern Lowlands. in N. Grube, Ed., *The Emergence of Lowland Maya Civilization*, pp. 87-101, Acta Mesoamericana No. 8, Berlin.

Chase, Arlen F. and Diane Z. Chase (1996a) A Mighty Maya Nation: How Caracol Built an Empire by Cultivating its 'Middle Class,' *Archaeology* 49(5):66-72.

Chase, Arlen F. and Diane Z. Chase (1996b) More than Kin and King: Centralized Political Organization Among the Ancient Maya. *Current Anthropology* 37(5):803-810.

Chase, Arlen F. and Diane Z. Chase (1996c) The Organization and Composition of Classic Lowland Maya Society: The View from Caracol, Belize. in M.G. Robertson, M. Macri, and J. McHargue, Eds., *Eighth Palenque Round Table, 1993*, pp. 213-222, Pre-Columbian Art Research Institute, San Francisco.

Chase, Arlen F. and Diane Z. Chase (1998a) Late Classic Maya Political Structure, Polity Size, and Warfare Arenas. in A. Ciudad Ruiz et al., Eds., *Anatomia de Una Civilizacion: Aproximaciones Interdisciplinarias a la Cultura Maya*, pp. 11-29, Sociedad Espanola de Estudios Mayas, Madrid.

Chase, Arlen F. and Diane Z. Chase (1998b) Scale and Intensity in Classic Period Maya Agriculture: Terracing and Settlement at the 'Garden City' of Caracol, Belize. *Culture and Agriculture* 20(2).

Chase, Arlen F. and Diane Z. Chase (in press) *Investigations at Caracol, Belize: 1988-1996*, Monograph 8, Precolumbian Art Research Institute, San Francisco.

Chase, Arlen F., Grube, Nikolai, and Diane Z. Chase (1991) *Three Terminal Classic Monuments from Caracol, Belize*, Research Reports on Ancient Maya Writing No. 36, Center for Maya Research, Washington, D.C.

Chase, Diane Z. and Arlen F. Chase (1994) Eds., *Studies in the Archaeology of Caracol, Belize*, Monograph 7, Pre-Columbian Art Research Institute, San Francisco.

(1997) La Guerra Clasico Maya Vista por Caracol, Belize. in Sylvia Trejo, Ed., *La Guerra Entre Los Antiguos Maya*, Primera Mesa Redonda de Palenque, INAH, Mexico.

Coe, William R. (1967) *Tikal: A Handbook of Ancient Maya Ruins*, The University Museum, University of Pennsylvania, Philadelphia.

Coe, William R. and William A. Haviland (1982) *Introduction to the Archaeology of Tikal, Guatemala*, Tikal Report No. 12, Monograph 46, The University Museum, University of Pennsylvania, Philadelphia.

Cook, Orator F. (1921) Milpa Agriculture: A Primative Tropical System. *Annual Report of the Smithsonian Institution, 1919*, pp. 307-326, Smithsonian Institution, Washington, D.C.

Cooke, C. Wythe (1931) Why the Mayan Cities of the Peten District, Guatemala were Abandoned. *Journal of the Washington Academy of Sciences* 21(13):283-287.

Dahlin, Bruce H. (1983) Climate and Prehistory on the Yucatan Peninsula. *Climatic Change* 5:245-263.

Demarest, Arthur A. and Antonia E. Foias (1993) Mesoamerican Horizons and the Cultural Transformations of Maya Civilization, in D.S. Rice, Ed., *Latin American Horizons*, pp. 147-191, Dumbarton Oaks, Washington, D.C.

Diehl, Richard A. and Janet C. Berlo (1989) Eds., *Mesoamerica After the Decline of Teotihuacan: A.D. 700-900*, Dumbarton Oaks, Washington, D.C.

Graham, Elizabeth (1987) Resource Diversity in Belize and Its Impications for Models of Lowland Trade. *American Antiquity* 52:753-767.

Graham, Ian (1997) Mission to La Corona. *Archaeology* 50(5):46.

Grube, Nikolai (1994) Epigraphic Research at Caracol, Belize. in D.Z. Chase and A.F. Chase, Eds., *Studies in the Archaeology of Caracol, Belize*, pp. 83-122, Pre-Columbian Art Research Institute, San Francisco.

Gunn, Joel and Richard E.W. Adams (1981) Climate Change, Culture, and Civilization in North America. *World Archaeology* 13(1):87-100.

Haviland, William (1972) A New Look at Classic Maya Social Organization at Tikal. *Ceramica de Cultura Maya* 8:1-16.

Henden, Julia (1991) Status and Power in Maya Society: An Archaeological Study. *American Anthropologist* 93:894-918.

Hodell, David A., Curtis, Jason H., and Mark Brenner (1995) Possible Role of Climate in the Collapse of Classic Maya Civilization. *Nature* 375:391-394.

Houston, Stephen D. (1987) Appendix II: Notes on Caracol Epigraphy and Its Significance. in A.F. Chase and D.Z. Chase, *Investigations at the Classic Maya City of Caracol, Belize: 1985-1987*, pp. 85-100, Pre-Columbian Art Research Institute, San Francisco.

Houston, Stephen D. (1991) Appendix: Caracol Altar 21. in M.G. Robertson and V.M. Fields, Eds., *Sixth Palenque Round Table, 1986*, pp. 38-42, University of Oklahoma Press, Norman.

Issac, Barry L. (1996) Approcches to Classic Maya Economies. *Research in Economic Anthropology* 17:297-334.

Jones, Christopher (1991) Cycles of Growth at Tikal. in T.P. Culbert, Ed., *Classic Maya Political History:Hieroglyphic and Archaeological Evidence*, pp. 102-127, Cambridge University Press, Cambridge.

Kelley, David H. (1983) The Maya Calendar Correlation Problem. in R.M. Leventhal and A.L. Kolata, Eds., *Civilization in the Ancient Americas*, pp. 157-208, University of New Mexico Press and Harvard University Press, Cambridge.

Kepecs, Susan, Feinman, Gary, and Sylviane Boucher (1994) Chichen Itza and Its Hinterland: A World-Systems Perspective. *Ancient Mesoamerica* 5:141-158.

Krejci, Estella and T. Patrick Culbert (1995) Preclassic and Classic Burials and Caches in the Maya Lowlands. in N. Grube, Ed., *The Emergence of Lowland Maya Civilization*, pp. 103-116, Acta Mesoamericana 8, Berlin.

Laporte, Juan Pedro and Vilma Fialko C. (1990) New Perspectives on Old Problems: Dynastic References for the Early Classic at Tikal. in F.S. Clancy and P.D. Harrison, Eds., *Vision and Revision in Maya Studies*, pp. 33-66, University of New Mexico Press, Albuquerque.

Laporte, Juan Pedro and Vilma Fialko C. (1995) Un Reencuentro con Mundo Perdido, Tikal, Guatemala. *Ancient Mesoamerica* 6:41-94.

Marcus, Joyce (1995) Where is Lowland Maya Archaeology Headed? *Journal of Archaeological Research* 3:3-53.

Martin, Simon and Nikolai Grube (1995) Maya Super-States. *Archaeology* 48(6):41-46.

Meggers, Betty J. (1954) Environmental Limitation on the Development of Culture. *American Anthropologist* 56:801-824.

Messenger, Lewis C., Jr. (1990) Ancient Winds of Change: Climatic Settings and Prehistoric Social Complexity in Mesoamerica. *Ancient Mesoamerica* 1(1):21-40.

Millon, Rene (1988) The Last Years of Teotihuacan Dominance. in N. Yoffee and G.L. Cowgill, Eds., *The Collapse of Ancient States and Civilizations*, pp. 102-164, University of Arizona Press, Tucson.

Nicholson, H.B. (1960) The Mixteca-Puebla Concept in Mesoamerican Archaeology: A Re-examination. in A. Wallace, Ed., *Men and Cultures*, pp. 612-617, International Congress of Anthropological and Ethnological Sciences, University of Pennsylvania, Philadelphia.

Proskouriakoff, Tatiana (1950) *A Study of Classic Maya Sculpture*, Publication 593, Carnegie Institution of Washington, Washington, D.C.

Puleston, Dennis E. (1979) An Epistemological Pathology and the Collapse, or Why the Maya Kept the Short Count. in N. Hammond and G.R. Willey, Eds., *Maya Archaeology and Ethnohistory*, pp. 63-71, University of Texas Press, Austin.

Puleston, Dennis E. (1983) *Tikal Report No. 13: The Settlement Survey of Tikal*, Monograph 48, The University Museum, Philadelphia.

Rathje, William L. (1971) The Origins and Development of Lowland Classic Maya Civilization. *American Antiquity* 36:275-285.

Rathje, William L. (1977) The Tikal Connection. in R.E.W. Adams, Ed., *The Origins of Maya Civilization*, pp. 373-382, University of New Mexico Press, Albuquerque.

Rice, Don S. (1993) Eighth-Century Physical Geography, Environment, and Natural Resources in the Maya Lowlands. in J.A. Sabloff and J.S. Henderson, Eds., *Lowland Maya Civilization in the Eighth Century, A.D.*, pp. 11-63, Dumbarton Oaks, Washington, D.C.

Sanders, William T. (1973) The Cultural Ecology of the Lowland Maya: A Reevaluation. in T.P. Culbert, Ed., *The Classic Maya Collapse*, pp. 325-365, University of New Mexico Press, Albuquerque.

Sanders, William T. (1977) Environmental Heterogeneity and the Evolution of Lowland Maya Civilization. in R.E.W. Adams, Ed., *The Origins of Maya Civilization*, pp. 287-297, University of New Mexico Press, Albuquerque.

Sanders, William T. and Barbara J. Price (1968) *Mesoamerica: The Evolution of a Civilization*, Random House, New York.

Satterthwaite, Linton (1958) The Problem of Abnormal Stela Placements at Tikal and Elsewhere. in E.M. Shook, W.R. Coe, V.L. Broman, and L Satterthwaite, *Tikal Reports Numbers 1-4*, pp. 61-83, The University Museum, Philadelphia.

Schele, Linda and David A. Freidel (1990) *A Forest of Kings: The Untold Story of the Ancient Maya*, William Morrow, New York.

Schuster, Angela M.H. (1997) The Search for Site Q. *Archaeology* 50(5):42-45.

Sharer, Robert J.

Schuster, Angela M.H. (1991) Diversity and Continuity in Maya Civilization: Quirigua as a Case Study. in T.P. Culbert, Ed., *Classic Maya Political History*, pp. 180-198, University of New Mexico Press, Albuquerque.

Stone-Miller, Rebecca (1993) An Overview of 'Horizon' and 'Horizon Style' in the Study of Ancient American Objects. in D.S. Rice, Ed., *Latin American Horizons*, pp. 15-39, Dumbarton Oaks, Washington, D.C.

Thompson, J. Eric S. (1950) *Maya Hieroglyphic Writing: Introduction*, Publication 539, Carnegie Institution of Washington, Washington, D.C.

Tozzer, Alfred M. (1930) Maya and Toltec Figures at Chichen Itza. *Proceedings, XXIII International Congress of Americanists*, pp. 155-164, New York.

Tozzer, Alfred M. (1957) *Chichen Itza and Its Cenote of Sacrifice: A Comparative Study of Contemporaneous Maya and Toltec*, Memoirs of the Peabody Museum of Archaeology and Ethnology, Nos. 11 and 12, Harvard University, Cambridge.

Vaillant, George C. (1935) Chronology and Stratigraphy in the Maya Area. *Maya Research* 2:119-143.

Valdes, Juan Antonio (1995) Desarrollo Cultural y Senales de Alarma entre los Mayas: El Preclasico Tardio y la Transicion hacia el Clasico Temprano. in N. Grube, Ed., *The Emergence of Lowland Maya Civilization*, pp. 71-85, Acta Mesoamericana 8, Verlag von Flemming, Berlin.

Valdes, Juan Antonio, Federico Fahsen, and Gaspar Munoz Cosme (1997) *Estela 40 de Tikal: Hallazgo y Lectura*, Instituto de Antropologia e Historia de Guatemala, Guatemala.

Wallerstein, Immanuel M. (1974) *The Modern World System*, 3 volumes, Academic Press, New York.

Webster, David (1988) Copan as a Classic Maya Center. in E.H. Boone and G.R. Willey, Eds., *The Southeast Classic Maya Zone*, pp. 5-30, Dumbarton Oaks, Washington, D.C.

Willey, Gordon R. (1974) The Classic Maya Hiatus: A Rehearsal for the Collapse. In *Mesoamerican Archaeology: New Approaches*, edited by N. Hammond, pp. 417-444, University of Texas Press, Austin.

Willey, Gordon R. (1982) Maya Archaeology. *Science* 215:260-267.

III. CHANGES IN DYNASTIES AND EMPIRES: FAR HORIZONS

-

In the far horizons concept set, we depart, for western preconceptions, the familiar historical venues of Europe and the Maya lowlands to examine sudden social changes in distant and less familiar places. For me these chapters clarified the affects of the A.D. 536 event. The "fall of Rome" and the "Maya Collapse" are all too familiar corridors of cultural change. Expanding to the distant horizons of Africa and east Asia somehow augments the range of perspectives sufficiently to place the stereotypes of Rome and the Maya in a new and broader context: a context that reveals the rich variety of human reactions to surprise circumstances; a context that almost incidentally reverses the Roman and Maya stereotypes. Should global warming precipitate sudden worldwide environmental change, these chapters open the doors of past human knowledge of how to cope with unusual and extreme circumstances. It is in these chapters that the derangement of empires and dynasties is most evident.

Margaret Houston's and Peter Schmidt's chapters hold in common physical separation from any other chapters' regional foci and from each other. In part, this departure from the pattern of closely spaced pairs reflects an effort to cast about for future fields of research. The chapters clearly demonstrate that tropical and subtropical Africa and the far east of Eurasia are fertile fields for this research. Both historical and archaeological records mark the passage of A.D. 536^{cal}. They also both demonstrate and support the contention that by A.D. 536^{cal}, the munificent global conditions of the Roman Optimum (200 B.C.-A.D. 200) were already well faded.

Houston's chapter delves into the first millennium history of China. During the globally cool Vandal Minimum (A.D. 350 to 750^{cal}), China experienced drier than usual climate. The sixth century in particular is noted for cooler and drier climate, frequent warfare, and state fragmentation. Both the Vandal Minimum and sixth century climate are considered against the backdrop of Chinese regional climates. The causes and affects of the A.D. 536^{cal} dry fog are investigated through Chinese historical records. Chinese records suggest comets that could match the event. Chinese sunspot observations show low solar activity during this period, which could have augmented the generally cooler conditions of the Vandal Minimum episode (see chapter 1 for more sources on this issue). Chinese dynastic histories note severe droughts, unseasonable frosts, and famine in years following A.D. 536^{cal}. The year presaged a half century of massive political instability in China with rival dynasties in the North and South and internecine warfare in the North. The unfavorable climate probably worsened that instability for both the expected economic reasons and ideological ones. Even before the A.D. 536 crisis, possibly since the first downturns of global temperatures in the third century, cultural adjustments to accommodate possible reduced crop yields included developing new institutions to store government grain and heightened knowledge of agricultural techniques suited to dry conditions. After the crisis the Chinese excavated a canal to facilitate transportation of emergency stores. Any incompetence in dealing with disasters or portents affected Chinese ideas of the legitimacy of the state. If the state cannot manage grain shortages, floods, and droughts, or explain comets and meteors, it risked losing the "mandate of heaven," its bid for legitimacy. Individual adaptations could have included sixth century retreats into other-worldly religions such as Buddhism and Daoism.

Houston's take on Chinese social history finds a fisioning of the Chinese government into eastern and western factions. One faction moved its capital to the more steppic northwestern China region and the other to the better-watered North China Plain. Houston's Chinese chapter, along with the Schmidt's, Jones's, and Young's chapters, bolsters the intriguing historical side of the issue and adds details to a catalogue of social coping mechanisms and local climatic variations.

I was intrigued by the Chinese comet data. One wonders if a comet of some provenience would be more or less visible from Europe than China? An interesting question is raised if a comet did strike the earth on the Chinese side of the earth. In the case of a severe blow to the earth, the sphere of the earth acts like a lens and focuses energy on the opposite side of the earth (XX). Earthquakes and eruptions in the Byzantine Empire (Baillie 1995) could suggest a comet/meteor strike on the opposite side of the earth.

Schmidt's chapter carries the issue to equatorial and subtropical climes. His chapter provides yet another combination of archaeological and historical comparison. He first relates his research in archaeologically known Lake Victoria area. Schmidt then turns to the ancient Empire of Axum for comparative historical material. Over the last several decades archaeological research west of Lake Victoria has shown that in the early first millennium A.D. an economically advanced culture based on an innovative iron technology grew to singular prominence on the African continent. With this growth came degradation of the environment. Gallery raid forest—used for fuel in smelting

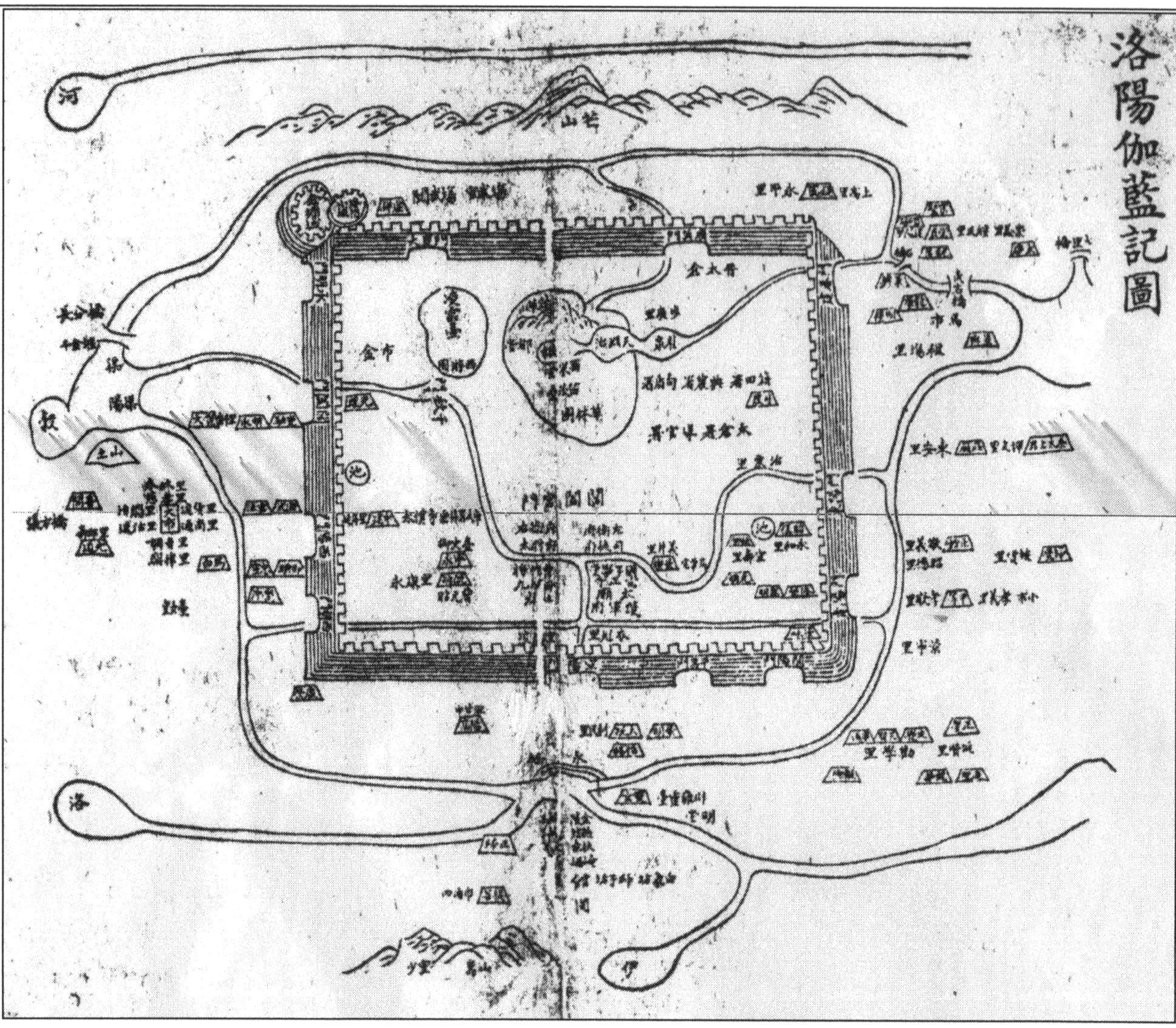

A woodcut from a Chinese book shows the city of Luoyang before its abandonment in A.D.534. The scholar laments losing Luoyang, probably around A.D. 555 (Hsuan-chih Yang, Lo-yang chieh lan chi. [A Record of Buddhist Monasteries in Luoyang], Shanghai, 1833. The book resides in the University of Hawaii's library.

and converted to agriculture—disappeared over a four hundred year period. With deforestation came soil erosion cycles. Yet by A.D. 600^{c14} iron smelting had disappeared along with most of the population. There is no documentation to explain why this region was so rapidly abandoned. However, historical linguistics confirms that groups left the region to seek a more pastoral life in the grasslands to the north. The A.D. 536 event, coming as it did at the nadir of environmental difficulties, could have added stress to an already marginal subsistence system. Historical evidence from the Axum Empire in Ethiopia and Eritrea indicates crop failures, famine, and ancillary diseases associated with A.D. 536^{cal}. In A.D. 524^{cal} the Axum Empire fielded and sustained 120,000 soldiers in a campaign on the Arabian Peninsula. By A.D. 536^{cal} Axum only manage 3,000 soldiers who were easily defeated in a campaign of self-defense. Within five years of A.D. 536^{cal}, historical sources indicate that disease and plague had ravaged "Ethiopia" as well as Egypt and the Middle East. Alternative but alternatively parallel accounts can be seen in Keys (1999).

Schmidt has been about documenting environmental degradation in tropical Africa during the Iron Age for some time and to good effect in raising consciousness of landscape evolution in tropical Africa. This discussion sorts out interaction between human-induced degradation and apparent A.D. 536^{cal} jarring of the Iron Age societies resulting in a cultural switch to pastoralism. Also, I was very pleased to see that there is a historical side of the story from the Axum Empire. It makes the chapter balance very well with the other chapters with historical components. While seasoned with history, the chapter is not a powerful historical account as the dry fog eye witness reports from Europe and China. As in Jones's chapter, it provides cross-references of history and archaeology and suggests means of reconciling the two to the advantage of both.

Global Change. The African and Chinese studies bring the question of the collapse/reformulation of civilization, ably dealt with in previous chapters into stark focus. This sharpened focus is a product of, more than anything else, an expanded number of closely examined cases. Until now the question of the collapse of civilization has hung somewhere in the doldrums between the supposed slow death of the Roman Empire in the early part of the first millennium and the presumed sudden and catastrophic demise of the Maya late in the millennium. All of the historical studies from the eastern hemisphere seem to suggest that while empires were generally not the well-toned social institutions of the Roman Optimum, they were, nevertheless, doing well. Empires were universally disturbed by the A.D. 536 event, some never to rise again, some to recover and rule another day. Oddly, it was the Roman Empire, or at least its subsequent form, that suffered grievous harm, while the Maya as a population rose relatively suddenly to formulate a new outlook and social form. Civilization in east Africa seems to have received a crushing blow, as did Celtic Roman Briton civilization. Chinese civilization shattered, but reformulated itself in about a half century. The central regions of France seem to have suffered patiently but continued with some amount of continuity.

These world wide snap-to-grid investigations, facilitated by the monumental character of the A.D. 536 event, minimally raise a number of questions about the collapse of civilizations, and maximally uncovers a broad range of responses ranging from civilization death, as in the Roman and African cases, to serious reformulation of society and institutions as in the Chinese and Maya cases.

Keys, David (1999) Catastrophe: An Investigation into the Origins of the Modern World. Ballatine Books, New York.

6. CHINESE CLIMATE, HISTORY, AND STATE STABILITY IN A.D. 536

Margaret Snow Houston

St. Andrews College, Laurinburg, North Carolina 28352

Global and Local Climate from Chinese Historical Records

[China, east, west, north, central; historical observation, eye-witness accounts, regional climate variation; drought, moisture, floods, solar emissions, state instability; 28^{cal} B.C. to A.D. 1638^{cal}, Vandal Minimum, Wei and Liang dynasties]

Chinese written records provide a trove of data on astronomical processes as well as local climate and sociocultural adaptations for the last 2,000 or more years. Following the model of Gunn (1994), this chapter will examine possible causes of the A.D. 536^{cal} event and its aftermath, how it interacted with regional meteorological processes, and whether it contributed to state instability in north and central China in the Eastern Wei, Western Wei, or Liang Dynasties. As noted by Baillie (1995) and other scholars, there were great disruptions of Chinese life in the years following A.D. 536^{cal} which I will discuss below. My specific interest, however, is in the impact on state stability.

The Vandal Minimum climate episode from approximately A.D. 350-750^{cal} has been noted as an episode of lower global average temperature (Gunn 1994; see Chapter 2). When seeking possible causes of climatic change, one component that can be investigated in the Chinese historical records is solar emissions. Chinese court astronomers recorded large sunspots visible to the naked eye since at least 28^{cal} B.C., often describing their size and shape. The record is not comparable to the modern weighted count of telescopically observed sunspots, but it could indicate periods of no sunspots, which are correlated with low solar emissions (Eddy 1994; Gunn et al. 1995; Lean et al. 1992). The French astronomer de Moidrey (1904:63-66) found 94 sunspot accounts in Chinese records between 28 B.C. and A.D. 1638^{cal}; between A.D. 350 to 750^{cal} he noted 23 sunspots, from an equal time period from A.D. 750 to 1150^{cal} he found 29 sunspots, not a significant difference. However, sunspots are absent from A.D. 513 to 567^{cal} and a larger gap occurs from A.D. 579 to 787^{cal}. The absence of sunspots hints that the Vandal minimum was a period of low solar activity.

Studies of China's past climates show a cool episode from A.D. 350 to 750^{cal} that was drier than normal. Numerous climatological investigations have compiled the flood and drought records from official Chinese histories, local chronicles, and other sources to detect precipitation patterns. Wang et al. (1987) analyzed records for the last 2,000 years by region. Before A.D. 951^{cal} they could only assess the most serious floods and droughts in the middle reaches of the Yangtze (Changjiang) and Yellow (Huanghe) rivers (Figure 6.1) by the relative frequency of droughts to floods (D/D+F). Dry periods cover the fourth to sixth centuries in the Yellow River and the fourth century in the Yangtze River (Wang et al. 1987:27). Li Zhaoyuan and Quan Xiaowei (1987:57-58) concentrated on the Chang-an region (modern Xi'an) analyzing climatic change at several time scales. The greater part the Vandal Minimum from A.D. 380-680^{cal} was dry; at their most resolved scale, A.D. 380-450^{cal} was dry, 460-500^{cal} was very dry, and 510-680^{cal} was again dry (see Figure 6.1). Thus the A.D. 536^{cal} event occurred in a China that had been drier than normal for several centuries, especially in the north.

Climatic Perturbation in A.D. 536^{cal}

[China, World; history, tree rings, ice cores; meteors and comets, aerosols, volcanoes; A.D. 530 to 545^{cal}]

The ultimate cause of the A.D. 536^{cal} weather crisis is not precisely delineated in the Chinese historical sources. Speculation by modern researchers centers on a volcanic eruption in the tropics, possibly Mt. Rabaul (Stothers 1984). Also being considered are comet impacts and meteor swarms. The ash from a volcano blocks sunlight from reaching the earth's surface; the effect of sulfuric acid aerosols persists longer and is responsible for most of the cooling. Chinese astronomical records display negative evidence for the specific year A.D. 536^{cal}. Ho Peng Yoke (1962:164-165) found comet sightings in A.D. 530, 533, 535, and 539^{cal}. Biot (1848:145) found records of meteor showers in A.D. 532^{cal}. The same meteor shower occurs in both the Chinese and Korean observations (Imoto and Hasegawa 1958:135). While there is no massive comet or meteor shower reported in the Chinese records in A.D. 536^{cal}, the sightings in the immediate years before should be considered as potential causes, especially A.D. 535^{cal} (see Baillie 1998 for similar antecedent phenomena). Also, any causative event did not have to be visible from China. Bright comets tend to be visible around the world at some point in their orbit; any large comet close enough to graze or hit the earth would probably be visually prominent. Meteors that glow when coming through the earth's atmosphere are commonly visible over a 500 mile radius depending on the altitude they begin to burn. Comets and meteors were portents of disaster for the Chinese; they could have been reported and would have influenced how people viewed associated climatic changes (see Young's chapter for similar beliefs in the occident). A possible historical record of a massive volcanic dust cloud would be several years of redder than usual sunsets or the reddish color of the moon during a partial eclipse as after El Chichon, Mexico erupted in 1982. An

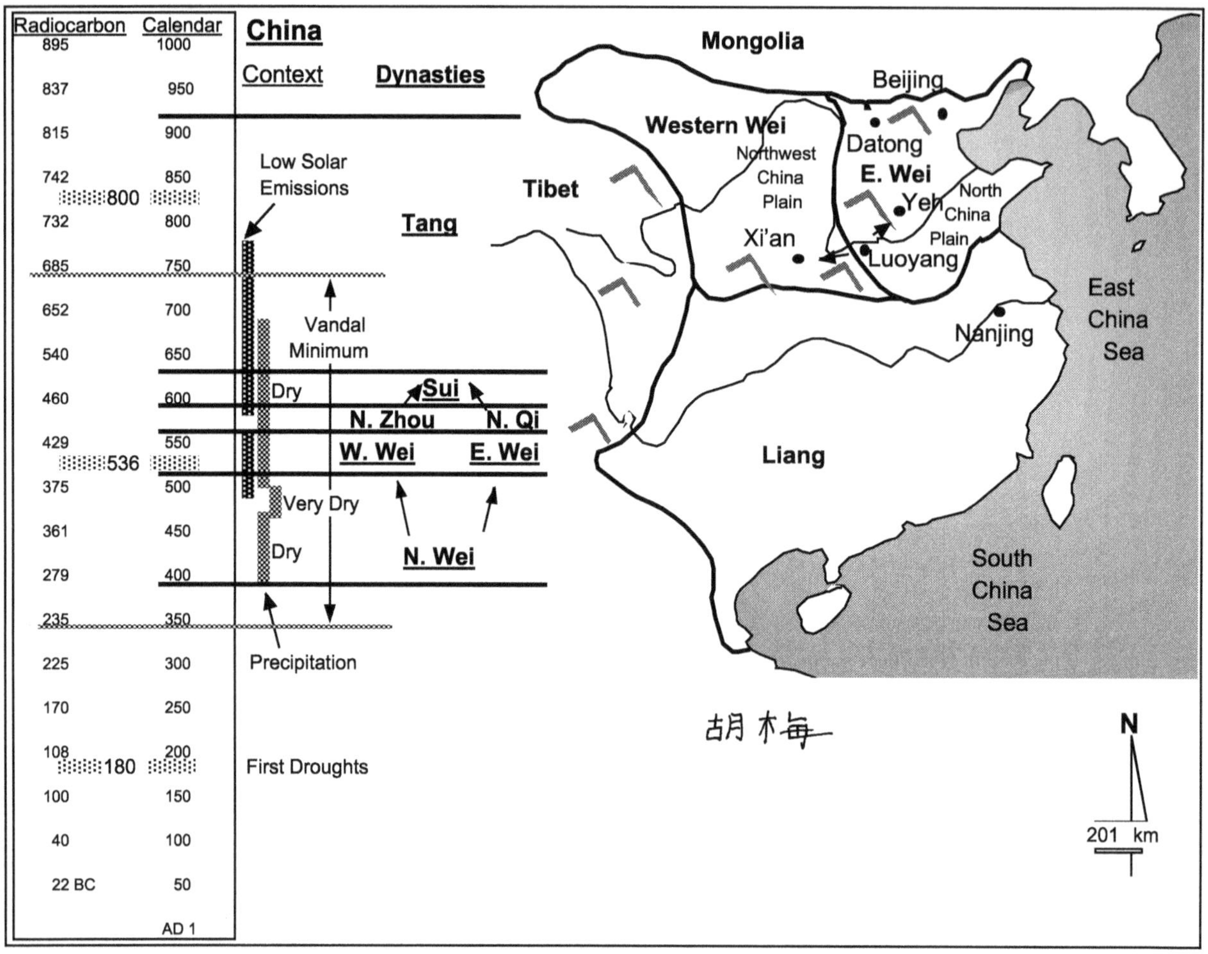

Figure 6.1. Map and Chronology of East Asia.

exhaustive search of the Chinese historical accounts might discover descriptions of such phenomena.

The A.D. 536^{cal} climatic event is mirrored in both tree-ring and ice core records. Cold years leave narrow tree-rings while frost can damage tissue. Sulfuric acid aerosols mark the Greenland ice core annual layers with increased acidity and directly implicate volcanic activity. Baillie (1995:70-71) notes the period from A.D. 536 to 545^{cal} as having the worst conditions for Irish oak trees in nearly thousands of years. One narrow ring occurs at A.D. 536^{cal} and the narrowest rings continue from A.D. 540 to 542. The sixth century is covered in both Greenland and Antarctic ice cores. The closest volcanic acid peak from the Byrd Station, Antarctic ice core dated to A.D. 505, although Hammer et al. (1997:6) note that not all northern hemisphere volcanos appear in the Antarctic ice record. From Greenland the Crete core only reaches back to A.D. 553, and the Camp Century core is unreliable after A. D. 43 (Hammer et al. 1980). Zielinski et al. (1994:949) measured sulfate concentrations in the Greenland GISP2 core and suggested which volcanos might be responsible; their A.D. 508 event is listed with volcanic eruptions from Oshima, Japan dated to A.D. 500+ 200 years and Sheveluch, Kamchatka, Russia dated to A.D.500+ 50 years. Clausen et al. (1997:26,714) measured the acidity in the southern Greenland GISP 1 (or Dye 3) ice core and compared the results with the Central Greenland GRIP Summit core. The GISP 1 or Dye 3 core shows acidic signals in A.D. 516, 530, and 534. The GRIP core closely matches at A.D. 514, 527, and 532 with the acidity lasting from one to two years. While not as precise as tree-ring dates, the last set is reasonably close to the A.D. 536 event.

China's Regional Climates

[China; meteorology.; shifting high pressure, monsoons South Pacific High, Mongolian High; modern]

China's regional climate is markedly continental with westerlies poleward of 30 degrees north and the Mongolian high pressure making most of central China cold in the winter. Easterlies and variable monsoons from both the Indian and Pacific Oceans bring rains gradually northward during the summer months (Zhao 1986:17-18). The westerlies must flow around the high Tibetan Plateau, which influences airflow in China. China's temperature is influenced by the position of the Mongolian High, while rainfall is determined by the position of the Southern Pacific High. As the Mongolian High moves south and reduces the strength of the Southern Pacific High, both China's temperature and precipitation decrease (Feng et al. 1993:175). What factors can change the normal temperature and rainfall patterns? Factors decreasing the temperature in Mongolia-Lake Baikal region, such as snow cover, can move cold air masses rapidly into eastern or southern China (Bao 1987:27). United Nations summaries indicate that a southward shift in the West Pacific High can affect the timing and position of the rainfall belt over eastern China, often leaving North China dry; at present this is due to decreasing solar radiation and lack of clear skies caused by modern pollution (WMO 1995:46). Presumably major volcanic dust and aerosols would generate similar weather patterns. Also, China has been burning coal since at least the fourth century A.D.(Fairbank and Reischauer 1989) and the North China plain suffers from dust storms from the Gobi Desert during the dry winter. Both would add to the burden of pollution from any other cause.

Chinese Historical Records of Climatic Disturbances

[China, northern, eastern; history; unusual weather and climate, frost, ice, snow, famine, drought, star visibility; sixth century, A.D. 530 to 544[cal]]

Extreme weather conditions characterize the A.D. 536 decade in the Chinese historical accounts. The Chinese from very early state periods sensed that writing a truthful history of a dynasty was a means to learn from the mistakes of the past. A yet older custom encouraged noting astronomical events, and floods and droughts as portents or disasters; if rulers did not cope with them efficiently, such portents could be a sign that the rulers had lost the "mandate of heaven" or legitimacy. Hence these events were deemed worthy of recording and inclusion in official dynastic histories. A dynastic history was written after its fall, usually by officials of the succeeding regime. Systematic use of all Chinese historical documents for research on many topics including climate is currently a major focus of Chinese scholarship.

Kevin Pang and Hung Hsiang Chou (in Weisburd 1985:93) searched for references to atmospheric conditions around A.D. 536[cal]. The star Canopus was not visible at the spring and fall equinoxes that year and in the Ching state just south of the Shandong peninsula "frost and snow in July and August that killed the seedling crop, causing a major famine the following autumn." Problems continued into A.D. 538[cal]. A map created by Pang and Chou shows early frost, drought, and famine in A.D. 536[cal], summer snow and famine in A.D. 537[cal], and famine in A.D. 538[cal]. They identified an account from a kingdom north of the Yellow River where 70 to 80% of the population perished (Spaceflight 1989). As indicated by Pang and Chou's research, cold and drought are associated in China because onslaughts of cold Siberian air block the northward progress of moist tropical air much as in eastern United States.

The systematic study of Chinese flood and drought records requires special analytical techniques. They do not measure rainfall; a drought is catalogued as less rain than expected, while floods can be caused variably by excess rainfall or a failure of the authorities to maintain irrigation or flood control works. Also there is no guarantee of uniform coverage over time as records may not be as carefully kept in periods of overthrown dynasties.

In addition to Pang and Chou, several scholars provide information on moisture conditions. Yao (1941:178) has analyzed flood and drought records from the encyclopedia compiled from the official histories called the *T'u-shu Chi-ch'eng*, and found severe drought in June A.D. 535[cal] and worse in A.D. 537[cal]. The A.D. 537 [cal] drought occurred during late April to late May in four provinces: Shansi (Shanxi), Shensi (Xi'an), Kansu (Gansu), and Honan (Henan). It is described as being severe with frost and famine. Frost in April and May is unusual now, even for elevations in Shanxi. Floods are not noted for these years (Table *.1), the next occurred in A.D. 538[cal]. A practical agricultural handbook, *Ch'i Min Yao Shu*, probably written from A.D. 533 to 544[cal], is full of hints to conserve moisture and has flowering dates several weeks later than today's (Shih 1982). A rebellious general and his army were able to ford the Yellow River north of Luoyang by surprise because the water only came up to the horses's knees (Yang 1984:37-38). This occurred after November A.D. 530[cal] and indicates extreme dryness in north China.

There are indications that China was generally colder in the sixth century[cal] than it is today. Chu Ko-chen (1973:234) notes that the southern Liang dynasty capital Cheinyeh (modern Nanjing) was supplied with an icehouse. Currently it is not cold enough in Nanjing to freeze ice for stocking an icehouse.

Well before A.D. 536 [al], moister conditions are indicated. A hurricane, and indication of warmer and moisture air moving northward, overturned houses and trees in the Northern Wei capital of Luoyang in A.D. 526[cal] (Yang 1984:21).

To summarize, then, the 530s were in general colder and drier than normal. The year A.D. 536[cal] was remarkable even in this context with rare phenomena including reduced atmospheric transparency, correlated extreme cold and drought, and resulting famine. The period stands out from the other meteorological disasters in sixth century[cal] China noted in historical accounts. The episode of extreme conditions may have commenced before A.D. 536[cal]. However, as Baillie (1995) points out, there could be no precise onset; multiple disturbances could be indicated with A.D. 536[cal] and the years after being be most extreme expression of the episode.

Sixth Century Chinese History and Culture

[China, north, east, south, west; history; court factions, elites, changing capitals, dynasties, climate, intrigue, Buddhism; A.D. 534-590[cal]]

A climatic problem such as that imposed around A.D. 536[cal] must be interpreted against the backdrop of Chinese history and ideas in the sixth century. Before A.D. 536[cal] China was split into competing northern and southern fragments. Turkish barbarians, the Toaba, shared rule in the north with Chinese elites over a Chinese peasantry. In the south, a Chinese elite ruled over Han Chinese peasants and assorted other ethnic groups. Chinese bureaucrats could work for the northern Wei dynasty or turn coat and find appointment to a position in the southern Liang court. Both empires fed on taxes in grain collected from peasants and used it to pay officials's salaries and the army, purchase court luxuries, or stocked it as relief supplies. Political infighting was rampant at both courts, but probably worse in the north.

In A.D. 534^{cal} infighting in the northern Wei empire erupted into outright division; the court split into separate empires (see Figure 6.1). The Eastern Wei moved from the old capital at Luoyang to Yeh near modern Anyang while the Western Wei abandoned Luoyang for Changan (modern Xi'an). Dynasties evolved separately briefly. Following A.D. 550^{cal}, the Western Wei became the Northern Zhou, the Eastern Wei became the Northern Qi, while the southern Liang dynasty became the Chen. North and south China, and the Eastern and Western Wei empire, were not reunited until A.D. 590^{cal} by the Sui Dynasty.

Some scholars are under the impression that the A.D. 534^{cal} split of the Wei Empire coincided with a climatic disaster. One is left in wonder at the parallel timing of war and social change processes in the Mediterranean basin and Justinian's War (see Young chapter), and those described here in the orient for the disintegration of the Wei Empire. However, causes are by the nature of human social organization complex. Most traditional historians of China attribute cyclical order and disorder to social causes; the old idea holds that the founding emperor of a dynasty is energetic and good while the last emperor is a lazy corrupt villain. The dynamics of dynastic evolution could very well have played a roll in social changes of the 530s. Gernet (1982:194) sees ethnic tensions between Toaba Turks and Han Chinese at the root of the northern Wei breakup. The Wei Empire was composed of an alliance of Turks from an Asian steppe culture and Han from moister agricultural lands. When the empire split, the Turks retreated to the drier and colder interior of Northwestern China, while the Han departed to the east for the moister conditions of the North China Plain. The umbrella of multi-causality can encompass practical and ideological components in interactions as it clearly does in the Wei case. In addition to political intrigue, ethnic volition, and environmental chaos in the early sixth century, religious evolution could have been involved in the socio-environmental interactions of the period. While Buddhism entered China as early as in the first century A.D., the sixth century with all of its political instability was one of the times that Buddhism flourished and became popular with the ordinary people (see Young [1995] for discussion of contemporary and similar evolution of Christianity).

In the sixth century in China, people generally sought other worldly religions such as Daoism and Pure Land Buddhism, either from spiritual longing, dissatisfaction with life, or some very practical advantages. Most historical accounts are written by Confucian scholars who have a bias against Buddhists; Buddhist monks and nuns do not work, or marry and carry on the family, or pay taxes and serve in the army. Luoyang supported Buddhism on a grand scale; a city of 190,000 households had 1367 temples and monasteries (Yang 1984:246). The southern Liang emperor twice retreated into a monastery and had to be ransomed, whether from religious motives or a way to transfer state money to the monastery. Probably poor people were pleased to avoid paying taxes and being subject to the draft with easy salvation as an incentive. The economic, legitimacy, and religious issues can interact to motivate individual responses.

If portents or disasters pointed toward the mandate of heaven, Luoyang before the Wei split had a plentiful supply of indications in anticipation of the A.D. 536 event. One was simply horrifying; a rebellious general called all the government officials into his presence after his coup in A.D. 528^{cal} and massacred 2,000 of them (Hurvitz 1956:98). Astronomical portents were also clearly negative. Unless carefully explained by officials, all the comets and meteor showers (catalogued above) were signs of disasters to come, as well as the fact of the droughts. Further portents were reported from regional religious centers; a statue of Buddha that wept tears before disasters, started in A.D. 528^{cal} (Yang 1984:98-99); another statue moved by itself in A.D. 533^{cal} (Yang 1984:109). In A.D.534^{cal}, one of the main stupas (temples) caught fire amid thunder, rain, and hail mixed with snow, and burned for months (Yang 1984:40). These disasters predate A.D. 536, and thus set a tone of public anxiety. Whether they actually predate the disaster (anxiety), or were created in the public mind in retrospect (post traumatic stress disorder), they measure the emotional magnitude of the disruptions of the period.

State Stability and Recovery

[China, northern, eastern, western; history, landscape; landscape development, state stability, factions, climate-political linkages, collapse, adaptation, cultural experience, causal chain, fragmentation to adapt to drought, centralization, monasteries, grain storage; A.D. 470-618^{cal}]

The Chinese historical records bear information on responses undertaken to adapt to cold and drought conditions of the fifth and sixth centuries and following A.D. 536^{cal}. Ideally we would like to have a comprehensive picture of Chinese adaptations to climatic change from actions at court, down through the peasant, and through the Chinese landscape footprint. While often insightful, the histories give only hints of broader behavioral patterns. By the sixth century A.D., the Chinese had made massive changes in their landscape: forests cut down, fragile loess soils eroded, flood control levees built, extensive irrigation systems watered fields. Although they had attempted the beginnings of accompanying social engineering to provide social stability in the face of environmental disturbance, much remained to be done to insure state stability.

The climatic disaster in the several years following A.D. 536^{cal} had direct ramifications for state stability through the economic system: crops yields diminished so less grain tax could be collected. There were, as a consequence fewer resources for government expenses and to control unruly nobles and factions. The north-south political split further circumscribed response options by limiting ability to make up shortfall from another ecological zone. While this causal chain from climate event to political breakup of the northern Wei begs further scholarly scrutiny, it provides a convincing first approximation hypothesis for this study.

The Northern Wei fragmentation into two empires may itself have been an adaptation to more limited resources. A common reaction to economic stress is to split into smaller social units. The first real military attack on the Western Wei by the Eastern Wei was pressed in A.D. 537^{cal} (Dien 1990:345), presumably under the shadow of the dry fog and still within the time of direct impact of the A.D. 536^{cal} event.

Some of the direct effects of unstable climate can be detected in historical records. The climate problems could have contributed to the Eastern Wei difficulties with establishing in their new capital of Yeh. In A.D 493^{cal}, when the Northern Wei decided to move their capital from extreme northwestern Datong to central Luoyang, the move and construction of the city was accomplished in two years (Yang 1984:7). After the A.D. 534^{cal} split, the Eastern Wei fled a short distance to Yeh. They were not really settled there until they built the city wall in A.D. 539^{cal}, five years later. Complicating circumstances are further suggested: in A.D. 538^{cal}, the Eastern Wei forbade additional building of temples in Yeh, perhaps fearing that resources were being diverted from more critical construction; later that year the decree was extended to the rest of their territory (Hurvitz 1956:100).

The next stable dynasty, the Sui, appear to have undertaken corrective measures, to have learned from the Chinese cultural experience with climate change. The corrective actions of the Sui Dynasty after A.D. 590^{cal} offer clues to the problems of state organization in the mid sixth century. The Sui increased centralization by eliminating local patronage posts in the provinces. This forced the aristocracy to move from their home base if they desired government service (Ebrey 1978:83). In effect the Sui reduced the dependence of the central government on grain taxes to maintain order among the nobility. Earlier, the Northern Wei tried to ensure that relief grain was managed well and surpluses stored by giving prisoners of war and relief grain to monasteries in a system of Sangha and Buddhist households (Ch'en 1964:154). This system was instituted between A.D. 470 and 476^{cal}. Peasants worked for monasteries supporting them by doing work monks were not permitted by orders's rules. In this system, however, government grain tended to get used, and bought and sold for profit, not stored for times of shortage. The Sui solution, however, proved to be a more enduring approach.

The Sui, who controlled both north and south China, also started the Grand Canal from southern to northern China to transport grain from the double-crop south to the single-crop north. This breached the old Wei-Liang political ecological barrier and allowed the Sui to respond to shortages by transport between ecological regions. Another practical response to the dry conditions was instituted in A.D. 525^{cal} by ordering immediate collection of land taxes for the subsequent six years (MacGowan 1973: 234). I mentioned above the practical responses to dryness issued in an agricultural handbook in the 530s and 540s.

Conclusions

[China; history; dry fog, other worldly religion, Buddhism, famine, volcanism, comets, legitimacy, disintegration of government; A.D. 535-536^{cal}]

China shows clear evidence of the dry fog event in A.D. 536^{cal}, its action was to lower the temperature and decrease rainfall causing great hardship and famine. It was more severe because it arrived in the middle of an already dry period. The Chinese records reveal an intersection with a comet in A.D. 535^{cal} just before A.D. 536^{cal}. Current data from Greenland ice cores shows the sulfuric acid signature of a volcanic eruption around A.D. 532 - 534, reasonably close to the A.D. 536 event. Drought combined with other portents led people to question the legitimacy of the three competing Chinese dynasties and in a practical sense it made the government, especially in the north, difficult to maintain. Ethnic origins, political circumscription of resources, and decentralized government appear to have contributed to tensions that precipitated the disintegration of the Wei Empire. The A.D. 536 event could have contributed to the disintegration of that government and the popularity of other worldly religions, especially Buddhism. The richness of the Chinese records beyond the official dynastic histories warrants a much more exhaustive search for phenomena related to the A.D. 536 crisis and Chinese responses to it.

ACKNOWLEDGMENTS: I would like to thank Joel Gunn for numerous stimulating discussions on climate change and constant encouragement on this project.

References

Baillie, M. G. L. (1995) Patrick, Comets, and Christianity. *Emania* 13:69-78.

Baillie, M. G. L. and M. A. R. Munro (1988) Irish Tree Rings, Santorini and Volcanic Dust Veils. *Nature* 332:344-346.

Bao Chenglan, editor (1987) *Synoptic Meteorology in China*. Ocean Press, Beijing.

Biot, Ed. (1848) Catalogue General des Etoiles Filantes et des autres Meteores Observes en Chine. *Memoires presentes par divers savants a l'Academie des Sciences* 10:129-415, Paris.

Ch'en, Kenneth K.S. (1964) *Buddhism in China: A Historical Survey*. Princeton University Press, Princeton, N.J.

Chu Ko-chen (1973) A Preliminary Study on the Climatic Fluctuations during the Last 5,000 Years in China. *Scientia Sinica* 16:2:226-256.

Clausen, Henrik B., Claus U. Hammer, Christine S. Hvidberg, Dorthe Dahl-Jensen, Jorgen P. Steffensen, Josef Kipfstuhl, and Michel Legrand (1997) A comparison of the volcanic records over the past 4000 years from the Greenland Ice Core Project and Dye 3 Greenland Ice Cores, *Journal of Geophysical Research* 102:26,707-26,723.

Dein, Albert E. (1990) *State and Society in Early Medieval China*. Stanford University Press, Stanford.

Eddy, John A. (1994) Solar History and Human Affairs. *Human Ecology* 22:23-36.

Ebrey, Patricia Buckley (1978) *The Aristocratic Families of Early Imperial China*. Cambridge University Press, Cambridge.

Fairbank, John and Edwin Reischauer (1989) *China*. Houghton Mifflin, New York.

Feng, Zhaodong, L.G. Thompson, E. Mosley-Thompson, and Tandong Yao (1993) Temporal and Spatial Variations of Climate in China during the last 10,000 Years. *The Holocene* 3:174-180.

Gernet, Jacques (1982) *A History of Chinese Civilization.* Cambridge University Press, Cambridge.

Gunn, Joel D. (1994) Introduction: A Perspective from the Humanities-Science Boundary. *Human Ecology* 22:1:1-22.

Gunn, Joel D., William J. Folan, and Hubert R. Robichaux (1995) A Landscape Analysis of the Candelaria Watershed in Mexico: Insights into Paleoclimates Affecting Upland Yucatan Peninsula Semi-karst. *Geoarchaeology: An International Journal* 10:1:3-42.

Hammer, C.U., H. B. Clausen, and W. Dansgaard (1980) Greenland Ice Sheet Evidence of Post-Glacial Volcanism and Its Climatic Impact. *Nature* 288:230-235.

Ho Peng Yoke (1962) Ancient and Mediaeval Observations of Comets and Novae in Chinese Sources. *Vistas in Astronomy* 5:127-225.

Hurvitz, Leon (1956) *Wei Shou on Buddhism and Taoism* (part of Wei Shu). Kyoto.

Imoto, Susumu, and Ichiro Hasegawa (1958) Historical Records of Meteor Showers in China, Korea, and Japan. *Smithsonian Contributions to Astrophysics* 2:131-144.

Lean, Judith, Andrew Skumanich, and Oran White (1992) Estimating the Sun's Radiative Output during the Maunder Minimum. *Geophysical Research Letters* 19:1591-1594.

Li Zhaoyuan and Quan Xiaowei (1987) The Climatic Changes of Drought-Wet in Ancient Chang-an Region of China during the Last 1604 Years. In *The Climate of China and Global Climate*, edited by Ye Duzheng, Fu Congbin, Chao Jiping, and M. Yoshino, pp. 57-62, China Ocean Press, Beijing.

Macgowan, J. (1973) The Imperial History of China:Being a History of the Empire as Compiled by the Chinese Historians. Curzon Press, London.

de Moidrey, J. (1904) Observations Anciennes de Taches Solaires en Chine. *Bulletin Astronomique* 21:1:59-69.

Shih Sheng-han (1982) A Preliminary Survey of the Book Ch'i Min Yao Shu: An Agricultural Encyclopaedia of the 6th Century. Second edition, Science Press, Peking, China.

Spaceflight (1989) Krakatoan-Class Volcanos 31:159.

Wang Shaowu, Zhao Zongci, Chen Zhenhua, and Tang Zhongxin (1987) Drought/Flood Variations for the last 2000 Years in China and Comparison with Global Climatic Change. In *The Climate of China and Global Climate*, edited by Ye Duzheng, Fu Congbin, Chao Jiping, and M. Yoshino, pp. 20-29, China Ocean Press, Beijing.

Weisburd, Stefi (1985) Excavating Words: A Geological Tool. *Science News* 127:91-94.

World Meteorological Organization (1995) The Global Climate System Review: Climate System Monitoring June 1991-November 1993.

Yang Hsuan-chih (1984) *A Record of Buddhist Monasteries in Lo-yang.* Princeton University Press, Princeton, N.J.

Yao, Shan-Yu (1941) Floods and Drought in Chinese History: Their Distribution and Correlation, 206 B.C. - A.D. 1911. Unpublished Ph.D. dissertation, University of Pennsylvania.

Zhao Songqiao (1986) *Physical Geography of China.* Science Press, Beijing.

Table 6.1. Calendar of Records from Chinese Histories.

Year A.D.	Event	Source
505	Byrd Station, Antarctica ice core acid peak	Hammer et al. 1997
508	Greenland GISP 2 ice core sulfate peak	Zielinski et al. 1994
514	Greenland GRIP ice core acid peak	Clausen et al. 1997
516	Dye 3 ice core acid peak (GISP 1)	Clausen et al. 1997
516	flood Xuzhou June-Oct., 10,000 drown	Yao 1941
517	flood Hebei Oct.	Yao 1941
520	flood Hiangsu, Anhui July	Yao 1941
521	flood Hebei May	Yao 1941
524	possible comet in China	de Visser in Baillie 1995
527	flood Luoyang Aug.	Yao 1941
527	GRIP ice core acid peak	Clausen et al. 1997
530	GISP 1 ice core acid peak	Clausen et al. 1997
532	flood Luoyang July, 300 houses gone	Yao 1941
532	meteor shower Aug. 28	Imoto & Hasegawa 1958
532	GRIP ice core acid peak	Clausen et al. 1997
533	flood Nanjing June	Yao 1941
533	comet Mar. 1, Liang	Ho Peng Yoke 1962
534	GISP ice core acid peak	Clausen et al. 1997
535	comet noted in Wei Shu	Ho Peng Yoke 1962
535	severe drought June-July	Yao 1941
536	dry fog Europe late Mar.	Rampino 1988
536-545	bog oaks Europe reduced growth	Baillie 1995
536	Canopus not seen at spring & fall equinoxes (Pang & Chou)	Weisburd 1985
536	frost & snow July & Aug. Ching state	" text
536	early frosts, drought, & famine, Shanxi	" map
536	frost southern Shanxi	" map
537	summer snow & famine, coast south of Shandong & Yangtze	" map
537	autumn famine in Ching state	" text
537	nova, Jan.-Feb.	Ho Peng Yoke 1962
537	severe drought, frost, famine in Shansi, Kansu, Shensi, Henan, Apr.-May	Yao 1941
538	famine south of Shandong	Weisburd 1985 map
538	severe weather, other kingdoms north of Yellow River, 70-80% starved	Weisburd 1985 text & Spaceflight 1989
538	flood Hebei & summer in Shandong	Yao 1941
539	comet	Ho Peng Yoke 1962
540	previous date acid peak Dye 3 ice core	Hammer 1980
541	narrow Irish tree ring, narrowest from 13 B.C. to A.D. 894	Baillie & Munro 1988
541	nova	Ho Peng Yoke 1962
542	flood Hebei	Yao 1941
542	drought Nov. to May	Yao 1941
543	drought Nov. to May	Yao 1941
545	meteor seen at siege of Yu-pi (Eastern Wei)	Biot 1841
547	drought Nov. to Feb. & Nov. to Apr.	Yao 1941
550	severe drought Feb. to Aug., Nanjing cannibalism during siege warfare	Yao 1941
550	flood Xi'an Aug. to Feb.	Yao 1941

7. ARE THERE AFRICAN TRACES OF THE A.D. 536 EVENT?

Peter R. Schmidt

Department of Anthropology, University of Florida, Gainesville Florida 32611

Similar Trends in East Central and Northeast Africa

[East Central Africa, Kagera Region, Northeast Africa, Ethiopia, Eritria, Arabian Peninsula; archaeology, paleobotany, ethnography; decline, erosion, empire; A.D. 1 to 600^{cal}, Iron Age]

Archaeological, paleobotanical, and ethnographic research have contributed over the last three decades to a deep-time perspective on the ecological history the Kagera Region of northwestern Tanzania (Figure 7.1). The development of iron technology and of settled agricultural life are fundamental to understanding the environmental changes that arose in the region over the last two millennia or more. This chapter provides an overview of the most salient observations derived from this research program and it addresses how documented changes in settlement and environment during the mid-first millennium A.D. in Kagera Region could be linked to a global event in A.D. 536^{cal}. I will begin with a short review of the history of Iron Age settlement, technology, and the major environmental changes that form the backdrop to a concluding discussion of the role of global climate i nfluences.

The second part of the chapter addresses what I see as possible parallel and complementary changes that occurred during the same period of time to the north in the northeastern Africa. This region was then in the Empire of Axum, which was located in what is today northern Ethiopia, Eritrea and parts of the Arabian peninsula. One of the most powerful states of eastern Africa and the Middle East in the first millennium A.D., the Axumite state depended on agricultural production of grain in the highlands of Ethiopia and Eritrea. As is readily evident from recent experience, the grain crops of highland Ethiopia are particularly susceptible to climatic variations. Environmental research at Axum shows that the first period of occupation (A.D. 1 to 350^{cal}) is marked by very wet and prosperous conditions (Butzer 1981). This corresponds to the period during which most of the immense granite stele that characterize classical Axum were erected. Starting in A.D. 350^{cal} rains became less regular or more subject to interruption (Butzer 1;81). Significant erosion and loss of soil quality resulted from this period of rapidly intensified agriculture and it probably set the scene for the diminished influence of the Axumite state in the sixth century A.D.

In this chapter, I want to explore the possibility that part of the diminished capacity in both east central Africa and the Axumite state in the sixth century was linked to the cosmic event of A.D. 536^{cal} and its subsequent climatic changes and other stresses that it introduced.

Eastern Central Africa in the First Millennium

[Africa east central, Lake Victoria, Kemondo Bay, Karagwe depresssion, Rwanda, Burundi; paleobotany, archaeology, linguistics; forest species diversity, agriculture, cattle breeding, charcoaling, iron smelting furnaces, Cushitic, Bantu, interaction; 600 B.C.-A.D. 600]

The eastern part of Kagera Region (see Figure 7.1) is called Buhaya and is occupied by the Bahaya people. They once practiced a mixed economy based on an iron industry, fishing, hunting, and an agricultural system based on bananas interplanted with beans, as well as cash cropping (coffee and tea) and some cattle keeping. Over the past several decades, culture change in Buhaya has led to the extinction or near extinction of many aspects of the traditional economy. The strongest sector of the productive economy, other than agriculture, was formerly steel production and its attendant industries such as forging and wire drawing. My environmental research in the region has been directed toward this industrial sector of the economy and its distinctive needs and impacts (Schmidt 1992, 1994, 1997a).

My research goals were to document when the most recognizable sectors of exploitative economy developed and when they influenced Buhaya landscape evolution. I also wanted to analyze what changes in technology, economy, and environment have occurred over the past three thousand years (Schmidt 1978, 1994, 1997a). Very dense populations now live in the zone of greatest historical interest, a 15 to 25-kilometer-wide strip to the west of Lake Victoria. Most of the Early Iron Age settlements and industrial locales (600 B.C. to A.D. 600^{c14}) in this region are overlain by contemporary villages, a condition that limits access to prehistoric remains as well as contributes to their destruction by deep hoe agriculture.

Little is known of iron production or settlement in the Buhaya area between 600 B.C.—when the first pioneer communities of iron workers are documented—and 200 B.C. The population densities of the pioneer iron workers may have been so insignificant that traces of their presence on the landscape are difficult to find. It is possible that initial settlement was short-lived and that the earliest populations, which by that time would have incorporated some grain agriculture from Sudanic speakers-mostly finger millet—relocated to more open environments. If the agricultural repertoire of these earlier settlers had come to depend on some cereals, then the wet heavily forested ridges of eastern Buhaya would not have been as hospitable as the more easily cleared patches of savanna woodlands of the central and eastern uplands in Burundi and Rwanda, forest not so

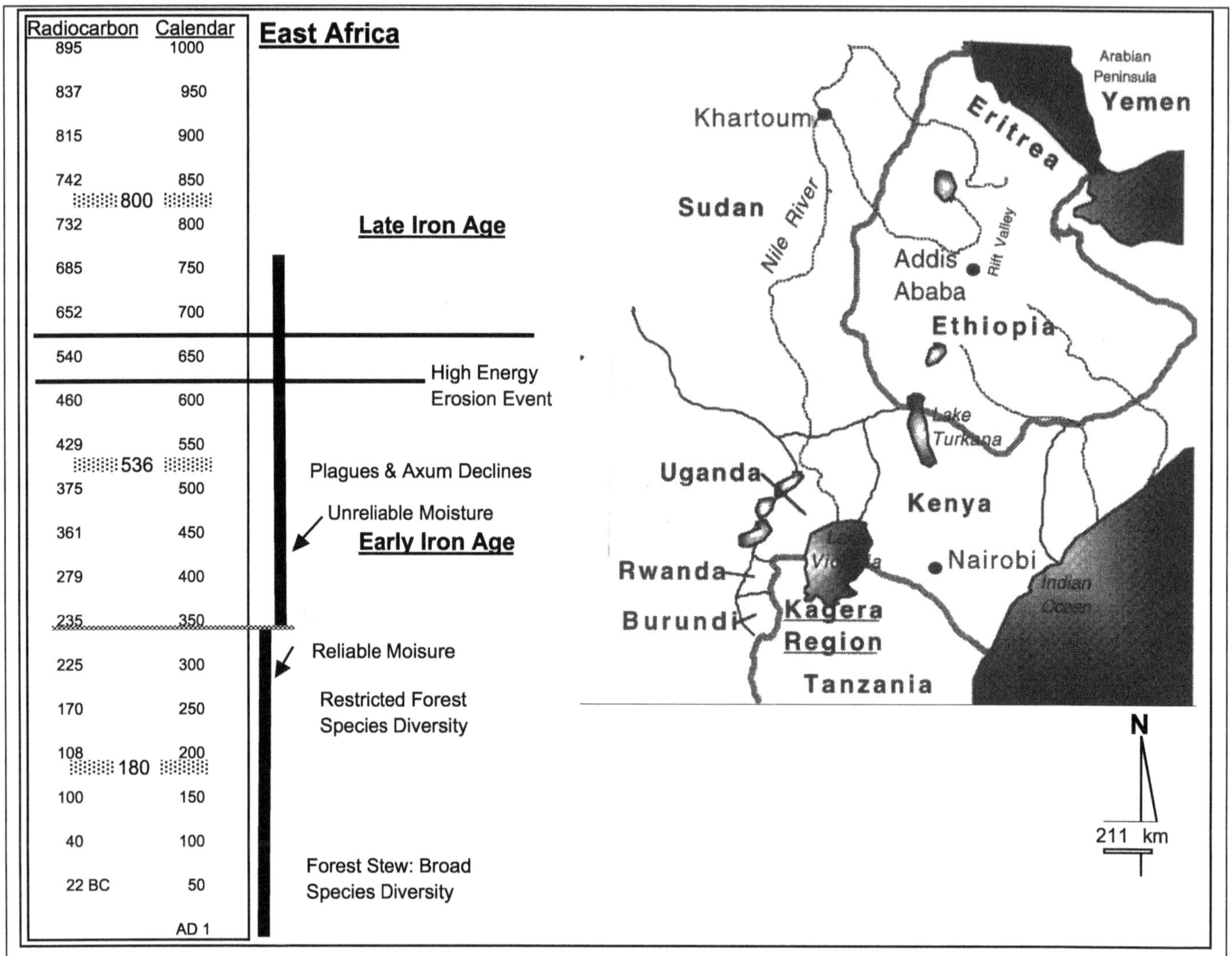

Figure 7.1. Map and Chronology of East Africa.

tropical and dense and where several communities dating to the mid-first millennium, and earlier, have been documented (Van Grunderbeek et al. 1983). By 200 B.C., however, we have the first strong evidence of settlement by Early Iron Age peoples along the shore of Lake Victoria (Schmidt and Childs 1985). Bantu-speaking groups engaged in iron production expanded along the lake shore after A.D. 1. This population growth and increase in community size followed a period during which the Bantu-speakers would have been in regular contact with Cushitic speakers living in the Karagwe depression immediately to the west of central Buhaya (Schoenbrun 1993a:15-21). Such interaction would have diversified Buhaya subsistence economy. Additional cereal crops such as sorghum and increased cattle breeding would have helped to meet the needs of a large population (Schoenbrun 1993a). Evidence for population growth and the proliferation of communities and technological locales is particularly strong in the area surrounding Kemondo Bayand its environs after A.D. 1 (Schmidt 1980). The Kemondo Bay communities were not isolated entities. The Bantu and Cushitic communities were in close proximity and in regular contact, undoubtedly with regular economic exchange as well as social and political interactions.

One of the best indices for environmental change that can be linked directly to the production economy is charcoal from prehistoric smelting furnaces. In Buhaya, such evidence provides an exceptional opportunity to study the type and location of forests used for iron-working purposes. Because charcoal had been excavated from precisely dated furnaces, these analysis provide a means to trace changes in forest ecology as well as variations in the range of forest species used in iron industry. The findings span the entire Early Iron Age. Similar research into species of trees identified from industrial charcoals in Rwanda and Burundi (see Figure 7.1) have shown clear ecological changes that can be linked to industrial production as well (Van Grunderbeek et al. 1983, 1984; Van Grunderbeek and Doutrelepont 1988).

A great variety of forest species are reported from the earliest iron-smelting furnaces at the KM3 site near Kemondo Bay. Analysis of its charcoal shows the use of at least twelve genera of forest trees (Schmidt 1994). This remarkable range of types derives from several microenvironments and suggests that by the early first century B.C. iron technologists were exploiting predominantly first-growth, mature, wet-forests. Some species, such as *Celtis durandi,* are huge; others, such as *Entandrophragma cylincrium* (mahogany) and *Chatacme microcarpa,* which grows on shallow ridge-top soils, are very hard and very large. At the same time, middle-story trees and larger swamp trees (such as *Uapaca,* which makes excellent charcoal) were also in use. At the nearby KM2 site, a furnace of similar date yielded charcoal made from sixteen different genera. The great diversity of genera and species in this case likewise points to a predominance of mature, moist-forest trees, including mahogany and ironwood (extremely difficult to cut), supplemented with middle-story trees and trees common to swampy and riparian environments. Several large forest trees with light, soft woods (such as *Canarium schweinfurthii)* are also represented at this time.

The woody contents of Early Iron Age furnaces in this earlier period appear to resemble a forest stew: bits and pieces of any available tree were tossed into the furnace, even though some species would in fact have had little value in smelting iron. The parsimonious explanation is that the first iron technologists along the lake shore were working in an environment where forest clearance for agriculture, perhaps under swidden conditions, was being practiced. The clearance of mature, moist forest would have left great quantities of readily available wood—much of it of high quality—lying about within new farm plots. The presence of charcoal from light and soft woods, as well as from wetland hardwoods, supports a picture of "omnivorous" wood use.

The margin of Lake Victoria appears by A.D. 100^{c14} to have been settled by social groups engaged in this form of "unselective" forest use from about 10 km north of today's regional headquarters, Bukoba, to about 30 km south of Kemondo Bay. These communities practiced iron production and a form of subsistence that probably drew significantly from the cereal diet and pastoralism borrowed from the earlier Southern Cushitic speakers in the region. These early communities probably continued to exploit the rich fish resources of Lake Victoria, along with hunting, gathering, root crops, and a gradual but growing use of beans (Schoenbrun 1993a:25). As forest clearance continued along the valleys and ridges near the lake, favorable conditions for cereal agriculture would have expanded.

The early communities were probably lineage and clan-based groups, some of which monopolized control of iron production. This interpretation suggests some social differentiation based on wealth. Even among the iron producing groups there appears to have been differences, with some groups following parallel but distinct lines of technological development. While each iron working group shared a common idea about how to build iron smelting furnaces, there were clearly differences among them in the use of materials used to smelt iron.

Rank and privilege can be detected in some iron smelting practices. Between A.D. 1 and 200^{c14} smelters working 200 m from Kemondo Bay at the KM2 site appear to have been privileged in the quality of clay they used in making heat-resistant tuyeres important in high temperature smelting. This certainly would have been one domain in which control over technologically advantageous resources would have led to different levels of technological achievement, quality of product, productive capacity, and wealth. Thus there appears to have been social and economic differentiation based on control of resources, perhaps the beginnings of social ranking and social stratification among the Early Iron Age peoples of Buhaya.

Iron-working and non-iron-working social groups both voraciously consumed the forested environment for agricultural, domestic, and industrial purposes. This had two immediate consequences: 1) diminished resources that were needed to sustain iron production; and 2) small groups departed to settle more interior areas, such as the high central plateau where competition for resources and productive agricultural land was not as intense. Those groups that continued to reside near the lake shore turned to more marginal forest resources as the wet gallery forest dwindled. By A.D. 200 to 300^{c14} communities had been established on the plateau, as well as at its base.

By A.D. 200^{c14} some social groups involved in iron working along Lake Victoria appear to have lost some of the characteristics that distinguished them during the first two centuries. The technology appears more standardized, with more universal access to critical resources such as heat-resistant clays. Even so, inequities between groups remained in evidence. Standardization of production may also point to the growth of centralization, where a central authority grew to control resources and the standards of production, an interpretation that agrees with indications that social stratification may have been developed somewhat earlier.

As our focus shifts to three to five centuries later, a different picture of forest utilization emerges. Fourth century evidence from the KM3 site shows a much more restricted fuel list: only two moist-forest species were used. The KM2 site, which near the shore of Lake Victoria, shows only slightly less confined list of fuel woods utilized during the fifth century. In one furnace pit only three species are represented: an older, wet-forest type; a secondary tree that colonizes the fringe of the forest; and a tree with soft, light wood that today is commonly used to make beer boats—large vessels like dugout boats used to brew beer. In a nearby furnace from the same period we found a similarly limited mix of mahogany, two other forest species now uncommon in East Africa, and two very common swamp species. The variety of forest species formerly common to the environment was no longer available. Large remnant trees were used, but the iron smelters were now consistently venturing into the swamps to obtain supplementary fuel. It suggests that central authority may not have been sufficient to ensure sustainable forests.

I conclude that over the centuries after the first practice of iron smelting along the coastal hills, the environmental plenty of moist forests diminished to the point that iron smelters were obliged to use only a few remnant moist-forest species along with some secondary regrowth and they were ranging farther afield to exploit the swamps for supplementary fuels. The human industrial record thus gives a clear picture of the spatial form that forest clearance took and its precise expression in the range of species eliminated from the landscape. From it I can also estimate the scale of early forest exploitation for both land clearance and industry, until forests were eliminated from the coastal valleys and ridge tops and ironworkers increasingly sought other suitable fuels.

With the passage of time and increased agricultural and industrial production, then, ecological stresses increased along the margins of Lake Victoria. By A.D. 300-500^{c14}, iron-producing groups had increased the tempo and scale of production and were forced to exploit more difficult-to-reach forests while others continued to seek new opportunities in the hinterland. By A.D. 300-400^{c14}, and possibly earlier, the area around Lake Ekimba had been settled. Its more open woodland environment and woody grasslands had probably already been utilized by southern Cushitic-speakers for cereal agriculture and cattle pastoralism, uses that may have made it more susceptible to environmental degradation when it was occupied by Bantu-speaking agriculturists. The 400-500 period in Buhaya, as well as in Rwanda, was a turning point for Early Iron Age populations (Van Grunderbeek et al. 1983:40-41).

Another perspective on the vegetational history of the central coastal zone of Buhaya was obtained from pollen and sedimentary histories derived from core samples. They were taken along several east-west transacts ranging from the drier central depression through the moist ridges next to Lake Victoria. One sample from the eastern part of Lake Ekimba (basal core deposits radiocarbon-dated to A.D. 265^{c14}) holds particularly important information: it shows an erosion horizon that began as a catastrophic environmental event (Schmidt 1994:109). The deposition in the lake bed of limonite nodules from the hillsides above and east of the lake suggests that land there had been severely degraded, perhaps even stripped of protective shrub and grass cover, which meant that high-velocity erosion carried nodules into the lake. This episode occurred at approximately A.D. 450-500^{c14} and provides a definitive environmental marker. Lower deposits—prior to this erosion event—show, first, a decrease in arboreal pollen, apparently related to deforestation, and then a decrease in nonarboreal pollen. This marked and rapid pattern of environmental degradation is particularly signaled by the fall in nonarboreal pollen, which indicates widespread alteration of low-lying vegetation and shrubs presumably owing to agriculture.

These changes, along with the severe erosion, resulted from Early Iron Age cultivation and settlement in this zone to the east of Lake Ikimba. The spread of Early Iron Age agriculture and technology to the woodlands of the central basin would have posed a different array of adaptive problems compared to the moist forested coastal ridges and the well-watered central plateau. Land clearance would have occurred at a much more rapid rate in the woodlands being further compounded by cattle grazing in a zone with a history of pastoralism, a phenomenon that may be reflected in the severe erosion episode of A.D. 450 to 500^{c14}. The more open woodlands of the Lake Ekimba area, if opened to cereal agriculture and intensive herding-initially by the Tale Southern Cushitic speakers and later by Bantu speaking agriculturists influence by interactions with the former—could well have experienced a more rapid and amplified environmental degradation. By the time the Early Iron Age ended in the interior of Buhaya, the southern margins of the unique "montane" swamp forests had been gradually cleared until woody shrubs and grasses prevailed. Other profound changes had occurred on the hilly slopes around Lake Ikimba. In that area initial agriculture, with a herding component, apparently had a severely degrading effect. The consequences can be read in a major erosion event and in the subsequent abandonment of the area, apparently during the late sixth century and seventh century A.D. The causes of human depopulation in this area cannot be definitively determined. We have no specific archaeological evidence that disease was the direct cause of population decline in Buhaya, but the region's history over the most recent century strongly suggests that human disease is a common response to conditions of ecological stress. It also appears that another response to such stress was the movement of farmers from west of the lake margins into the drier grasslands to the northwest, where the cereal and cattle-keeping parts of their food systems would have allowed a fairly rapid and easy transition (Schoenbrun 1993a:28).

Faced with erosion difficulties, diminished forests, and hundreds of communities of significant size—some up to 50 ha or more in size—these peoples witnessed a failure of technology to keep pace with a degraded resource base. During the sixth century A.D. the production of iron stopped in the once prosperous factory sites around Kemondo Bay and in the interior. The prosperity of the first half of the millennium had passed. The most poignant testimony to this decline was the rapid diminution of communities in the interior, away from the lake margins.

What happened to these populations? There is linguistic evidence that suggests that many of these "Rutaran-speaking" peoples left the region and moved into the area that we know today in the north as Ankole, western Buganda, and Bunyoro (Schoenbrun 1993b:48). The increasing popularity of pastoralism—perhaps linked to more grasslands resulting from deforestation—undoubtedly contributed to the movement north into open grasslands where pastoral peoples seem to have established either ritual centers or centralized polities by the tenth to eleventh century^{c14} A.D. at sites such as Ntusi (Reid 1990, 1991).

There is also the distinct possibility that adverse climate conditions added to the stress induced by deforestation and soil erosion. As noted earlier, the industrial production on which the region depended for much of its prosperity had fallen on hard times. An analysis of radiocarbon dates from well dated sites along the littoral of Lake Victoria shows that only a few furnaces at best can be fixed to the sixth century at a confidence level of 67% (Schmidt and Childs 1985). Even then, it is clear that iron smelting was not practiced beyond the second half of the sixth century A.D. Why was there such an abrupt cessation of production and settlement in this once very prosperous region? Why was there not a gradual decline, with the last vestiges of production and settlement in the interior spread out over several centuries?

As is apparent in other chapters in this volume, we now know that a global event of major negative impact occurred in A.D. 536^{cal}—either the explosion of the most powerful volcano during the last three millennia—a contested interpretation—or a cosmic event that resulted in significantly diminished sunlight for over a year (Stothers 1984; Baillie 1994). Historical records explicitly mention the loss of productivity, famine, and disease that ensued over the next decade in Europe, China, the Middle East, and elsewhere (Gunn, this volume; others, this volume). Given the global probable impact of this event on agriculture, it is reasonable to suggest that the negative affects in Africa may also have been profound. If environmental conditions had already reached stressful levels and subsistence activities based on agricultural had become increasingly risky to the west of Lake Victoria, then the conditions during the decade immediately following A.D. 536^{cal} could have precipitated the abrupt termination of occupation that evident in human settlement in Buhaya. The A.D. 536^{cal} event could be the proverbial "extra stick that broke the camel's back" in Buhaya during a period when populations were still making a living, but under increasingly adverse environmental conditions that threatened their way of life. A shift to increasing pastoralism with a corresponding movement of populations into the grasslands in the northern Karagwe depression seems to have been one response to conditions in which agriculture in Buhaya was no longer as reliable a form of subsistence.

As populations from the Buhaya area sought out new opportunities to the north in the upper extension of the Karagwe Depression—in what today is south central Uganda—they appear to have been attracted to open grasslands suitable for their herds as well as to areas where there were ancient religious and ritual centers, perhaps established by earlier Rutaran speakers who had also made their way into this area. One of those was Mubende Hill, which is well known as a sacred shrine for Ndahura, the first 'ruler' of the Bacwezi line. Archaeological inquiries at Mubende Hill have documented an Early Iron Age occupation at this site (Robertshaw 1988:38). The presence of a significant Early Iron Age occupation at Mubende Hill, an important ritual site associated with cults of affliction, suggests that this site may resemble sites such as the prominent and very ancient Kaiija shrine in Buhaya. It appears that it is likely that the stresses experienced by these early 'Rutaran' populations may have been played out at important ritual centers, where cults of affliction may have been headquartered in ancient times, as well as up through this millennium (Schmidt 1997b). The location of several major ritual centers precisely in the area where the refugee populations ended up suggest further that such centers could have grown up around a culture of poverty, famine, and disease that contributed to the demise of one of central Africa's most significant civilizations in the first millennium A. D.

Axum and Northeastern Africa

[Africa northeast, Ethiopia, Eritrea, Arabian Peninsual, Red Sea, navy; history; trade, war, soil degradation, soil erosion, precipitation change; A.D. 100-600]

There is reason to believe that Africa, and perhaps Buhaya, were not spared from the A.D. 536 phenomenon. In northeast Africa, the Axumite empire rose during the first half of the first millennium A.D. Located mostly in northern Ethiopia and in Eritrea (see Figure 7.1), and noted for its trade with the Mediterranean world, interior Africa and Asia, Axum is one of the great empires of ancient Africa. The reasons for the beginning of its decline in the sixth and seventh centuries^{cal} A.D. have been traditionally linked to its waning political influence over the Arabian peninsula and its subsequent failure to control the Red Sea trade, which was central to its prosperity. Examination of historical records, however, suggest the strong possibility of different precipitating causes for declining political and economic influence, namely, that famine and pestilence gripped Axum at a key moment in its florescence, unexpectedly weakening it at a time when it was deeply involved in empire building. If the event of A.D. 536^{cal} affected Axum as severely as other regions, then it may hold the key to understanding why Axum suddenly was unable to maintain its territorial and economic interests on the Arabian peninsula, thus loosing the trade networks that helped to sustain the empire and setting into train a sgries of other processes over the coming two hundred years that led to its economic and political demise.

The agricultural prosperity of Axum from the second to fourth centuries was based on a moister climate in which two crops a year were possible to sustain a growing population (Butzer 1981). After the fourth century, rainfall was more episodic while remaining predominantly moist. The heavier rainfalls, when combined with accelerated land clearance, led to increased soil erosion, but not initially to degraded soil productivity (Butzer 1981:487). Thereafter, however, soil degradation increased with much devegetation of the areas surrounding the capital. During this subsequent period, Axum remained dependent upon cereal agriculture, and it appears that the potential for stresses during times of drought was beginning to take shape in the sixth and seventh centuries with a rapidly intensifying agriculture on soils of decreasing productivity.

By the sixth century A.D. the Axumite state was at full imperial extension. In A.D. 518^{cal} Axum mounted a campaign to subdue the Jewish Hiryamite kingdom in the southern Arabian peninsula. That venture resulted in the eventual loss of the garrison posted there and, possibly because of internal political problems, Axum did not respond until six years later. In A.D. 524^{cal}, according to a Greek text *The Martyrdom of St. Arethas,* the Axumite king Kaleb transported some 120,000 men across the Red Sea using 60 vessels (Pankhurst 1961). Other accounts say 70 vessels (see Kabishchanov 1979). It is not clear if the vessels were used repeatedly to transport the troops, or if they were moved in one wave. If the former, then the numbers may well have been significant. If the latter, then the large numbers are most certainly exaggerated. While this seems an exaggerated number of troops, and some contend that the numbers were perhaps a tenth (or less) of that number (Kobishchanov 1979:123), it does point to the capacity of the Axumite state at this time to marshal significant resources and manpower for overseas expeditions. By A.D. 536^{cal}, however, Axum fielded only 3,000 soldiers to the same area, and these troops turned, discontented, on their leader and went over to the enemy (Kobishchanov 1979). Another campaign shortly after also ended in defeat. The failure of the Axumite state to mount effective campaigns and to retain the loyalty of its troops speaks to internal disorder and a seriously reduced capacity to maintain its empire.

It could be, given the other effects of the A.D. 536 event in the Middle East and North Africa, that Axumite agricultural output was significantly diminished along with the state's ability to sustain its central functions, including the successful governance of outlying tributaries. This position is supported by deleterious health reported in Axum. Axum could have suffered from the plague that reached Egypt in A.D. 541^{cal} (Procopius, ed. Dewing 1914: 451 ff.; Pankhurst 1961). It is not clear where this disease originated, but some have claimed that it came from Ethiopia—which could be Axum or other areas of Africa that fell under that generic reference (Munro-Hay 1991). One of the immediate effects on Axum could have been its failure to field sufficient troops to maintain its foreign interests. In the years that followed, Axum continued to experience weaknesses, a condition that restricted trade and made it susceptible to loss of regional influence during the rise of Islam during the seventh century. To be sure, there were many other geopolitical and environmental influences at work in the region and in greater Axum that diminished Axumite influence, but A.D. 536^{cal} appears to have been a major watershed, and the A.D. 536 event marks Axum's apex as well as its decline. As is discussed in several other chapters in this volume (Houston, Robichaux, Young), this pattern of sudden decline conforms logically to an event-forced change of political status.

Decline in Buhaya and Axum

[Africa east, central and northeast; history, inference; Muslims, capital relocation, abrupt decline from apex, recovery; first millennium]

We see, then, significant stresses in two major civilizations in Africa simultaneously during the sixth and seventh centuries—both of which appear to be linked to the A.D. 536 event and its immediate aftermath. In Buhaya, by the end of the millennium the only permanent communities were located on the shores of Lake Victoria, where there would have been still-reliable rains for cereal agriculture and root crops as well as the ever-reliable fish resources of the lake. Otherwise, the landscape appears to have been mostly empty of humans. Over time the land healed and forests again partly took hold until the end of the first quarter of the second millennium, when once more the interior of Buhaya wcs occupied, at least as far as the eastern margin of the Lake Ikimba basin. In Axum, the story was one of incremental and steady loss of regional influence and prosperity exacerbated by accelerating environmental degradation. By the end of the millennium most of the buildings in the capital city had collapsed, the major port of Adulis had been sacked and destroyed by Muslim armies several centuries before, and the state capital had relocated several hundred kilometers to the south in a much attenuated form.

References

Baille, M. G. L. (1994) Dendrochronology Raises Questions about the Nature of the AD 536 Dust-veil Event. *The Holocene* 4:212-217.

Butzer, Karl W. (1981) Rise and Fall of Axum, Ethiopia: A Geo-Archaeological Interpretation. *American Antiquity* 46:471-495.

Kobishchanov, Yuri M. (1979) *Axum*. The Pennsylvania State University Press, University Park and London.

Munro-Hay, Stuart (1991) *Aksum: An African Civilisation of Late Antiquity*. Edinburgh University Press, Edinburgh.

Pankhurst, Richard. (1961) An Introduction to the Economic History of Ethiopia from Early Times to 1800. Staplgs, Rochester.

Procopius. (1914) *History of the Wars*. Edited and translated by H. B. Dewing. Loeb, London.

Reid, Andrew (1990) Ntusi and Its Hinterland: Further Investigations of the Later Iron Age and Pastoral Ecology in Southern Uganda. *Nyame Akuma* 33:26-28.

Reid, Andrew (1991) The Role of Cattle in the Later Iron Age Communities of Southern Uganda. Ph.D. dissertation, University of Cambridge.

Robertshaw, Peter (1988) The Interlacustrine Region: A Progress Report. *Nyame Akuma*. 30:37-38.

Schoenbrun, David (1993a) We Are What We Eat: Ancient Agriculture between the Great Lakes. *Journal of African History* 34:1-31.

Schoenbrun, David (1993b) Cattle Herds and Banana Gardens: The Historical Geography of the Western Great Lakes Region, ca AD 800-1500. *The African Archaeological Review* 11: 39-72.

Schmidt, Peter R. (1978) Historical Archaeology: a Structural Approach in an African Culture. Greenwood Press, Westport.

Schmidt, Peter R. (1980) Early Iron Age Communities and Locales in West Lake. *Tanzania Notes and Records* 84-85:77-94.

Schmidt, Peter R. (1992) An Interpretative Model for Sustainability Based on the Use and Transformation of a Forested Landscape in Eastern Equatorial Africa. In *Environment and Archaeology: Emerging Trends and New Techniques for Heritage Management & Sustainable Development in Tropical Forest Areas,* edited by A. Pantel, K. Schneider, and G. Loyola-Black, pp. 227-234, United States Forest Service, Atlanta.

Schmidt, Peter R.(1994) Historical Ecology and Landscape Transformation in Eastern Equatorial Africa. In *Historical Ecology: Cultural Knowledge and Changing Landscapes*, edited by C. Crumley, pp. 99-125. SAR Press, Santa Fe.

Schmidt, Peter R. (1997a) Archaeological Views on a History of Landscape Change in East Africa. *Journal of African History* 38:?.

Schmidt, Peter R. (1997b) *Iron Technology in East Africa: Symbolism, Science, and Archaeology*. Indiana University Press, Bloomington.

Schmidt, Peter R. and S. Terry Childs (1985) Innovation and Industry during the Early Iron Age in East Africa: The KM2 and KM3 sites of Northwest Tanzania. *The African Archaeological Review* 3:53-94.

Stothers, R. B. 1984. "Mystery Cloud of AD 536." *Nature* 307 (5949): 344-46.

Van Grunderbeek, M.-C., H. Doutrelepont, and E. Roche (1984) Influence humaine sur le milieu au Rwanda et au Burundi a l'age du fer ancien (220-665 A.D.): Apports de la palynologie et de l'etude des charbons de bois," *Revue de paleobiologie* (Avril): 221-29.

Van Grunderbeek, M.-C., E. Roche, and H. Doutrelepont (1983) *Le Premier age du fer au Rwanda et Burundi: archeologie et envirnoment*. Brussels.

Van Grunderbeek and H. Doutrelepont (1988) Etude de charbons de bois provenent des sites metallurgiques de l'Age du Fer Ancien au Rwanda et au Burundi. *PACT:Journal of the European Study Group on Physical, Chemical, Mathematical and Biological Techniques Applied to Archaeology* 22: 281-295.

IV. CHANGES IN TIME AND TEMPERATURE: ARCHAEOLOGICAL TIME AND GLOBAL TEMPERATURE

The chapters in this concept set examine two issues important to understanding climate in a non-annual framework, archaeological time and global temperature. The familiar passage of years in previous chapters marks a historical perspective on time. In prehistory, time floats measured in the passage of cultural generations whose age is estimated by various indirect means of measurement such as radiocarbon years and relative vertical position in archaeological site layers. Understanding these indirect measures of time, and temperature as well, are necessary because the following concept sets are located entirely within a part of the world that left no written records as of A.D. 536^{cal} and therefore no direct time. The concept-set topics were selected to study subregional variation within the southeastern United States Atlantic Slope, the Appalachian Mountains, Piedmont, and Atlantic Coastal Plain. In addition to providing information on subregional responses to global conditions in the southeastern United States, the study may suggest approaches to subregional variation in other parts of the world.

The research regions of the archaeologist authors are located in a range of physiographies and habitats. The question they attempt to address is: how does one comprehend the consequences of an apparently world wide event at the time scales that archaeologists study the past? Those time scales are necessarily super-annual except in the rare cases where tree rings provide insight into annual environments (see Anderson chapter). Tree ring times are generally late in prehistory and do not shed information on the temporally remote and potentially interesting climate episodes such as the Middle Holocene when global temperatures exceeded those of the present. However, one tree ring chronology from North Carolina encompasses the A.D. 536 event providing something of an anchor for both time and climate in most of the region.

Archaeologists' measures of time arise primarily from radiocarbon dating of excavated remains of past cultures. The accuracy of these dates is an ongoing problem of considerable proportions. Standard statistical estimates of accuracy, given in standard deviations, or sigmas, typically have broad, i.e., century scale spans in which radiocarbon daters regard the dates as being accurate. However, since archaeologists have been dating sites for 50 years now, most century scale problems have been resolved by 1000s of radiocarbon dates. This broad threshold threatens the utility of the technique in the future, and saddles archaeologists with 1000s of dates that may be of little use for future studies such of this one. In other words, the interpretation of radiocarbon dates needs to be seriously rethought.

One possible avenue of rethinking radiocarbon dates is to consider the theoretical and practical aspects of accuracy. Radiocarbon scientists supply theoretical probability estimates of the resolution of dates. However, it is unclear what those probabilities mean in practical terms. An analogy can be drawn to early applications of statistical methods in the social sciences. When social scientists began using statistical procedures extensively in the 1950s, they initially adopted the same probability assumptions, or rules of thumb, used by physical scientists. Probabilities are mathematical estimates of the accuracy of measurement of phenomena. Needless to say, since the statistical phenomena studied by physical scientist are very systematic, these standards were very rigorous; in relatively common language terms, the rule of thumb was that a conclusion would be accepted if less than one chance in 1000 existed that the conclusion would be incorrect. The phenomenology of these estimates is such that the applicability of the probability estimate to the situation in question depends on the judgment of the investigator.

In our 1950s example, the statistical standards of the physical sciences were soon found by social scientists to be unworkable in most situations and misleading in others where patterned human action was clearly present but obscured by intervening variables. To adapt statistical methods to human situations, new and less stringent rules of thumb were developed to accommodate the less physically constrained context of social data collection and analysis.

In the chapter on radiocarbon dates from eastern United States, we suggest that the practical limits of radiocarbon date resolution probably lie somewhat inside the one standard deviation probabilities calculated for dates, and they certainly are more accurate than the more conservative two sigma standard. As judged by comparison with tree rings, the data suggest that radiocarbon dates are measuring time at a resolution of about 50-100 years.

The other side of the archaeological time coin is; how resolved is the cultural data that archaeologists recover from sites? This is also worked out to be about 50 years based on a theory of human generations. Thus a practical perspective on radiocarbon dating and human generations converge at about a half century of time. The conditions of the convergence considers both the inherent limitations on archaeological data imposed by site destruction, obscure artifact-to-person and -site relationships, etc., and the practical radiocarbon interpretation model. Taken together, the inherent limitations suggest about a human lifetime as the lower limit on archaeological dating, which is about 50 years.

In William Tanner's chapter, the question of estimating global temperatures at a scale compatible with a lifetime temporal scale is discussed. Tanner uses land surface rebound following the melting of the Scandinavian Ice sheet to provide a stage for measuring changes in seal level. The setting of his study is the Baltic Coast of Denmark. Here low energy waves have marked sea level against a slowly rising coastline since the glaciers retreated. Because of the rising beaches, beach ridges left by the Baltic waves were lifted out of the range of subsequent sea level rises preserving their record for present day reference. Interpreting the beach ridge history is neither simple nor easy, but Tanner works through the geological interpretive process to provide about 150 data points since 8,000 years ago. These approximately 50-year interval measurements of sea level fall at approximately the lifetime time scale of archaeological time developed above.

The importance of sea level changes to our studies of both historic and prehistoric regions becomes apparent when one realizes that sea level is an indirect measure of the global average temperature, and thus global change. As global temperatures rise, glaciers worldwide retreat and sea level rebounds. As the world cools, more ice is locked into glaciers and sea level retreats. Thus, measuring sea level at any given time estimates global temperature. Fortunately, Tanner's 50-year sea level measurements fall at the correct time resolution to provide archaeological-scale lifetime data on global environmental change. How these global measurements are translated into regional climates is a much larger issue than can be addressed in this book. Techniques have been developed to translate global scale changes into local climate chronologies and are published in various peer reviewed journal articles (Gunn et al. 1995, Gunn and Crumley 1991 and in Woodall's chapter in this book).

Together the concepts of archaeological time and global temperature change provide a backdrop for the first millennium A.D. archaeological studies of the southeastern United States that follow. I have to emphasize that this study is a mere down payment on the many studies that need to be conducted to come to a useful understanding of global change at archaeological time scales. Of course, the overriding question is: how do archaeologists see time-limited events such a 15-year period of global cold? The answer in general terms is that they do not; they see its aftermath, which is at the more appropriate lifetime scale. Being able to date the event to the year, however, provides a global time boundary, or grid (see chapter 1), for the event-induced episode following. Through the principles of thermodynamics and earth system processes such as glaciation, researchers can infer the most likely consequences of an event. The cooling of the A.D. 536 event, for example, would be expected to reduce sea level through increased glaciation.

Gunn, Joel, and Carole L. Crumley (1991) Global Energy Balance and Regional Hydrology: A Burgundian Case Study. *Earth Surface Processes and Landforms* 16:579-592.

Gunn, J., W. Folan, and H. Robichaux (1995) A Landscape Analysis of the Candelaria Watershed in Mexico: Insights into Paleoclimates Affecting Upland Horticulture in the Southern Yucatan Peninsula Semi-Karst. *Geoarchaeology: An International Journal* 10:3-42.

8. BEACH RIDGE HISTORY, SEA LEVEL CHANGE, AND THE A.D. 536 EVENT

William F. Tanner

Regents Professor, Geology Department, Florida State University, Tallahassee, Florida 32306

Sea Level and Global Climate

[globe, North Atlantic, Florida, Denmark; beach ridge granulometry, 50 year resolution; global climate, Little Ice Age, sea level cycles; Holocene, Late Holocene]

Long beach ridge sequences provide information at 30-50 year intervals about sea level changes through much of Holocene time. The longest known sequence (northern Denmark) extends to about 9,550^{c14} B.C., with a small gap near 6,050^{c14} B.C. It can be checked against other sequences, as in Florida (U.S.A.), reaching about 1,150^{c14} B.C., and against still others for shorter intervals. The "Little Ice Age" (Lamb 1981) appears clearly as a sea level low stand, and other widely-recognized global episodes are reflect in sea level low or high stands.

The use of beach ridge plains on both sides of the Atlantic Ocean assures that the resulting history is global rather than regional. Precision is 30-50 years, and accuracy may be close to 50 years, perhaps 75-100.

This history includes three important sea level declines (1-4 m each; hence global cooling episodes) in the last 3,000^{c14} years. Of these three, the first was the most severe, and the last, the "Little Ice Age," was the mildest. The second of the three, from A.D. 500-550 to 800-850^{c14}, was intermediate in severity. The "dry fog" of A.D. 536^{cal} appeared at the beginning of, or early in, this decline; therefore they could have been related.

The "dry fog" could have been due to volcanic ash, cosmic dust, wind-blown terrigenous dust or perhaps some presently unknown cause. There is little or no evidence for the first two of these, or a fourth. Zhao et al. (1996) reported *no* unusual cosmic dust concentrations in loess profiles, in China, since 8,000^{c14} B.C. The associated sea level drop also requires a duration (more than three centuries), which is not compatible with minor volcanism or with a minor comet or meteorite.

This leaves terrigenous dust as the likely cause. Wind-blown dust events were common in the Pleistocene, and one such episode was observed in recent history in the American west in the 1930s: the so-called "dust bowl."

Beach Ridge Studies

[Europe, South America, North America; sand granulometry; Holocene]

This chapter uses long beach ridge sequences in the time sense to estimate a history of sea level changes during the Holocene. Many beach ridge plains extend only to 1,000-3,000^{c14} years ago, but the very wide beach ridge plain in extreme northern Denmark provides detailed data—at intervals of 30-to-50 years—across essentially all of the Holocene, the most recent 10,000 years.

Other beach ridge systems that have been studied are located in Germany (on The Darss; at Heiligenhafen; and northeast of Schleswig), in Venezuela (near Rio Chico), in Brazil (on and near Cabo São Tomé), in Honduras (two different systems), and in the U.S.A. (on Mesa del Gavilán east of Brownsville, Texas; in southwestern Louisiana; and at several locations in Florida).

The Danish study area is south of the town of Skagen (located close to the northern tip of Denmark; Figure 8.1). The town of Jerup is near the eastern edge of the study area, and the town of Tversted is near the western edge.

I undertook field work on the Jerup beach ridge system in the late spring months of 1989, 1990 and 1991. The granulometric analysis of the Jerup system ridge sands was based on procedures that were described previously in a series of articles (Tanner 1991a, 1991b, 1995a), and a detailed report on the Jerup beach ridge system (including a list of the granulometric moment measures) has been published (Tanner 1993). Field and laboratory work have been completed on the Tversted ridges, but the written report is not yet in print.

Jerup Beach Ridge Sets and Systems

[Jerup, Denmark; variability range since 5,800 B.C.; Tversted, Jerup, 9,500 B.C. maximum]

On wide beach ridge plains, individual ridges occur in sets with typically five to 25 ridges in one set. Several sets may make up a system that might include as many as 100 or 150 ridges. In northern Denmark, the older Tversted system contains more than 120 ridges (the western edge is covered by dunes, so the count is incomplete); the younger Jerup system contains 154 well-defined ridges with no significant eolian cover. The two systems are separated by a creek that flows northward into a relict depression. The latter is drained by another creek flowing eastward (see Figure 8.1).

The Jerup system dates to about 5,800^{c14} B.C. with reasonably good radiocarbon control back to 3,600^{c14} B.C., and the Tversted system falls before it at about 9,500 and 6,200^{c14} B.C. The interval between the systems (6,200-5,800^{c14} B.C.) contains no beach ridges at all. It is thought to represent a sharp sea level rise of about 14 or 15 m, perhaps more, followed by an equally sharp decline, both together encompassing roughly 400 years.

The Tversted ridge sands are characterized by a significant amount of dark organic matter, but the Jerup ridges are composed of clean buff-colored quartz sand. The sharp contrast in color is taken as one line of evidence, among others, that the Tversted ridges were submerged to a depth of perhaps 14-15 m or more, *after* the deposition of this system was completed on the beach but before sea level fell again and the Jerup system was initiated.

Since 5,800^{c14} B.C., mean sea level has varied within a few meters of its present elevation. Therefore, the most important factor responsible for the present elevated position of individual Jerup ridges, is glacio-isostatic rebound. This is the slow upward movement of the land surface due to melting and removal of the Wisconsinan (= Saalian) ice sheet from Scandinavia. For the last 6,000^{c14} years, the rebound rate in the study area has been on the average 2 mm/yr, 20 cm/century, or about 15-16 m in the last 8,000^{c14} years.

This rebound has been large enough that individual beach ridges have been raised so they are no longer subject to wave attack. Hence they are well preserved. On the other hand, it has been slow enough (only about 3 micrometers per tidal cycle) that it has not affected wave dynamics in the surf and swash zones. The isostatic rebound, therefore, does not appear in the granulometric data.

Jerup Granulometry

[Jerup; granulometry, sample frequency, sieves, kurtosis; low energy beaches; Holocene]

The uses of granulometric procedures for identifying agents and locales of deposition have been discussed in detail elsewhere ("Suite statistics;" Tanner 1991a, 1991b, 1995a, and references therein). Each of the 154 ridges in the Jerup system has been sampled on the mid-point of the face according to strictly defined procedures. Each sample for each ridge was 40-50 grams. There is no historical advantage to taking additional samples from one ridge, but it is necessary to adhere rigidly to the prescribed sampling pattern so that within-ridge variability does not warp the historical results.

Each sample was cleaned, dried, weighed and sieved for 30 minutes in a nest of quarter-phi sieves. The resulting fractions were weighed to 0.001 grams. The computer program GRAN-7 uses the raw weights to produce a histogram, a probability plot, and a listing of various parameters, including the first six moment measures. The moment measures are much more useful and reliable than graphic measures, which depend on very little of the data set.

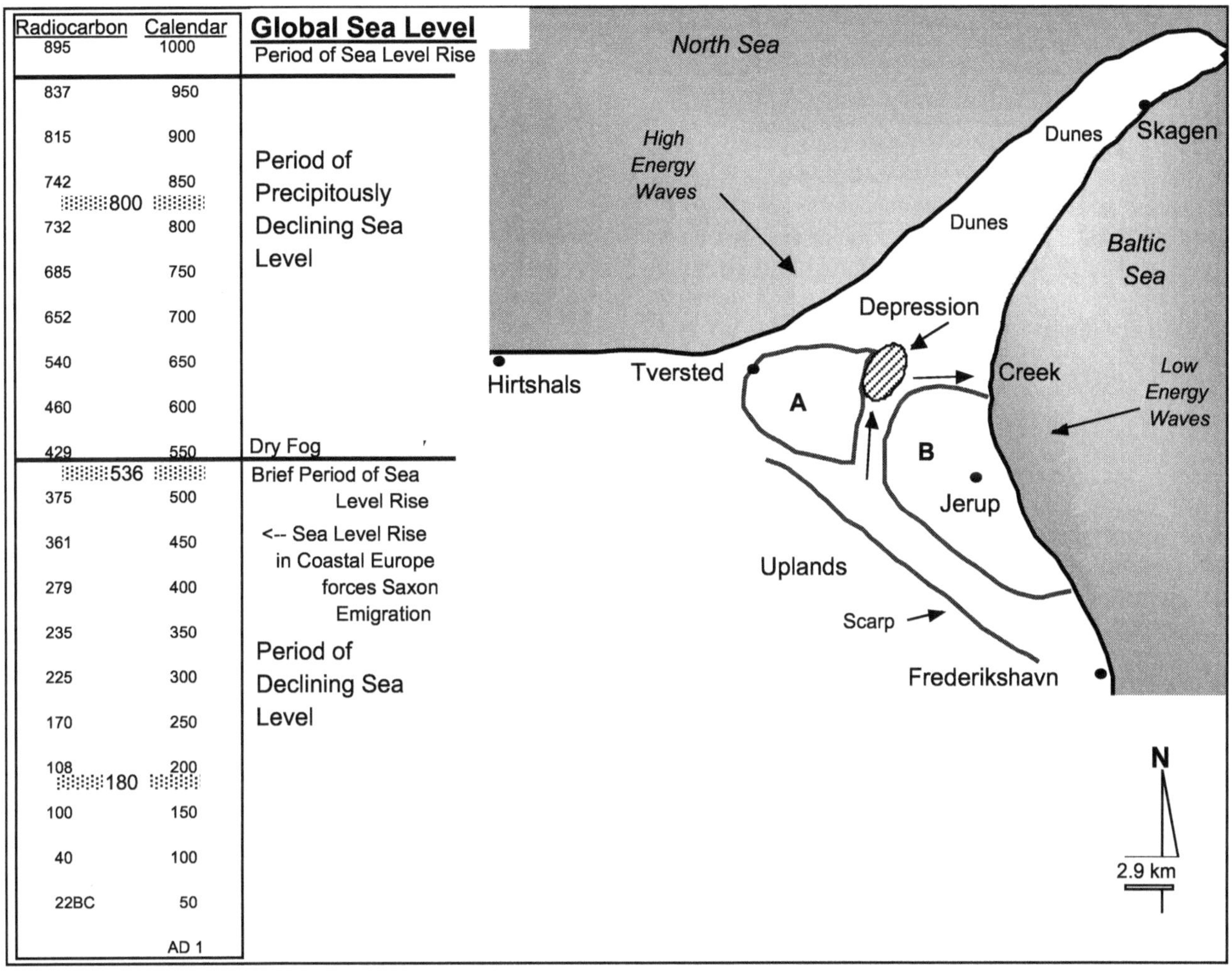

Figure 8.1. Map of extreme northern Denmark showing five towns and the two main low-energy beach ridge systems: Tversted ridges at A, and Jerup ridges at B. The Skagen high-energy ridge system, close to the town of that name, is not shown. Within the areas A and B, no individual ridges are indicated, but there are at least 120 in A, and precisely 154 in B. The ridges in B are, in general, parallel with the modern low-energy beach to the east of Jerup. A creek, draining northward into the depression, separates A and B; it occupies a residual, or relict, valley, but is much too small to have formed that valley by erosion.

These moment measures for the full set of samples are then analyzed by the computer program SUITES-7 to identify the agent of deposition. This procedure relies heavily upon the calculated values of the moments of the moments, e.g., the standard deviation of the kurtosis. It therefore reflects important ideas such as the variability of any one moment measure from sample to sample within the suite. The method has been calibrated by many thousands of samples from known localities.

The findings in northern Denmark are unambiguous; the sand in the Jerup ridges was deposited on a low-energy beach, and cannot be attributed to high-energy waves, wind, river, ice or deep-water processes (Tanner 1993). The modern beach east of Jerup (facing Sweden) is in fact now a low energy beach, and represents accurately the conditions that prevailed in the area during the time of accumulation of the Tversted and Jerup systems of ridges.

A series of experiments (Tanner 1992a) showed that the fourth moment measure of the size distribution, the kurtosis, when inverted, is an excellent index to the local wave energy level. The kurtosis value within any one beach ridge *set* (requiring hundreds of years to form) is typically almost constant from earliest ridge to latest. This demonstrates that the near-shore wave energy level was almost constant for that time period or episode.

However, at ridge set boundaries, with an elevation differential of 1-4 m, the kurtosis changes markedly. This shows that initiation of a new ridge set, at a different elevation, was associated with a change in long-term wave heights. Therefore a marked change in kurtosis indicates a change in sea level (see the next section). I developed this in detail on St. Vincent Island, Florida, U.S.A. (Tanner et al. 1989; Tanner 1992b). The hydrodynamic explanation for this relationship is given below.

Beach Ridge Hydrodynamics

[Jerup; kurtosis, kurtosis/time plot; sea level; 7800^{c14} years, 2000^{c14} years]

The near-shore transverse profile on sandy coasts is gently, upwardly concave, with the maximum concavity near the beach. When sea level rises by a small amount (about one meter), the water close to the beach is significantly deeper than that prior to the rise. Therefore less wave energy is subtracted in the shoaling process, and more energy is left for the surf zone. This is not a matter of a storm tide, or even of a few storms: the mean near-shore wave energy level is larger until a new transverse profile is established, some few centuries later. When sea level drops by a small amount, the reverse effect is obtained: the energy level in the surf zone is reduced, until a new profile can be established at a later date.

The result is that K (kurtosis) is a function of 1/E (wave energy density), from an hydrodynamic point of view; or E is a function of 1/K, from an interpretative point of view. A calculation of dK/dT (where T is time, and d is a difference) shows sharply-defined peaks at those times when sea level changed by more than 10-40 cm.

Therefore 1/K can be plotted against time, and this plot can be read in terms of sea level changes: peaks on the plot show high sea level positions and troughs show low sea level positions. The sea level changes that are indicated in this fashion were typically 1-4 m in the last few thousand years.

The changes in kurtosis (Figure 8.2) for the last 2,000^{c14} years represent the latter part of the Jerup beach ridge system, and hence the recent history of sea level change. I have published a plot of sea level changes since 5,800^{c14} B.C. (Tanner 1993), and a redrafted part of a figure in that article, in this case covering the last 2,000^{c14} years, is used in the present discussion. Two different curves are given for the Jerup ridges: one for the raw data (1/K), and the other for smoothed data (moving averages, with a window of W = 3, which reduces single-point spikiness in the data).

Beach Ridge Origin

[Jerup; Sea level change couplets, rise and fall, ridge and swale; Holocene]

It is necessary to keep in mind that no one of these beach ridges was built by, or during, a storm, despite the wide popularity of this poorly grounded hypothesis. The basic idea (no storm origin) was sketched in broad form, with supporting evidence, by Douglas Johnson (1919:404-457) in a long and useful discussion that subsequently has received very little attention. However, he did not suggest sea level changes.

Instead, each beach ridge was the product of a couplet of small sea level changes: one rise, then one fall. Alternatively, one can say that each swale in the system was the product of one fall plus one rise. This addition to and clarification of the work of Johnson was developed in detail by me (Tanner 1995b). The necessary sea level changes for one ridge were only a few centimeters.

The present scheme includes the following:

1. Sea level change couplet, 5-30 cm, decades apart: One ridge-and-swale.

2. Two sea level changes of 1-4 m, some centuries apart: One ridge set. Of course there have been larger sea level changes than these, such as the sea level rise, from about -140 m, between 20,000 and 10,000^{c14} years ago, but such large changes do not produce individual ridges, ridge sets, or ridge systems.

Dating

[Jerup, St. Vincent Island; radiocarbon, survey; interval resolution 50.5 years; glacio-isostatic rebound; Holocene, Late Holocene]

The granulometric data from northern Denmark have been plotted in a diagram that also shows a radiocarbon calendar (not corrected). This calendar was based on radiocarbon dating of nine peat samples from beach ridge swales, which approximate sea level at the time of formation. They were obtained by Poul Hauerbach and surveyed precisely to map location and altitude. The oldest date is about 3,650^{c14} B.C. at an altitude of almost 14 meters, and the youngest is about A.D. 550^{c14} at an altitude of about 3.3 m. There is a tenth point, as well: modern sea level.

My plot of Hauerbach's data (Tanner 1993:Figure 4) duplicates in a general way the sea level history that had been derived from St. Vincent Island, Florida (Tanner et al. 1989; Tanner 1992b), previously, showing clearly: (a) the low-stand that bracketed 2,250^{c14} B.C. (about -1.6 m), (b) the important sea level rise (2-3 m) that took place around 1,250-1,150^{c14} B.C., and (c) the high stand that bracketed A.D. 50^{c14} or 150^{c14} (about +1.0 m). Hauerbach did not have enough data points to show other details, nor did his set of swale-peats extend far enough in time to show changes in the earliest part of the Holocene.

Unlike shell dates in the same area, these swale-peat dates are internally consistent, and also match the St. Vincent Island, Florida, data (Tanner et al. 1989).

In each case, a precise location and elevation was obtained by Hauerbach, who is a registered and certified surveyor. The peat sites are only some 10-15 km north of the Jerup beach ridge plain, which had been surveyed by the Danish government in great detail; tilting between the two areas, due to glacio-isostatic rebound, has been negligible. Therefore, dates were transferred to the Jerup beach ridges on the basis of elevation. This must be done in a time sequence, because Hauerbach's data show clearly that there have been both rises and falls (in agreement with results from other areas), hence a given elevation does not necessarily match a unique date, unless one has determined the proper position in the sequence.

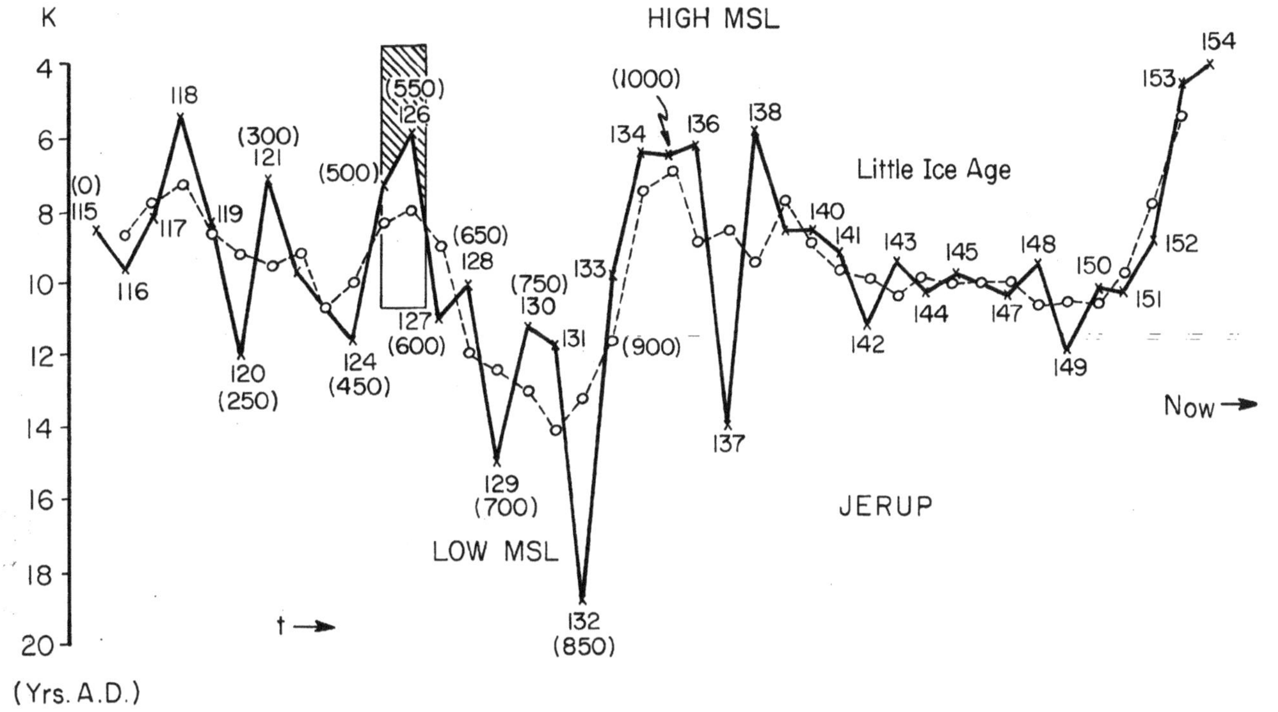

Figure 8.2. A plot of 1/Kurtosis for Jerup system ridges No. 115 through No. 154 (hence the most recent 2,000 years of history). The continuous line shows the actual kurtosis values of the grain size distribution of each sample (one per ridge), and the dashed line shows moving averages (window is W = 3, hence smoothing is minimal). The high K values (low positions on the plot) show low sea level positions, hence by inference cooler climates; the low K values, high sea level positions, and warmer climates. The Little Ice Age appears between Sample No. 138 and Sample No. 153. The drop in sea level, in each case, was 1-3 or 1-4 meters. The cooling event that followed A.D. 525-550 (see box in this figure) was more severe than the Little Ice Age, as can be seen here. Dates are given in parentheses. The uncertainty of each date is thought to be about 50 years, but perhaps is as much as 75-100 years. The K values are precise to three decimal places, although this precision cannot be shown in the figure. Data for the entire set of 154 Jerup beach ridges were given in table form, as well as in diagram form, in an earlier paper (Tanner, 1993).

The radiocarbon dates provide an average interval of 50.5 years between ridges, to $3,650^{c14}$ B.C. This average was employed uniformly on all ridges in the Jerup system. A different, and more involved, procedure is available, giving an estimate of unique time intervals between each two successive ridges, rather than a mean value; this work is summarized in the next section.

I think that the error for the date of any given ridge is perhaps not more than 50 years. Nevertheless, individual ridges were not dated directly, and in fact there are no uncontaminated dates from the ridges and swales in either the Tversted system or the Jerup system; for the time being, at least, one must depend on Hauerbach's dates, locations, and elevations.

Sea level *changes* were deduced from the granulometry, but *not* from the present vertical position of any ridge, and *not* from the vertical differences between ridges. These results are not subject to surveying error (if there was any). Therefore, any errors in the resulting plot, however large or small they may be, are in the dates. For this and other reasons, it has been deemed appropriate to calculate moving averages, using various windows (from W = 3 to W = 9), and also using, in one case, two successive passes (W = 5; W = 2 but in the latter instance without overlap).

Sea Level Change Sequence

[Jerup; kurtosis, standard deviation; smoothing; calibration; A.D. 536^{cal}]

The basic time schedule that has been worked out in the previous section was based on the use of a 50.5-year average spacing between ridges. This figure was derived from Hauerbach's dates, and was applied systematically across all of the Jerup system. The radiocarbon control on the Tversted system is not nearly as refined; the average spacing between those ridges is something like 30-35 years, but there is little precision. However, the application of a rigidly-fixed mean value to each and every ridge in either system is not realistic.

An alternate method of creating *estimates* of interpolated time intervals for Jerup ridges was based on an equation using the granulometric standard deviation combined with the granulometric kurtosis:

Ln time = 6.27 - (5.4 * Standard Deviation) - (0.244 * Ln Kurtosis).

Because the value of 50.5 years was used as a fundamental constraint (hence total elapsed time), the total time for the system is the same as that obtained directly from Hauerbach's dates, but the ridge-by-ridge dates differ slightly.

The logic behind this equation is that sorting should improve with time (smaller standard deviation), and that kurtosis will reflect these same changes, also, although much less clearly. Therefore, those samples, or ridges, having low values for the standard deviation (good sorting), should have been built more slowly. Calibration was done in part with data from the St. Vincent Island beach ridge system in Florida.

The maximum *computed* interval between ridges near Jerup was 92.2 years, the minimum, 10.2 yrs. A three-point moving-average produced a maximum of 68 years and a minimum of 22 years revealing a great deal of spikiness in the calculated results. Whether the dates based on the mean value, or the computed results, are better, cannot now be posited with assurance.

One of the corollaries of this result is that spacing in time between ridges was relatively small during low stands of sea level: that is, sorting of near-shore sands was improved more rapidly when near-shore water depths were relatively small, rather than when they were 1-4 m deeper during high stands.

The resulting plot of relative sea level positions (based on inverted kurtosis) versus age (based on the equation, above), is very much like what can be seen in Figure 8.2, although not exactly. On the new plot, a minimum sea level position was reached at about 2,950^{c14} B.C. (perhaps 2,750^{c14} B.C. in Figure 8.2), which was followed by an almost steady climb to a sea level high stand at about A.D. 500-540^{c14}. This in turn was followed by a sea level drop, into the Little Ice Age, with a low stand at about A.D. 1,700^{c14}. Subsequently, there has been a more-or-less steady rise. This rising trend is now about 300 years old.

If global temperatures are assumed to be related to sea level, the sea level numbers suggest an important change in global climate at about A.D. 500-540^{c14}. Another important observation is that the years A.D. 500-540^{c14}. marked the beginning of an important cooling episode. This result was obtained from either method of estimating spacing in years between ridges.

Verifying Beach Ridge Formation Processes and Links to Global Climate

[All locations; cross-checking ridge systems; ; Holocene]

The Tversted-Jerup beach ridge plain provides a fairly detailed history of sea level change for almost all of Holocene time. This history can be verified by cross-checking with other beach ridges in the most recent three millennia; because no other beach ridges are known in an isostatic rebounding situation, no other sandy beach ridge plain has been found that extends so far into the past.

The St. Vincent Island, Florida, curve (Tanner et al. 1989) extends to about 1,250-1,150^{c14} B.C. For this length of time, the Jerup and St. Vincent histories are almost identical, although they were formed on opposite sides of the Atlantic Ocean, and at quite different latitudes (57° North vs 29° North, or a difference of about 28°).

The Little Ice Age (Lamb 1981) was recent enough to show up at many study locations, and therefore provides a good cross-check of processes between ridge systems, and on sea level response to global temperatures. If each beach ridge plain history does in fact record changes in global sea level, then this event should be seen at all, or almost all, study locations. The associated fall in sea level appears at Jerup (Denmark), St. Vincent Island (Florida), Dog Island (Florida), and Mesa del Gavilán (Texas), as well as elsewhere, as is discussed immediately below.

A lack of radiometric or other dates at some locations prevents a detailed comparison, although visual inspection suggests that the Little Ice Age was recorded in those ridge systems as well. For example, near Rio Chico (Venezuela), the ridges show a pattern of drowning by sea water much like what can be seen on St. Vincent Island (Florida) (Tanner et al. 1989).

A relevant match exists between dates of ridge-set boundaries in northern Denmark and the dates of initiation of Mississippi River sub-deltas (Tanner 1991c).

The examples given in this section indicate that the history of sea level change obtained in extreme northern Denmark was global rather than regional or local. It further indicates that glacio-isostatic rebound in northern Denmark was slow enough that it did not affect granulometric characteristics on the beach near Jerup.

Cross-checking is also possible on a spotty basis with historical records and other paleoclimatic data series. Here a major concern is whether any given curve senses global conditions at the episode level (smoothed; hundreds of years) or sub-episode level (jagged; tens of years).

East Asia

[Asia; cross-checking; climate; sixth century]

Many of the examples mentioned here—such as Denmark, Germany and Florida—are on the shores of the north Atlantic basin, the northernmost part of which was more-or-less encircled by the major continental ice sheets (except for Antarctica). Therefore changes in climate and changes in sea level are expected to be synchronized: cooler climate, sea level drop.

Although important changes in sea level and in climate may be global, the types of climate change do not have to be. It is well known that a shift to wetter conditions in one region may be accompanied by a shift to drier conditions in another region at about the same time. A recent example of this was reported in a popular summary by Jung-Hüttl (1996).

For this reason, in eastern Asia, specifically China, we may find evidence for marked climate change, but not necessarily in what appears at first glance to be the right direction. Thus

Zhang et al. (1996) reported an historical climate shift from dry to wet at A.D. 540^{cal}, or, alternatively, from warm to cool at A.D. 530^{c14} (p. 171; but note that text matter is quoted here, and likewise the table on p. 172, whereas the more generalized dates in the table on p. 171 are not used here). Unfortunately this article has no bibliography. It contains the single statement (p. 170) that the climate record was "rebuilt" by Kezhen Zhu from archeological data, studies of dates of animal migrations and plant flowering, and "historic records."

The A.D. 536 Event

[global; curve comparison; dry fog, sea level change; A.D. 1-present]

The original numerical data from the Jerup beach ridge plain have a mean apparent spacing in time of 50.5 years. The curve summarizing these data has been published in a smoothed form having one point every century (Tanner 1993) (also see Figure 8.2 of this chapter). The curve given in this chapter has a data point every 50 years (these values are based on uncorrected radiocarbon dates). The pertinent part of this history can be summarized as follows:

1. Peak at A.D. 300^{c14}.
2. Trough at A.D. 450^{c14}.
3. Peak at A.D. 550^{c14}.
4. Trough at A.D. 600^{c14}.
5. Small peak at A.D. 650^{c14}.
6. Trough at A.D. 700^{c14}.
7. Small peak at A.D. 750^{c14}.
8. Trough at A.D. 850^{c14}.
9. Peak at roughly A.D. 1000 or 1100^{c14}.
10. Drop, into the Little Ice Age. The years are given in multiples of 50; I do not think that greater precision can be obtained from the data. It is possible that uncertainty for any given point might be as great as 75-100 years.

The A.D. 536^{cal} event appears in the box in Figure 8.2. This is close to the initiation of a major (although zig-zag) drop and low-stand, covering three centuries (A.D. 550 to 850^{c14}). If the A.D. 536^{cal} event, and the start of a sea level drop, have a cause-and-effect relationship, then the "dry fog" reported at that time could mark an atmospheric event having serious climatic consequences, as is discussed below. The total drop, in this 300-year interval, apparently was something like 1-3 m, therefore the rate was perhaps as much as 1 cm/yr.

Use of lightly-smoothed averages (W = 3; Figure 8.2) simplifies the history somewhat, and gives a more-or-less straight drop from roughly A.D. 550 to 850^{c14}. As discussed above, however, there is reason to believe that the sample-to-sample sub-episode variations are meaningful as well.

Questions about Generalization

[global; research questions; dry fog, cross checking; 800 B.C.-A.D. 1150]

We should now ask the question, Did the "dry fog" of A.D. 536^{cal} take place at the beginning of a sea level drop (hence climate change), or was the close approximation in time merely a coincidence, perhaps based on poor data? The agreement of the Jerup beach ridge record with the St. Vincent Island record (Tanner et al. 1989; Tanner 1992b) supports the contention that the Jerup-dry fog coincidence was not a mere artifact of the Jerup data. There may be a half century of intervening lag when radiocarbon corrections are considered, but it is not unreasonable to expect that the oceans will respond to changes in global termperatures after a period of adjustment.

A further pertinent question is whether the record of dry-fog initiation at the beginning of a sea level decline can be generalized to other declines of sea level. A sharp fall initiated around A.D. 1000 and 1150^{cal} (a century later in calendar years) coincides with the beginning of the "Little Ice Age" (Lamb 1981). Record keeping at that time was presumably somewhat better than it had been five or six centuries earlier; the written records of the "Little Ice Age" with its cold weather, crop shortages, and advancing glaciers are reasonably detailed and complete. Was a "dry fog" reported then? Gill (1994) cited sources documenting various eruptions, including that of El Chichón at about this time; perhaps dry fog-like phenomena do initiate sea level changes.

Another sharp decline in sea level took place in the interval 800-500^{c14} B.C. (not shown in Figure 8.2 which does not extend that far, but it is clearly visible in an earlier report [Tanner 1993:fig. 6]). Is there evidence for a "dry fog" at that time? Discovering a dry-fog phenomenon at that time would do much to support generalization of the statement.

Causes

[global; meteorite, volcano, dust; A.D. 536^{cal} and later]

An attractive hypothesis to explain the "dry fog," especially in connection with a general recorded cooling, is a volcanic eruption. However, volcanic ash or acid aerosols from a single eruption probably would not persist in the atmosphere in significant amounts for more than two or three years. Therefore, to explain the full 300 years of the associated cool spell, a string of volcanic eruptions might be invoked. Unfortunately, no such eruptions are known. In the Great Plains of the United States, many volcanic ash falls were recorded in the last two million years, and these have been used in many efforts to explain the cold-warm-cold pulsations of Pleistocene time (and hence the waxing and waning of the North American ice sheet). No significant ash falls from the interval in question has been reported.

It has also been suggested that the "dry fog" may have been cosmic dust, perhaps part of a comet tail or debris from a meteorite impact somewhere on the face of the planet. Again there is now no evidence to support this conjecture, and the duration of the sea level drop (about 300 years) appears to have been much too long for a comet, or a meteorite impact, to have been the cause.

A third possibility is terrigenous dust, such as was blown from western United States (the American "dust bowl") in the 1930s and was observed as far away as Europe and the Mediterranean Sea. Such dust deposits are also known from earlier times where at least one example may have been Saharan dust (Herwitz et al. [1996]; however, a comparison with materials from the Gobi desert was not undertaken by them). Wind-blown dust is a strong contender as a cause because production of dust does not have to be limited to one short event, or even a few such events; instead, in itself, it indicates an important climate change. Another candidate is low solar emission (see Robichaux, Houston, this volume; Lean et al. 1992). Although in itself low solar emissions do not produce a "dry fog," cooler global conditions could have acted in a supporting role with terrigenous dust.

Volcanic ash, cosmic dust, and terrestrial dust are all rational hypotheses. Each would provide something that—at the proper distance from the source—could have been considered to be a "dry fog," and each is more-or-less a reasonably good candidate as a cause for cooling (hence sea level decline). Unfortunately, historical and/or geological details are not sufficient at the present time for us to identify any such event, except to comment that the duration does not seem to have been appropriate for the first two.

Cause and Comparisons of A.D. 536^{cal} Sea Level Decline

[global; granulometry; middle range decline; A.D. 536-800]

Granulometry of individual ridges in a very wide beach ridge plain in extreme northern Denmark, provides a fairly detailed history of sea level change for essentially all of Holocene time. This history has been duplicated on St. Vincent Island, Florida., and in various other areas, as far back as the record goes in each place. It includes, among many other changes, a sea level drop beginning about A.D. 550^{c14}., and continuing for approximately 300 years. It is inferred that there was a general global cooling associated with this fall in sea level.

The "dry fog" of A.D. 536^{cal}, described in written sources, may have been due to volcanic ash, cosmic dust, or terrestrial dust, and perhaps this is not a complete list of possible sources. At the moment there is no information available to help us select, with confidence, a "most likely" candidate from the three or more possibilities, except that the duration of the associated sea level drop seems to have been too great (roughly 300 years) for the first two to have been the sole cause.

The sea level drop that started about A.D. 550^{c14} (give or take 50 years), and the following low-stand, were spread over about 300 years, and appear to have reached a magnitude of 1-3 meters. It was not as severe as a drop about 1,000 years earlier, but appears to have been sharper (faster changes), and to have represented a greater change in global temperature, than was observed in the Little Ice Age, starting five or six centuries later. That is, the three main climate deteriorations in the last 3,000 years appear to have become progressively milder, culminating in the Little Ice Age, which was the mildest of the three.

The Danish sea level curve can be drawn back to about 9,000-9,500^{c14} B.C., with the latter half under much better time control than the earlier part. This curve shows eight or so sea level declines, three of which occurred in the last 3,000 years. Each represents a cooling of the global climate. If the "dry fog" of written records coincided with only one of them, one should question the idea of generalizing "cause and effect" until other such relationships have been demonstrated.

In any case, it seems clear that more information is needed before we can relate the "dry fog" of history, in a direct and dependable way, with various evidences of climate deterioration as seen here and there, such as in the Danish sea level curve.

ACKNOWLEDGMENTS: I wish to express my deep appreciation to Dr. Per Bruun, who was the first to suggest that the study area might well be worth investigating in detail; to Prof. N. Kingo Jacobsen, of the University at Copenhagen, Denmark, for introducing me to the area in the first summer that I worked there; and to Mr. Poul Hauerbach, registered surveyor from Randers, Denmark, who provided field logistics, field guidance, radiocarbon dates, precise locations and elevations, and much fascinating commentary on the written history of the Skagen-Jerup-Tversted area. Without their encouragement and assistance, this project would never have been started, much less completed.

References

Gill, R.B. (1994) *The Great Maya Droughts*. Unpublished Dissertation, The University of Texas at Austin.

Herwitz, S. R., D.R.Muhs, J.M.Prospero, S.Mahan and B. Vaughn (1996) Origin of Bermuda's Clay-Rich Quaternary Paleosols and their Paleoclimatic Significance. *Journal Geophysical Research* 101: 23,389-23,400.

Johnson, Douglas (1919) (reprinted 1938) *Shore Processes and Shoreline Development*. John Wiley, New York.

Jung-Hüttl, A. (1996) Die Ausreißer. *Bild der Wissenschaft (Stuttgart)* 12: 16-21.

Lamb, H.H. (1981) An Approach to the Study of the Development of Climate and Its Impact in Human Affairs. In *Climate and History*; edited by T.M.L.Wigley, M.J. Ingram and G.Farmer; Cambridge University Press, pp. 291-309, Cambridge.

Lean, Judith, Andrew Skumanich, and Oran White (1992) Estimating the Sun's Radiative Output during the Maunder Minimum. *Geophysical Research Letters* 19:1591-1594.

Paine, Jeffrey G. (1987) The Swan Lake Site (41AS16) and the Holocene Highstand Hypothesis. In *National Register Assessments of the Swan Lake Site, 41AS16, on Copano Bay, Aransas County*, Texas, Elton R.

Prewitt, Susan V. Lisk and Margaret Ann Howard, Prewitt & Associates, Inc., Reports of Investigations, Number 56, Austin. Submitted to the Galveston District Corps of Engineers.

Prewitt, Elton R., and Jeffrey G. Paine (1988) The Swan Lake Site (41AS16) on Copano Bay, Aransas County, Texas: Settlement, Subsistence, and Sea Level. *Bulletin of the Texas Archeological Society* 58:147-174.

Tanner, W.F., S.Demirpolat, F.W.Stapor and L.Alvarez (1989) The "Gulf of Mexico" Late Holocene Sea Level Curve. *Transactions Gulf Coast Association of Geological Societies* 39: 553-562.

Tanner, W.F. (1991a) Suite Statistics: The Hydrodynamic Evolution of the Sediment Pool. In *Principles, Methods and Applications of Particle Size Analysis*, edited by J.P.M.Syvitski; Cambridge Univ. Press, pp. 225-236, Cambridge.

Tanner, W.F. (1991b) Application of Suite Statistics to Stratigraphy and Sea Level Changes. In *Principles, Methods and Applications of Particle Size Analysis*, edited by J.P.M.Syvitski; Cambridge Univ. Press, pp. 283-292, Cambridge.

Tanner, W.F. (1991c) The "Gulf of Mexico" Late Holocene Sea Level Curve and River Delta History. *Transactions, Gulf Coast Association of Geological Societies* 41: 583-589.

Tanner, W.F. (1992a) 3,000 Years of Sea Level Change. *Bulletin American Meteorological Society* 73: 297-304.

Tanner, W.F. (1992b) Late Holocene Sea Level Changes from Grain-Size Data: Evidence from the Gulf of Mexico. *The Holocene* 2: 249-254.

Tanner, W.F. (1993) An 8000-Year Record of Sea-Level Change from Grain-Size Parameters: Data from Beach Ridges in Denmark. *The Holocene* 3: 220-231.

Tanner, W.F. (1995a) *Environmental Clastic Granulometry*. Florida Geological Survey, Tallahassee, Special Publication No. 40.

Tanner, W.F. (1995b) Origin of Beach Ridges and Swales. *Marine Geology* 129: 149-162.

Zhang, Chunshan, Yecheng Zhan and Jingjiang Hu (1996) Climatic Evolution during the Historic Period in the North China Plain. *Bulletin, Chinese Academy of Geological Sciences*, Special Issue 1996 pp. 170-175 (in English).

Zhao, Jixiang, Yonglin Xiao, Zhisheng An and Huayue Lu (1996) The Cosmic Dust in Loess Since Late Pleistocene, Shaanxi. *Bulletin, Chinese Academy of Geological Sciences*, Special Issue 1996 pp. 176-188 (in English).

9. SUB-REGIONAL PERSPECTIVES FROM RADIOCARBON DATES: SPATIO-TEMPORAL DISTRIBUTIONS OF FIRST MILLENNIUM A.D. RADIOCARBON DATES IN THE EASTERN UNITED STATES

Joel D. Gunn and C14 Consortium
(Jane Eastman, Keith Egloff, Robert Maslowski, Ken Sassaman)
Department of Anthropology, University of North Carolina, Chapel Hill North Carolina 27514

Radiocarbon Dates and Rapid Climate Change

[eastern United States; radiocarbon dating; population density; first millennium]

Over 2,700 Radiocarbon dates are available from states between Georgia and Ohio. The first millennial-scale period during which dates appear in statistically interesting numbers is the first millennium A.D. This high density of dates over a topographically varied region such as Georgia to Ohio suggests the examination of numerous hypotheses relating to cultural change, and population density and movement. This chapter opens the question of the impact of the A.D. 536 event and the subsequent 300 years on high and low elevation populations in the study region. It is an effort to construct the skeleton of a grid of date frequency densities across the region. It lends some insight to subsequent chapters and also raises many questions and research topics that could be pursued by analysis of radiocarbon date distributions in time and space. The most basic of these questions is the social and historical, as opposed to the statistical, resolutions of radiocarbon dates. Do the statistical deviations calculated by and for the physics of radiocarbon dates have equivalent meaning from the point of view of their practical application? We suggest that the answer to this question is probably more a matter of experimental application that physical theory.

The first work on globally oriented environmental inferences derived from radiocarbon dates was published in 1974 by Wendland and Bryson. They accumulated the then-extant corpus of dates on cultural and biological boundaries and produced a credible description of Holocene discontinuities. The results were used to support Bryson's long-held position that global climate changes occur over short periods of time, i.e., events. Events are followed by relatively extended periods, or episodes, of internally similar if not constant climate. This position has received some attention among archaeologists (Barreis and Bryson 1965, 1966; Custer 1989; Gunn 1994a) because it appears to coincide with the observed tendencies in the archaeological record for relatively sudden onset of cultural phases that are followed by decades to centuries of internally similar though not unchanging cultural manifestations.

As archaeologists turn to the linking mechanisms between environment and culture, and to the links between global climate change and cultural change (Gunn and Crumley 1991; Gunn et al. 1995), their attention will inevitably focus on highly resolved climatic and cultural chronologies. For more recent centuries, tree rings provide a satisfying, if not unlimited, record of climate relative to culture during one of the critical seasons of the year, late spring/early summer when temperate zone vegetation requires an adequate supply of moisture for its spring growth surge. Anderson et al.'s (1995) study utilizing baldcypress tree rings to evaluate relationships between Atlantic Slope groups and annual spring moisture is an ideal example of a cultural change study. It uses a concept of storage potential to link spring moisture with archaeological and historical episodes of complex and non-complex social organization. This approach allows for the sorting out of at least some of the environmental influences on cultural milieus. This technique is suited to the environmental data base and cultural record in the Southeastern United States, its own subregional set of biases imposed by the nature of the archaeological record, and the environmental configuration of the area.

For more remote time, essentially before the second millennium A.D., students of culture and climate must turn to radiocarbon. The average sigma (1 standard deviation) of the over 2,700 Holocene radiocarbon dates from states between Georgia and Ohio is 110 years. This suggests that sub-century statistically studies of cultural change are possible, and that when the statistical properties of large numbers of dates are taken into account, as in Wendland and Bryson's study, considerably less than century time intervals are accessible. As can be seen in Figure 9.1, the one sigma mode for the data set is 60 years. For a large portion of radiocarbon dates on archaeological sites, the age of the phenomenon is known within a one sigma deviation of 120 years. This approaches what we believe to be a minimal potential temporal resolution for archaeological cultural data of ≈100-150 years and the global scale climate resolution of Tanner's sea level curve when smoothed with a three point filter (see Tanner chapter). In other words, the two data sets have been subjected to similar temporal filters. The cultural data set has been filtered by the inherent characteristics of radiocarbon dating and the tendency of cultures to smudge over their immediate temporal milieu with living in place, reliving in place, and reworking of deposits. To this can be added bioturbation. However, perhaps one of the advantages of the radiocarbon method is that consolidated chunks of carbon are extremely fragile and thus a date immediately implies a generally, if not totally, undisturbed context. Tanner's data (see Figure 1.4 and 8.2) are filtered by the time required for glaciers and oceans to respond to changes in the global average temperatures (see Tanner chapter).

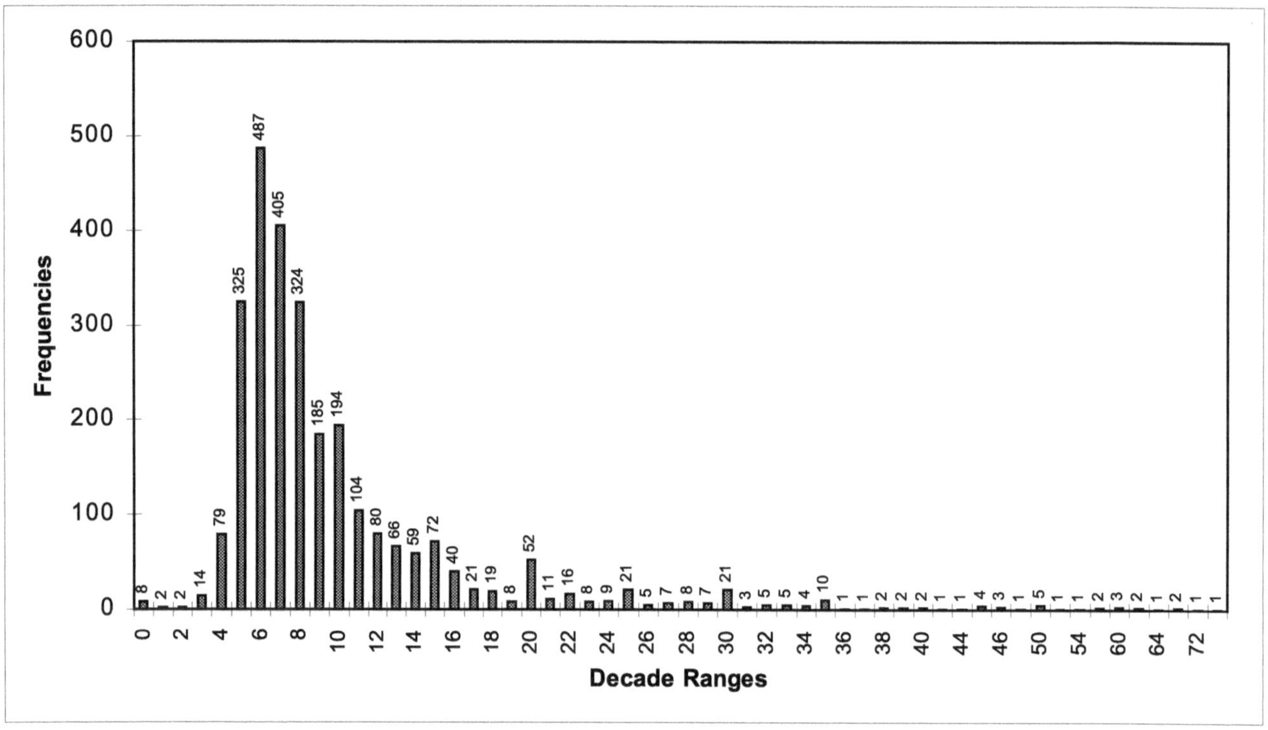

Figure 9.1. Frequencies of Radiocarbon Date Sigmas by Decades.

Data Set Structure

The current data set of radiocarbon dates is composed of 2724 dates that were assembled from the states of Georgia, South Carolina, North Carolina, Tennessee, Virginia, Kentucky, West Virginia, and Ohio. Numbers of dates from Georgia and Tennessee are small because they were included incidental to data gathering efforts in adjoining states such as at the Savannah River Plant in South Carolina. The largest number of dates is from Ohio, followed by Kentucky (Figure 9.2). In overview, the dates form a band from South Carolina extending northwestward over the southern and northern Appalachians, and down into the northern Mid South and eastern Midwest.

The data set used for the chapter is limited compared to data the C14 consortium hopes to accumulate. The assembled data consisted of variables that could be gathered from existing spreadsheets without extensive additional data collecting to augment states in which variables have not yet been coded. The variables that were consistent across all data bases were:

1. Site Trinomial designation; this provided for assembling the state, county and site number.

2. Radiocarbon age

3. Sigma; one standard deviation

4. Corrected age; most of the dates were associated with a corrected age. In this variable, the convention was used of treating B.C. dates as negative integers and A.D. dates as positive integers. This allows the dates to be used as a time series for purposes of statistical analysis and graphing.

In the future we hope to accumulate additional data on each date. These data will include:

1. UTMs of counties or site dates;

2. Elevation

3. Relief; normal differences between valley bottom and upland flats in the region.

4. It would also be useful to gather some of the variables that have traditionally proven useful in the analysis of settlement

patterns such as slope, aspect, distance and elevation to water, etc.

5. code for material

1 = charcoal

2 = shell, marine

3 = shell, fresh water; this will facilitate correcting dates from some of the obvious error factors and thus refining the resolution of the radiocarbon time series.

Assumptions

The primary assumption upon which the radiocarbon analyses are based is that the greater the population at given times in the past, the more opportunities they create for archaeologists to find hearths, smudge pits, post holes, etc., for archaeologists to date.

Eventually we should attempt to address archaeological biases and environmental biases that controvert this basic assumption. For example, as Bryson and Goodman (1980) point out while creating an index of global scale volcanism from radiocarbon dates, there is a tendency over time for older radiocarbon dates to be destroyed by erosion. In the case of archaeological dates, there will also be destructive processes such as reworking of deposits by later inhabitants of sites. Other biasing influence are various favored topics of archaeologists. Paleoindian remains are almost universally dated because of their rarity; later potential dates

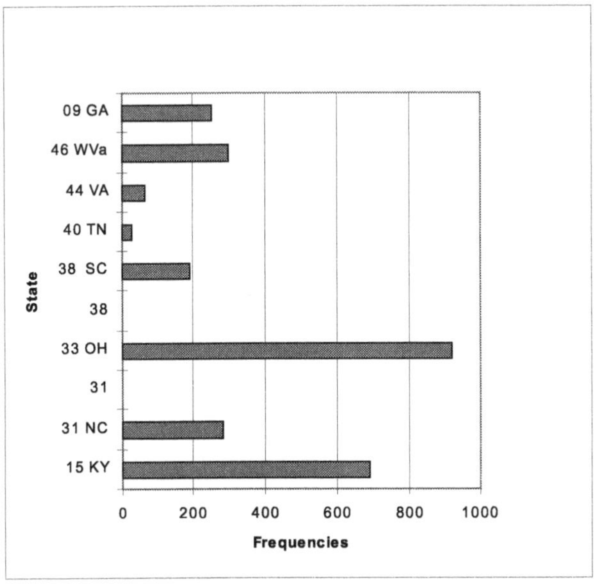

Figure 9.2. Frequencies of Radiocarbon Dates Per State.

are prioritized because they appear in greater numbers than budgetary constraints will bear. This has the obvious favorable effect of counteracting the date distortion caused by time's destruction of older dates, but it also must produce biases relative to later dates because a narrower range of phenomena are being dated more intensely. The effects of this process need to be evaluated.

Problem Orientations

Several problems have been identified to which the eastern radiocarbon data set could be applied.

1. Do people move out of high country during globally cold episodes, maybe because of shortened growing season? Following the Mt. Tambora eruption in 1815 there was a resettlement by many Americans. These resettlements include initial movement to the Ohio Valley from Maine (Stommel and Stommel 1979, 1983), and occupation of the U.S.-Mexican borderlands then along the Sabine River in present day western Louisiana and Texas (Gunn 1982). It could be that the large number of people at the Alamo (1836) from east Tennessee and western North Carolina can be attributed to unfavorable conditions at high altitudes during the 20 years of globally cold years following the Mt. Tambora eruption. Is there an evacuation of the mountains in the 300 years following the A.D. 536 event?

2. From Coe (1964) and Phelps (1964) onward, there have been numerous proponents of migration being an important process of population redistribution on the Atlantic Slope. Examples include Guilford, Savannah River, Algonkiaos (see Mathis chapter). Do radiocarbon dates shed light on this issue?

3. There were changes around 3700 B.P. Population appears to move inland and/or increase at that time: was this because of change in sea level?

Tanner's sea level and Lamb's (1977:Figure 16.23; Crumley and Marquardt 1987:238) European temperatures both show declines after approximately A.D. 536.

Space-Time Dates in North Carolina and Kentucky

Without precise spatial control, i.e., at the site or county level, the ability to address the elevational problem is limited. However it can be assumed, for example, that states such as South Carolina and Ohio generally represent lower elevation sites while those from West Virginia represent higher elevation sites. North Carolina, Virginia, Tennessee, and Kentucky clearly have a mix of elevations and can only be analyzed realistically when UTM coordinates are available. Of these states, only the North Carolina and Kentucky UTMs were available at this writing. Thus, in the following analysis we will endeavor to examine the change of altitude question by two cross checking methodologies. In the first, the North Carolina and Kentucky dates will be divided between physiographic zones—coastal zone, coastal plain, piedmont, mountains, and plateau—to examine the implications for impact of the A.D. 536 event and its subsequent episode at varying elevations. In the second methodology, the dates of South Carolina representing Atlantic Slope conditions, Ohio representative of the Midwest, and West Virginia, characteristic of the northern Appalachians, will be examined.

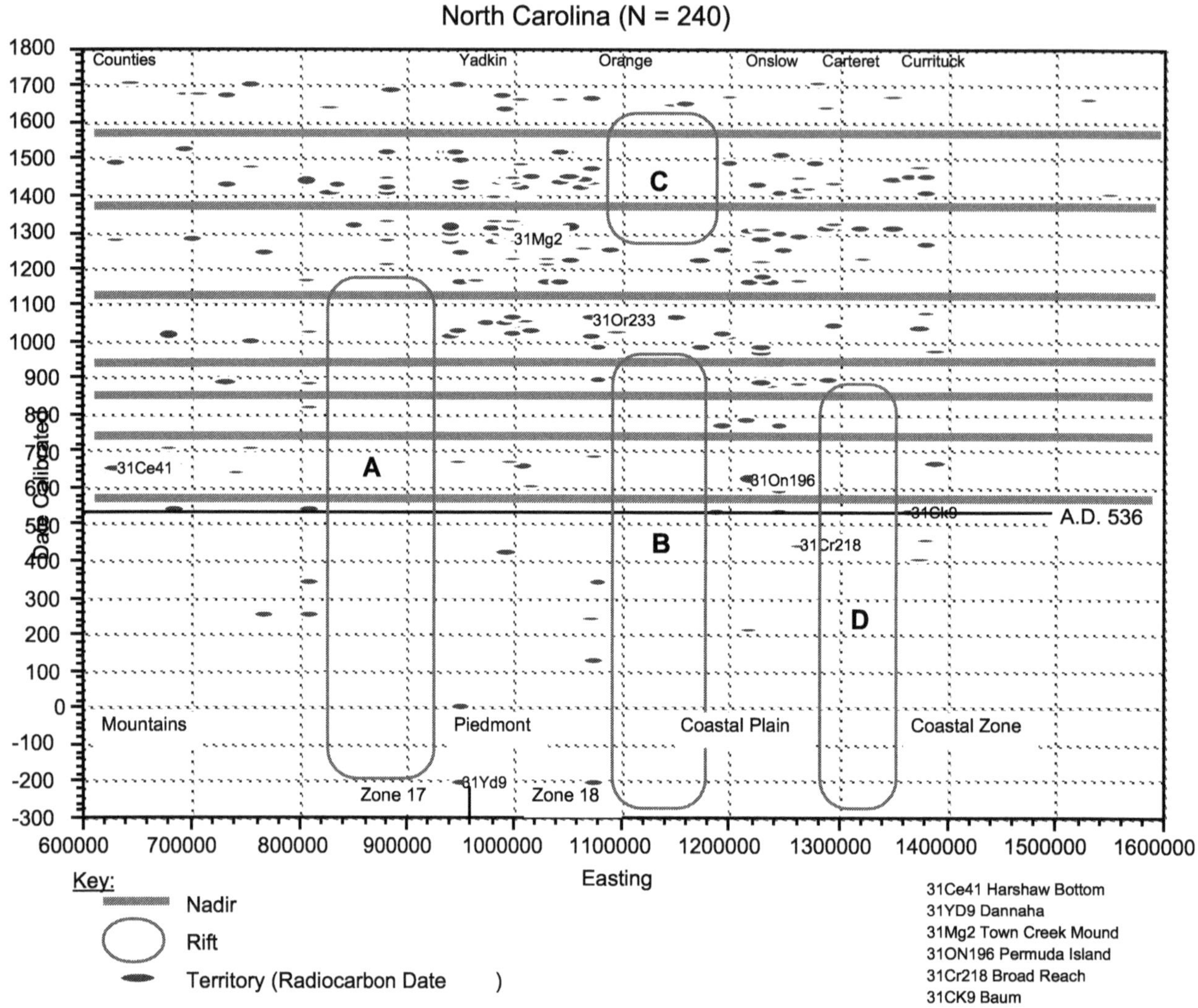

Figure 9.3. Calibrated Space-Time Distribution for North Carolina.

In the illustrations, dates are represented by disks that are about 10 miles in diameter and of 15 years duration. They roughly correspond to the exploitation range of a hunting and gathering group. Thus, one radiocarbon date, or fire hearth or its accouterments, represents the activities of a group around that time of the radiocarbon calibrated age. In effect, the represent a territory or some similar partial of land.

Plotting the North Carolina calibrated dates against easting UTM suggests several interesting possibilities (Figure 9.3). While not all of them are directly relevant to the A.D. 536 problem, they do, as an aggregate, add credibility to the argument that they represent real phenomena and provide insights into cultural change processes.

1. There is a distinctive rhythm to the appearance of dates with clusters and intervening absences or *nadirs* of dates. Where dates are relatively dense, i.e., after A.D. 800, nadirs appear at five times: 850, 940, 1120, 1370, 1580. The lapse between nadirs hovers below 200 years (Table *. 1). At this juncture we do not know if the intervals of date nadirs are an artifact of radiocarbon dating, perhaps the calibration technique, or real manifestations of cultural activity frequencies and perhaps interactions between humans and a climate mechanism. However, there are known climate phenomena that have such quasi-periods and could advantage and disadvantage human activities through atmospheric links to local climate. Whatever their cause, they tend to be synchronous across all four subregions of the state. As a whole, they suggest approximately century and a half intervals in which dates occur, with approximately half century date nadirs.

2. During the Woodland, there is a *rift* between the piedmont and the mountains (A), or an area of abandonment from a radiocarbon perspective. During the Mississippian, this rift is erased. The erasing of the rift brings to mind the Mississippian mounds along the mountain-piedmont interface such as the Town Creek mound (31Mg2; Coe 1995) and a mound reported in the upper Yadkin valley (Gunn et al. 1991), and several others (May 1996). Many of the post-rift A dates are from the Hardin site in Gaston County (Levy 1987).

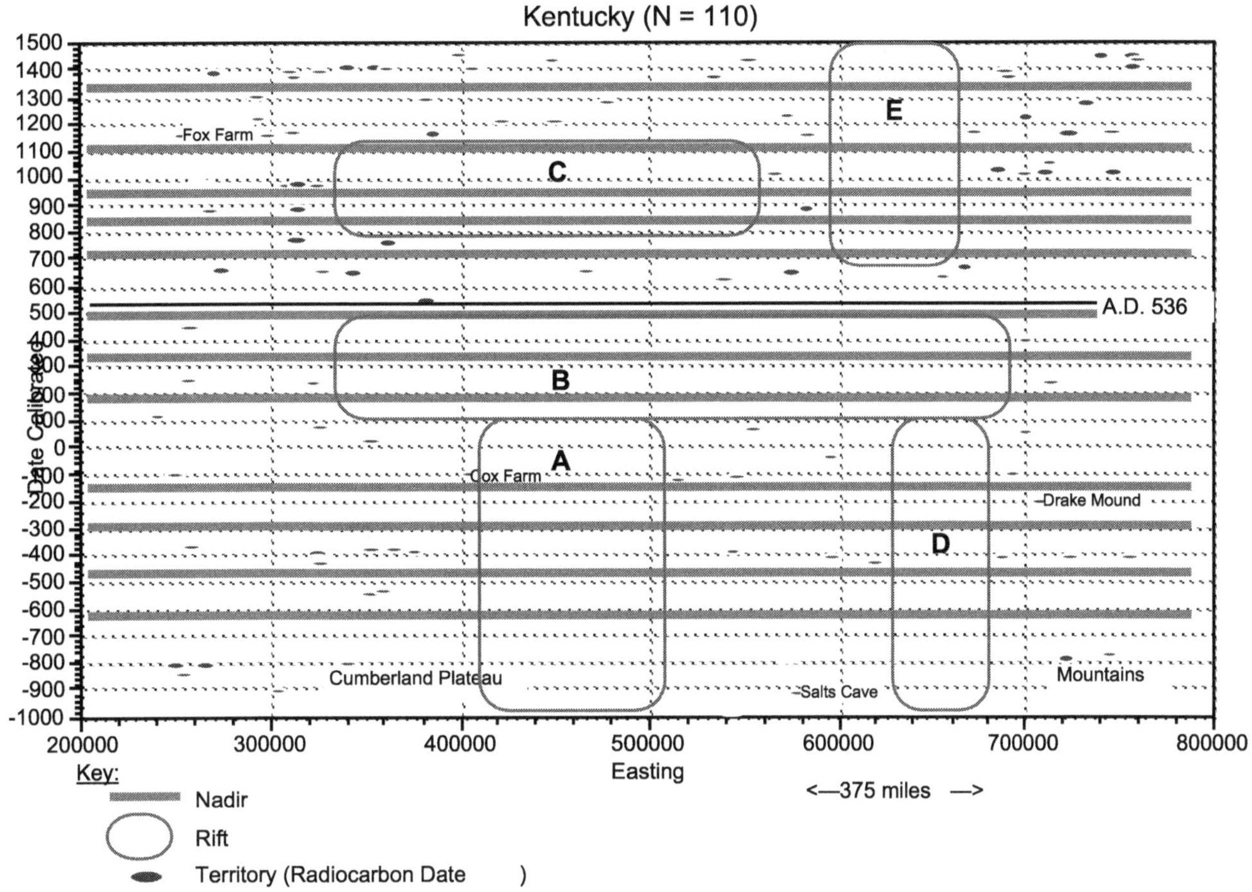

Figure 9.4. Calibrated Space-Time Distribution for Kentucky.

3. There is also a rift between the coastal plain and the piedmont (B). This rift disappears briefly during the early late Woodland, but then reappears by A.D. 1100 (C) and is not fused again until the 1600s. A similar rift appears between the coastal zone and the coastal plain (D). It, like the piedmont-mountain rift, seems to fuse during the Mississippian, though not as conclusively.

4. A date nadir could appear before and after A.D. 536 in the piedmont. Based on very few dates, the nadir seems to begin around A.D. 400 and continues to A.D. 600. Then, after a century of activity represented by five dates, the piedmont again finds itself unrepresented in the radiocarbon record. Although A.D. 536 is clearly an event involving cold global temperatures, there was considerable global cooling following A.D. 200, which was the most extensive episode of global volcanism in the Holocene according to Bryson and Goodman (1980). It could be that, from the perspective of piedmont peoples, the A.D. 536 event was just more trouble in an already troubled time, as in other areas of the world (see Houston chapter). On the other hand, the seventh century, judging by historical records in Europe, was a relatively pleasant century. Following a similar line of reasoning from cross-Atlantic parallels, the period A.D. 760 to 860 witnessed a distinctive run of cold winters signaling general adverse conditions (Lamb 1977). The piedmont appears to reflect a negative reaction to this as well. During the globally warmer years of the Medieval Maximum, however, the piedmont is abundantly occupied. (This interpretation needs to be tempered with forthcoming findings on dating of ceramics from Piedmont sites in Jane Eastman's research on pits at Ochenechee.)

5. In contrast to the piedmont, the mountains appear to be rather continuously dated though the post-A.D. 536 period. That they begin immediately after the event and continue to the 700s almost suggests that someone could have been escaping the piedmont to the mountains (see Wetmore, et al. chapter for possible explanations). The coastal plain dates imply a similar scenario to that of the mountains with dates bridging the A.D. 536 event and apparently increasing (see Mathis for possible explanation).

In brief and tentative summary, then, the distribution of dates among the North Carolina coastal plain, piedmont and mountains suggests a diminution of piedmont occupation while conditions improved in the mountains and on the coastal plain.

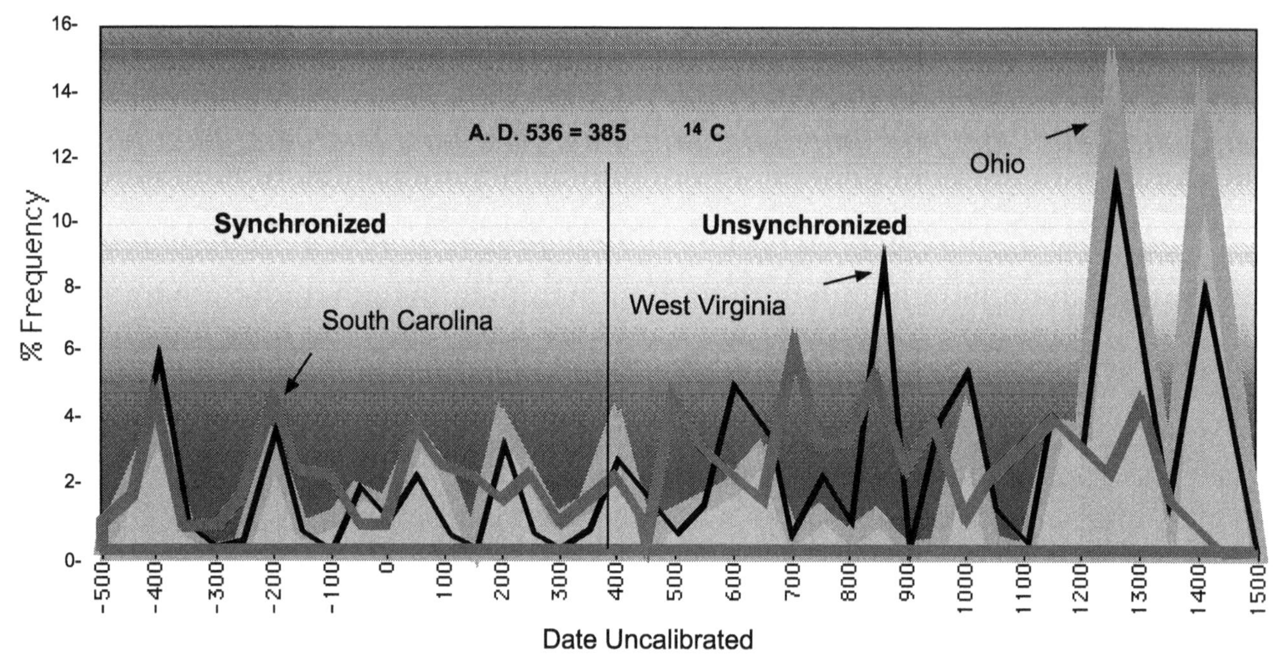

Figure 9.5. Ohio, West Virginia, and South Carolina Percentile Frequencies from 500 B.C. to A.D. 1500. .

A parallel analysis of Kentucky dates (Maslowski et al. 1995) on the west of the Appalachian Mountains is less conclusive relative to rifts and nadirs. There are less than half the dates in this suite (N=110) as the North Carolina dates (N=240). The lesser number of dates lends inherently less credibility to the inspection of date distributions (Figure 9.4). Where nadir lines can be threaded through the radiocarbon territories, they respond remarkably closely to those on the east side of the mountains in North Carolina (see Table 9.1) differing only by a decade or two in most instances. While the low number of dates leaves one hesitant to impute too much credibility to the parallel nadirs, they do suggest further investigation of the phenomena. Also, as will be seen in a subsequent analysis, the nadir phenomenon can be replicated in much larger suites of dates. As among the dates from east of the mountains, there are also rifts in the distributions. During the early centuries of the first millennium, the rifting of dates is widespread through the center of the state (B). In the century following A.D. 536, dates are widespread; then broad rifts appear again (C and E). The temporal and spatial distributions of dates in Kentucky, while awaiting further verification of rifts and nadirs, suggests much viz the North Carolina dates and the rift and nadir patterns need to be given special attention.

One could anticipate that eventually the rifts will disappear per se as more dates accumulate. At that time, the discussion will have to pass from one of presence and absence to one of frequency densities. However, barring substantial discrepancies between frequency densities and rifts and nadirs, it must stand as an hypothesis that there are subregions and episodes that are unfavorable to human occupation under certain conditions. Nothing in this analysis suggests what the linkages to global climate change are, or even if they exist. However, the broad temporal synchroneity of the nadirs implicates more than local conditions.

Comparisons of States

Turning to the broader but less resolved perspective of statewide date comparisons, we apply the assumption suggested above, that states with relatively uniform elevations can be used as aggregates of dates (Ohio low and midwest, West Virginia high northern Appalachian [Maslowski et al. 1995], and South Carolina low and Atlantic Slope [source: Sassaman files]). To evaluate the relative half century frequencies of radiocarbon dates, the frequencies of dates were counted by 50-year periods per state (Table 9.2). The period 500 B.C. to A.D. 1500 is included so that the dates can be converted to percentiles in long runs, visually eliminating the differences in date sample sizes between states.

The general appearance of the graphs (Figure 9.5) for the three states is clearly reminiscent of that cyclicity observed in the North Carolina and Kentucky date distributions, i.e., the century plus duration of high frequencies with intervening nadirs.

The relationships between state's radiocarbon frequencies change as time crosses the A.D. 536 line. Before the A.D. 500s, the three states exhibit relatively synchronous and, proportionally, relatively similar frequencies of dates. After A.D. 536, the states's frequencies follow differing courses. In the century following A.D. 536 the relationships appear to be chaotic; West Virginia date frequencies plunge, South Carolina date frequencies rise and tend to remain high for 500 years. West Virginia dates behave more synchronously but absolute frequencies vary wildly. It could be that this pattern would be anticipated should higher altitudes become less habitable. West Virginia dates, in the highest altitudes, vary between the highest and lowest percentiles.

We anticipated that populations would decrease in the mountains and increase at lower altitudes under the cooler post-A.D. 536 regime. South Carolina dates in fact display a general increase during the 300 years subsequent to A.D. 536 (see Anderson et al. 1995 for a discussion of related processes, also Woodall chapter). At the same time West Virginia dates vary radically from half century to half century; this suggests episodic favorable and unfavorable conditions; such is the supposed nature of colder world conditions. Following A.D. 900 when sea level and global temperature recover to warmer conditions, relative frequencies of South Carolina dates decline while West Virginia dates experience a general increase. Does this indicate a reversal of the trends initiated by the A.D. 536 cooling?

Through all of this, Ohio dates are even more depressed than those of West Virginia. As in West Virginia, there is a notable increase following the sea level/global warming of the 900s.

Conclusion

While much work remains to be done on discovering the precise mechanisms to which the movements of radiocarbon dates can be attributed, it is clear that the potential for using them as a discovery tool for conditions of cultural change exists. As is duly noted in the philosophy of archaeology literature (Salmon 1986), such correlations of presumed climate and culture changes are of little value unless undergirded by causal explanations. In Campeche, Mexico (Gunn et al. 1995), and Burgundy, France (Gunn and Crumley 1991), developing such explanations has required interdisciplinary research into ethnography, modern agricultural practices, ancient agricultural practices, hydrology, and seasonal slippage with global climate change. While such work has begun on the Atlantic Slope (see Woodall and Gunn 1993, Woodall chapter), a substantial amount remains to be accomplished relative to evincing reasonable causal mechanisms and accumulating sufficient information relative to the radiocarbon chronology. Nevertheless, without such background climatological work, any inferences relative to other influences on cultural change, such as charismatic leadership or cultural diffusion, are suspect.

References

Anderson, David G., David W. Stahle, and Malcom K. Cleaveland (1995) Paleoclimate and the Potential Food Reserves of Mississippian Societies: A Case Study from the Savannah River Valley. *American Antiquity* 60:258-286.

Baerreis, David, Reid A. Bryson (1965) Historical Climatology and the Southern Plains: A Preliminary Statement. *Bulletin of the Oklahoma Anthropological Society* 13:69-75.

Baerreis, David, Reid A. Bryson (1966) Dating the Panhandle Aspect Cultures. *Bulletin of the Oklahoma Anthropological Society* 14:105-116.

Bryson, R. A. and B. M. Goodman (1980) Volcanic Activity and Climatic Changes. *Science* 207:1041-1044.

Coe, Joffre (1964) The Formative Cultures of the Carolina Piedmont. *Transactions of the American Philosophical Society* 54. Philadelphia.

Coe, Joffre (1995) *Town Creek Indian Mound: A Native American Legacy*. University of North Carolina Press, Chapel Hill.

Crumley, Carole L., and William H. Marquardt, eds. (1987) *Regional Dynamics: Burgundian Landscapes in Historical Perspective*. Academic Press, San Diego.

Custer, Jay F. (1989) Prehistoric Cultures of the Delmarva Peninsula: an Archaeological Study. University of Delaware Press, Newark.

Gunn, Joel D. (1982) *Form and Environment: Historical Landscape Study of the San Antonio Missions*. Prepared by ECS Corp. Submitted to National Park Service, San Antonio.

Gunn, Joel D. (1994a) Global Climate—Human Life: Physical Contexts of Historic Landscapes. Special Issue of *Human Ecology* 22.

Gunn, Joel, and Carole L. Crumley (1991) Global Energy Balance and Regional Hydrology: A Burgundian Case Study. *Earth Surface Processes and Landforms* 16:579-592.

Gunn, Joel D., Marian D. Roberts, Barbara Lucas, Carol J. Poplin, and Eric C. Poplin (1991) Historical and Archaeological Survey and Historic Properties Management Plan for the W. Kerr Scott Reservoir, Wilkes County, North Carolina. Brockington and Associates, Inc., Atlanta. Submitted to the U.S. Army Corps of Engineers, Wilmington District.

Gunn, Joel, William J. Folan, and Hubert R. Robichaux (1995) A Landscape Analysis of the Candelaria Watershed in Mexico: Insights into Paleoclimates affecting Upland Horticulture in the Southern Yucatan Peninsula Semi-Karst. *Geoarchaeology* 10:3-42.

Lamb, H. H. (1977) Climate: Present, Past and Future: Climatic History and the Future. Vol 2, Methuen and Co. Ltd., London.

Maslowski, Robert F., Charles M. Niquette and Derek M. Wingfield (1995) The Kentucky, Ohio and West V9irginia Radiocarbon Data Bases. *West Virginia Archaeologist* 47(1&2).

Phelps, David S. (1964) *The Final Phases of the Eastern Archaic*. Ph.D. dissertation, Tulane University, New Orleans.

Salmon, Merrilee H. (1982) *Philosophy Of Archaeology.* Academic Press, New York.

Stommel, Henry, and Elizabeth Stommel (1979) The Year Without a Summer. *Scientific American* 240:176-186.

Stommel, Henry, and Elizabeth Stommel (1983) *Volcano Weather: the Story of 1816, the Year without Summer.* Seven Seas Press, Newport, Rhode Island.

Wendland, W. M., and Reid A. Bryson (1974) Dating Climatic Episodes of the Holocene. *Quaternary Research* 4:9-24.

Woodall, J. Ned and Joel D. Gunn (1993) Population Dynamics on the Yadkin River Following the Medieval Optimum Global Climate Episode. Paper presented to the 50th annual meeting, Southeastern Archaeological Conference, Raleigh, North Carolina.

Table *.1. Nadirs and Intervening Lapse Times.

North Carolina		Kentucky	
Nadir Date	Lapse	Nadir Date	Lapse
1580	210		
1370	250	1330	220
1120	180	1110	170
940	90	940	100
850		840	130
		710	210
		500	
Average	182.5		166

Table *.2. Frequencies and Percentiles of Dates for Ohio, West Virginia, and South Carolina by Half Centuries.

Date	OH		SC		WVa	
Lower bound	Freq	%	Freq	%	Freq	%
-500	0	0.0	1	0.8	1	0.5
-450	13	2.0	2	1.6	4	1.9
-400	26	4.0	6	4.8	13	6.1
-350	5	0.8	1	0.8	1	0.5
-300	0	0.0	1	0.8	0	0.0
-250	0	0.0	2	1.6	0	0.0
-200	22	3.4	6	4.8	8	3.8
-150	3	0.5	3	2.4	1	0.5
-100	7	1.1	3	2.4	0	0.0
-50	13	2.0	1	0.8	4	1.9
0	10	1.5	1	0.8	2	0.9
50	23	3.5	5	4.0	5	2.4
100	18	2.7	3	2.4	1	0.5
150	0	0.0	3	2.4	0	0.0
200	27	4.1	2	1.6	7	3.3
250	17	2.6	3	2.4	1	0.5
300	5	0.8	1	0.8	0	0.0
350	9	1.4	2	1.6	1	0.5
400	26	4.0	3	2.4	6	2.8
450	1	0.2	1	0.8	3	1.4
500	7	1.1	6	4.8	1	0.5
550	10	1.5	4	3.2	3	1.4
600	14	2.1	3	2.4	11	5.2
650	21	3.2	2	1.6	8	3.8
700	2	0.3	8	6.4	1	0.5
750	8	1.2	4	3.2	5	2.4
800	2	0.3	4	3.2	2	0.9
850	7	1.1	7	5.6	19	9.0
900	0	0.0	3	2.4	0	0.0
950	0	0.0	5	4.0	9	4.2
1000	27	4.1	1	0.8	12	5.7
1050	2	0.3	3	2.4	3	1.4
1100	0	0.0	4	3.2	0	0.0
1150	26	4.0	5	4.0	9	4.2
1200	24	3.7	4	3.2	7	3.3
1250	94	14.3	3	2.4	25	11.8
1300	47	7.2	6	4.8	10	4.7
1350	9	1.4	2	1.6	3	1.4
1400	88	13.4	1	0.8	18	8.5
1450	42	6.4	0	0.0	8	3.8
1500	1	0.2	0	0.0	0	0.0
Total	656		125		212	

V. CHANGES IN PLACE: MOVEMENT IN PREHISTORIC POPULATIONS

The theme that pervades archaeological investigations along the coastlines of the southeast United States is that people move with major changes in global climate and accompanying sea level adjustments. Reading Mark Mathis's and Karen Walker's A.D. 536 case studies leaves one with the impression that coastal residents either reeled or reveled in the double impacts of sea level change and changes in atmospheric temperature. The evident comparisons are with contemporary movements along the coasts of Europe (see Jones chapter).

Mathis reveals that between A.D. 600 and 900, substantial changes occurred in the cultural landscape of the southern coast of North Carolina. These changes are manifest in a variety of archaeologically identified cultural elements, including settlement size and distribution, architecture, ceramic technology, and mortuary practices. The current interpretive model holds that the changes are the consequence of an immigration of Algonquian-speaking societies from the north and a corresponding displacement of the indigenous societies, possibly Siouan or proto-Siouan speakers southward. The displacement appears to have primarily been towards the south, although some western and even northern movement can be detected. Comparable cultural materials are known to the south, where they appear to have continued in use two or more centuries longer. Contemporary population movements have been documented or suspected elsewhere along the Atlantic seaboard, some of which could be directly related. How rapidly the Algonquians spread into the south is unknown, as are the initial stimuli for the movement. The possibility of large-scale climatological change, such as might have occurred following the A.D. 536 event, offers at least a temporally consistent explanation.

Karen Walker also finds evidence of people moving in southwest Florida. She explores archaeological evidence for a rapid fifth–through eighth–century climatic cooling preceded and followed in rapid episodes. Built by the complex and sedentary Calusa people and their predecessors, coastal southwest Florida's shell middens and mounds contain stratigraphic sequences that, when topographically related to mean sea level, radiocarbon-dated, and zooarchaeologically quantified, suggest that sea level fell from a high over a meter above the present to as low as 60 cm below by A.D. 550^{cor}, and again rose to above-present levels by the ninth and tenth centuries. Middens 60 cm below today's sea level date to the sixth and seventh centuries. Mollusk populations represented in those middens and in others of the fifth and eighth centuries changed in their relative abundances and overall diversities, suggesting paleosalinity and elevation change in the regional estuarine gradient. In addition, wintering migratory ducks occur in great numbers during this time compared to earlier and later occupations. This suggesting a period of harsh winters in the Southeast, if not also in south Florida. During the earlier sea–level rise, people moved landward off the coastline, possibly as far as interior south Florida; during the sea–level fall, the coastline was reoccupied, this time by great numbers of people. A period of cooler temperatures–such as Europe's Vandal Minimum–for this time might explain the paleotopographical, chronological, paleoecological, and cultural patterns exhibited in southwest Florida's archaeological record. Walker's highly resolved study of sea level change compares favorably with Tanner's world wide sea level change measurements (see Tanner chapter). The two studies were developed in parallel circumstances.

Arguing Population Movements

Mathis achieves, in spite of relatively sparse dating of Middle Woodland evidence along the Atlantic seaboard, a convincing argument for southward population movement during the A.D. 536^{cal} aftermath. Such arguments are considered by archaeologists to be the most difficult of cultural change arguments to construct and substantiate. Mathis searches for substantiation of his hypothesis in a broad range of data ranging from ceramic types through mortuary practices and architecture. However, his studies are buttressed by similar findings to the north and south (see Anderson chapter).

As in Mathis's case study, Walker's draws on a broad range of paleoenvironmental indicators to discover evidence for ecologically significant changes in sea level. Thanks to a long-term project in Southwest Florida, Walker's inferences are supported by a surfeit of habitat-specific, cross-checked evidence. The evidence emanates from animal bones and shells recovered from several archaeological sites located at several distances from the open ocean. Through the perspectives of estuarine salinity gradients, predator-prey relationships, and abundance of shell fish and bird species, and the relationship of low lying archaeological sites to inundation, Walker documents powerful evidence for changes in sea level of perhaps 2 meters or more over a few hundred years.

Both authors detect population movements, but at vastly different scales. The difference in scale of the movements suggests that Walker's are local adjustments from coastline to uplands to compensate for loss of living space and changing soil moisture due to sea level change. Coastal movement could result from ±1m changes in sea level being amplified by coastal soils conditions, fresh water domes, and changed evaporation in coastal shallows. These are important issues relative to global change. Mark Brooks et al. (1986) see floodplain effect 100 miles inland, and at the Osprey Marsh site (Gunn et al. 1995) a sea level rise of 1 meter increased soil moisture at about 3 m above current sea level. Mathis's population movements are of a distinctly grander scale, the movement and replacement of linguistic groups along hundreds of miles of coastline.

The population movements on American coasts seem to compare favorably in scope with the historically recorded movements in Europe. Elizabeth Jones's discussion of sixth century Britain shows the more northerly people, Scandinavians, migrated over seas and hundreds of miles while the more southerly population, the Roman Britains seem to, in effect, change habitats. On the North American Atlantic seaboard populations move hundred of miles and a whole linguistic group is displaced. Baillie's (1995) observation that the intensity and duration of the post-A.D. 536 cold is greater in the north than the south could be relevant here.

Huge overwash events reported by Walker on the Florida Gulf Coast during the fifth century are intriguing. Are they a single unique happening in the history of the sites? Is there any reason to think that it was in the mid sixth century? If it is unique, is there any reason to suspect that it is a tsunami? Can one tell a tsunami from a storm? If it is a storm, there is reason to suspect that there would have been very big storms after a sudden cooling. The oceans seem to have been very high and hot. Sudden cooling, especially in the high northern latitudes would have caused very high rates of heat exchange between hot tropical oceans and cold northern air mass. Large, brutal hurricanes would undoubtedly have been a part of this heat exchange process. They would have continued at least as long as the upper 100 meters, the so-called mixing layer, contained an excess of heat out of equilibrium with the colder northern atmosphere.

The combination of sea level indicators Walker compiled seem to point an arrow right at A.D. 536^{cal} for the initiation of a sea level decline. They also support the validity of the Tanner sea level curve for the Gulf.

Walker has a very well-dated rapid sea level change associated with an important site on the southwest Florida coast in the A.D. 536 time frame. This site and sea level situation has great potential to illuminate the culture-sea level interaction as well as the climate-culture effects. The coastal sites have great potential for storing retrievable information on environmental change (Gunn et al. 1995).

Global Change. The results of the coastal zone analyses look in two important directions relative to global-local change and human dimensions issues. First, the changes in sea level were sufficient during a brief period of time to force abandonment and allow reoccupation of low lying archaeological sites. Second, Walker's analysis corroborates in one seashore location, the relationship between sea level and atmospheric temperature; because of its location in eastern North America, local temperatures generally correlate positively with global temperatures. The decline in sea level indicated by Tanner's analysis (see Tanner chapter) is reflected in the rise and fall of sea level in southwest Florida. In the southwest Florida sites, an abundance of ducks forced south from Maryland is the indicator of cooled global and regional climate. The evidence from oyster shells further indicates that the rises and falls were episodic or "jerky" rather than smooth upward and downward movements.

Of particular note is that southwest Florida culture and environment, particularly in the coastal zone, was greatly enriched by the global cooling associated with the A.D. 536 aftermath. As she so poignantly noted, "Why wouldn't you expect an increase in population if the summers were cooler in south Florida?" People moved coastward in numbers and organized themselves to build monumental architecture in the form of ceremonial mounds. Conversely, global warming impoverished the region. Sea level rise reduced the richés of the estuaries; storms lashed the coast line overwashing archaeological sites, making them uninhabitable.

References

Baillie, M. G. L. (1995) Patrick, Comets, and Christianity. *Emania* 13:69-78.

Brooks, Mark J., Peter A. Stone, Donald J. Colquhoun, Janice G. Brown, and Kathy B. Steel (1986) Geoarchaeological Research in the Coastal Plain Portion of the Savannah River Valley. *Geoarchaeology* 1:293-307.

Gunn, Joel D., Thomas G. Lilly, Cheryl Claassen, John Byrd, and Andrea Brewer Shea (1995) Archaeological Data Recovery Investigations at Sites 38BU905 and 38BU921 Along the Hilton Head Cross Island Expressway, Beaufort County, South Carolina. Garrow & Associates, Inc., Raleigh. Submitted to South Carolina Department of Transportation, Columbia.

10. THE MIDDLE TO LATE WOODLAND SHIFT ON THE CENTRAL COAST OF NORTH CAROLINA

Mark A. Mathis

Office of State Archaeology, Division of Archives and History, Raleigh, North Carolina 27601

Culture Change on the Atlantic Seaboard

[North Carolina, Atlantic Seaboard; chronometric dating, cultural change; sea level regression ; Middle Woodland 300 B.C.-A.D. 800^{c14}]

In the year A.D. 536^{cal}, a significant atmospheric event occurred in the eastern hemisphere. Though the exact location and cause are still being debated (see Introduction), it appears to have had locally severe and sustained climatological impacts. That the event occurred is documented by several contemporary European, African and Chinese historians, and supported by various paleoenvironmental and archaeological data. As discussed in other chapters of this volume, the event was followed by substantial socio-cultural and economic instability and change across much of the eastern hemisphere and, potentially, across the globe.

Could or did the A.D. 536 event affect the prehistoric cultural and environmental landscape as far afield as the central Atlantic Seaboard of the North American continent, and more specifically, the central coastal region of present-day North Carolina? Direct paleoenvironmental and archaeological data is lacking or insufficiently resolved chronometrically to make definitive statements for or against the proposition. We lack the type and depth of archaeological and geological research required to isolate and date specific episodes of environmental change, such as the period of sea level regression projected to have occurred following A.D. 536^{cal}. There is, however, a growing body of circumstantial and temporally coincidental evidence that supports a model of significant cultural change in the region during the two or three centuries following the event.

The year A.D. 536^{cal} falls in the latter end of the Middle Woodland cultural period, as traditionally defined for the coastal region of North Carolina (Figure 10.1) (South 1976; Loftfield 1976; Phelps 1983; Ward and Davis 1999). The Middle Woodland is usually assumed to span a period from about 300 B.C. to A.D. 800^{CAL}, with the Late Woodland lasting from that point until European contact between the sixteenth and eighteenth centuries. In fact, both the lower and upper range limits for the Middle Woodland, and the beginnings of the Late Woodland, are no more than *best guess* approximations, given the paucity of corroborating radiometric and archaeological data. Furthermore, other than an overall increase in the numbers and sizes of sites over time, the distinction between the Early and Middle Woodland, at least along the central and southern coast, is based almost exclusively on changes in ceramic technology, rather than in other sociocultural or economic characteristics, including settlement and subsistence patterns. The data to this point suggests an extraordinary level of cultural and adaptive continuity over several thousands of years, with only occasional additions and modifications to local technologies.

For much of the region, the Middle to Late Woodland transition or *shift* is equally vague, it, too, being based more on changes in ceramics than any significant or obvious changes in lifestyle and economy. The exception to this is along the outer reaches of the northern and central coast. It is particularly apparent in the central region.

In preface to the following discussions, it is necessary to note that archaeological research along the coast of North Carolina is still, in many respects, in it's proverbial *infancy*, with only two relatively limited site surveys prior to the 1970's (Haag 1958; South 1960). Only since the early 1970s has there been focused research on the coast, principally by David S. Phelps, at East Carolina University (Phelps 1983), and Thomas C. Loftfield, at the University of North Carolina at Wilmington (Loftfield 1977). Combined with data generated by a few theses and dissertations, and a growing body of cultural resource management studies, we have a reasonable cultural and taxonomic framework for coastal North Carolina prehistory, but not without the expected and unavoidable measure of ambiguity and ongoing debate. This is no more evident than in the ceramic typologies and cultural phase definitions variously employed in the central and southern coastal region of the state (Herbert and Mathis 1996; Herbert 1999; Mathis 1999). Consequently, and for heuristic purposes only, certain liberties are taken below in using ceramic *type* names interchangeably with and in lieu of cultural *phase* designations. Future research will, no doubt, render this exercise unnecessary.

Archaeological and Cultural Context

[Atlantic Seaboard; ceramic analysis, linguistics; grog temper, shell temper, fabric impression, population change, linguistic groups, Mockley, Hanover, Colington, White Oak, Algonkian; A.D. 300-1000^{C14}]

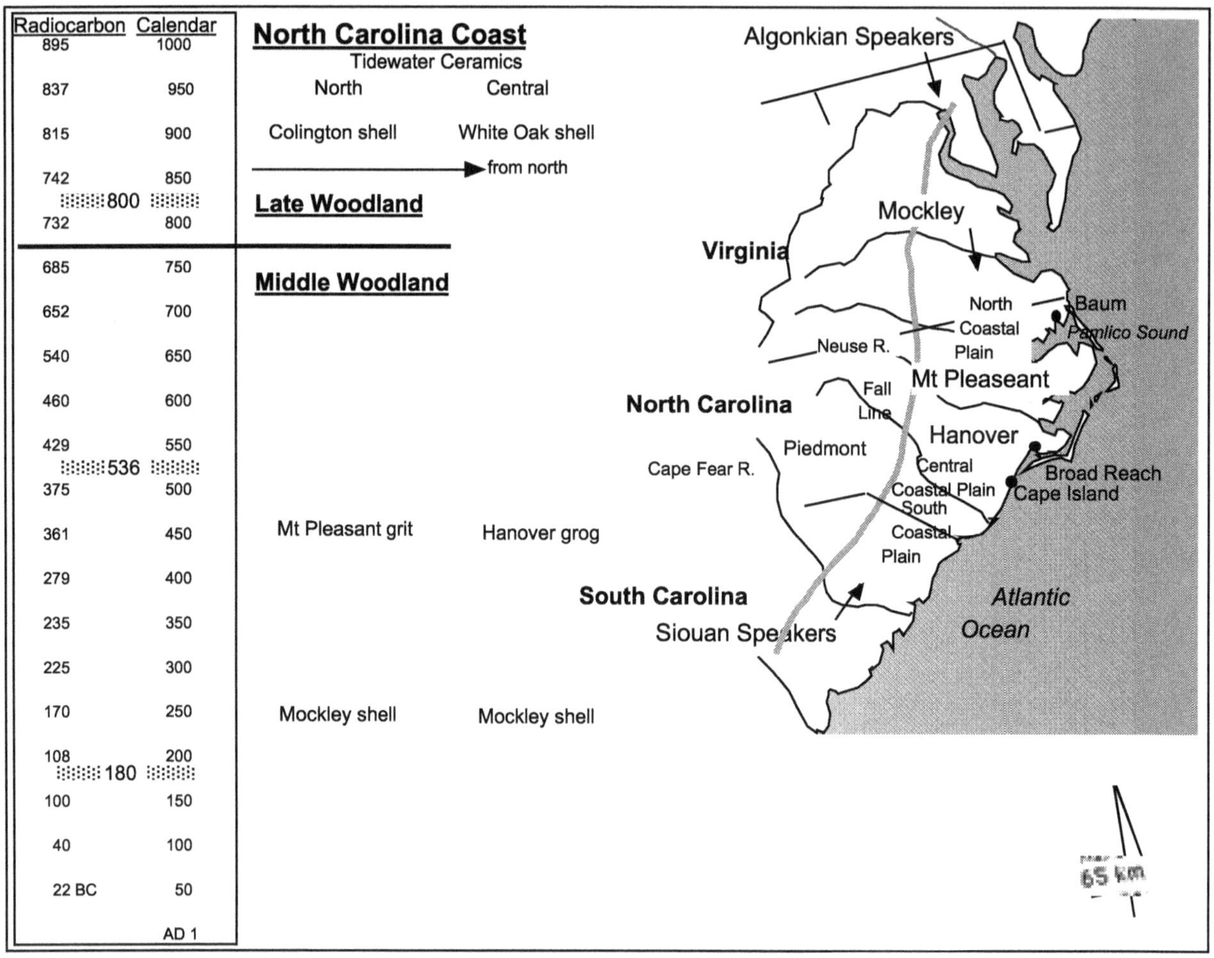

Figure 10.1. Map and Chronology of the North Carolina Seaboard.

By at least the third century^{C14} A.D., and probably earlier, the central and southern Tidewater region of North Carolina was occupied by people who manufactured a distinctive clay- or grog-tempered pottery referred to as *Hanover* (South 1976). Hanover ceramics are found throughout the coastal plain below the Neuse River, and occasionally above the Neuse (see Figure 10.1). While there is also evidence for a potentially contemporary sand-tempered ware in the region, often identified as *Cape Fear* (South 1976; see also Herbert 1999), Hanover is quantitatively predominate along the outer coast and is assumed to be the principle ceramic type of the period.

Radiocarbon dates indicate that the clay/grog-tempered Hanover ceramics were manufactured between A.D. 200 (if not earlier) and A.D. 1300 (if not a little later) (e.g., Wilde-Ramsing 1984; Mathis 1993a; Hargrove 1993; Eastman 1994a, 1994b; Reid and Simpson 1997; Botwick and Neville 1998; Daniel 1999; Shumate and Evans-Shumate 2000; Thomas H. Hargrove, personal communication 1998). However, as discussed below, the distribution of Hanover components dating after ca. A.D. 700 is not continuous or homogenous throughout the region. Along the estuarine zone of the central coast, clay- and grog-tempered pottery was "replaced" by a shell-tempered ware, called *White Oak*, around A.D. 700^{c14} (Loftfield 1976; see also Herbert and Mathis 1996). In the northern Tidewater, above the Neuse River, shell-tempered ceramics, called *Colington*, replaced the Middle Woodland sand- and grit-tempered ware known as *Mount Pleasant* (Phelps 1983). Mount Pleasant and Hanover appear to have been more or less contemporary, and have been found in potentially associated contexts (Phelps 1977; Robinson 1995, 1996).

Taxonomically, Colington and White Oak ceramics are almost identical, with only minor differences noted in the relative frequency of specific surface treatments (Herbert and Mathis 1996; see also Marshall 1999). Both are apparent derivatives of a northern or Middle Atlantic tradition of shell tempering that probably originated with the Middle Woodland Mockley phase (about A.D. 200 to 900^{c14}) of the lower Potomac River Valley (Handsman and McNett 1974; Egloff and Potter 1983; Potter 1993; cf. Byrd 1999). Mockley has been linked to the initial appearance of Algonkian-speaking groups who migrated into the Potomac Valley from the homelands further to the north (Potter 1993; Luckenbach et al. 1987). Colington and White Oak ceramics are considered *hallmark* characteristics of the Late Woodland period Carolina Algonkians, and continued in use until European contact (Loftfield 1976; Phelps 1983; Mathis 1995).

Precisely when shell tempering came into use on the coast of North Carolina is open to debate, even though we have casually accepted A.D. 800[C14] as a round figure (Phelps 1983). In fact, a number of radiocarbon dates from the region suggest it could have been a century or two earlier (Loftfield 1979; Eastman 1994a, 1994b; Mathis 1999; Daniel 1999; Shumate and Evans-Shumate 2000), although some of those dates have not been regarded as reliable by some authors (Phelps 1983; Herbert and Mathis 1996). Nevertheless, based on the limited presence of Mockley ceramics on the coast of North Carolina, a range of between A.D. 600 and 800[C14] appears reasonable for the initial, albeit tenuous and dispersed, appearance of shell tempering in the region.

The maximum southern distribution of shell-tempered ceramics appears to be around the southern end of Topsail Island (see Figure 10.1). To the north of this point, the coastal sounds are characterized by extensive estuarine and open water habitat supporting dense concentrations of shellfish and schooling fish, both of which were central to the prehistoric Algonkian settlement and subsistence strategies. To the south of Topsail, the sounds narrow, with salt marsh covering proportionately greater area than open water. Under the lower sea stand conditions of the past there was even less open water and, consequently, less fish and shellfish habitat. For an estuarine-oriented people, therefore, it seems a logical terminus in their expansion towards the south. For purposes of these discussions, this "end point" also serves as the southern delimiter of a "central" coastal region, bounded to the north by the Neuse River.

White Oak and Hanover pottery share two common characteristics—fabric-impressed surfaces and fabric impressing on the interior of the rim. It has been suggested that this could indicate cultural continuity between the Middle and Late Woodland periods, implying that the change in tempering agent, from clay/grog to shell, was an *in situ* adaptation of a new technology (Loftfield 1976; 1984). However, given the widespread nature of fabric impressing along the eastern seaboard as a whole, beginning in the latter part of the Early Woodland period, it appears more likely that it was an element of technology that transcended cultural, temporal and spatial boundaries. The associated Hanover and White Oak cultural assemblages differ in far too many other respects, including site size and distribution (a reflection of settlement and subsistence strategies), mortuary practices and, possibly, architecture.

Settlement and Subsistence Patterns

[Atlantic Seaboard; settlement patterns, fauna, flora; estuary, range of settings, shellfish, horticulture, maize; Hanover, White Oak]

Sites with Hanover ceramics are found in a wide range of environmental settings: on the estuarine shorelines of the sounds, along all classes of freshwater streams and rivers, and adjacent to inland swamps and pocosins. In contrast, White Oak ceramics occur only at sites within the estuarine zone, adjacent to contemporary brackish or saltwater habitats. In fact, White Oak pottery is only rarely found beyond a few hundred meters of saltwater.

Based on the distribution of ceramics alone, the Hanover and White Oak settlement patterns overlap only at the estuarine shorelines. Given its widespread distribution, Hanover subsistence strategies appear to have involved seasonal movement into other environmental zones, possibly requiring a fissioning of social units for extended periods. In this sense, the Hanover strategy appears to have been a continuation of earlier hunter-gather, seasonal rounds, though with an apparent increased emphasis on the exploitation of marine resources, particularly shellfish. Unfortunately, along the central coastal zone, undisturbed and discrete cultural deposits unequivocally associated with Hanover occupations are rare, and have provided little direct evidence of the Middle Woodland diet and, consequently, seasonality. Part of the problem stems from the fact that Hanover ceramics are found at virtually every shell midden site, in and among (and usually disturbed by) the much larger and usually later concentrations of the White Oak occupations. Furthermore, the use of subterranean storage or cooking pits, that might contain preserved floral and faunal remains, does not appear to have been a common practice during the Middle Woodland, at least by comparison to the Late Woodland.

While ethnobotanical and zooarchaeological data are slight for Hanover, it is substantial for White Oak. During the Late Woodland, the use of storage, cooking and roasting pits was routine, and hundreds have been excavated on the central coast. These pits yield evidence for the exploitation a wide variety of marine and terrestrial resources, though heavily weighted towards fish and shellfish (Scarry and Scarry 1997). Evidence of deer, turkey and other terrestrial fauna are found in the assemblages, but are and were available within a short distance of the estuaries. Ethnobotanical remains tend to be relatively limited, but include the range of available nuts and seed plants, as well as small amounts of cultigens (e.g., maize, squash), although dates for the latter appear to post-date A.D. 1000[c14] (e.g., Davis and Child 1995; Mathis 1993a).

The overall and combined evidence from both shell midden deposits and pit features indicate that the White Oak occupations were permanent and year-round (Loftfield 1979, 1987; cf. Daniel 1999), although analyses of isolated features and midden deposits often yield season-specific data (Claassen 1982; Lawrence 1987). The subsistence strategy was heavily oriented towards estuarine resources, with little direct evidence thus far for significant inland or upland resource gathering. This is not to say that hunting or gathering forays did not occur, only that sites with diagnostic artifacts (i.e., White Oak ceramics) are unknown in upland areas.

Mortuary Practices

[Atlantic Seaboard, North Carolina; mortuary analysis; flexed, semi-flexed, bundle, cremation, ossuary, proto-Souian; Hanover, White Oak]

Hanover-related mortuary practices are poorly documented, but appear to have included flexed and semi-flexed primary interments and cremation. Bundle burials are reported for the region, primarily to the south and west, in what could be temporally consistent contexts. In two instances on the central coast, primary interments included two individuals within the same burial pit (Mathis 1993c; Thomas C. Loftfield, personal communication). A "mass" cremation, containing up to 10 separate but commingled cremations, also has been recorded (Mathis n.d.). A cluster of possible Hanover-related cremations was recently reported on the White Oak River, on the central coast (Shumate and Evans-Shumate 2000). With these two exceptions, each of the documented burials appears to have been an isolated occurrence, and not a component of an organized cemetery.

White Oak burial practices, by comparison, reflect both the more sedentary nature of the settlement pattern, and a complex, community-based ceremonial system involving the ritual treatment of the dead. White Oak burial types include primary, secondary bundle, cremation and ossuary interments. Ossuaries, containing the redeposited (but occasionally primary) remains of up to 150 individuals, are another "hallmark" characteristic of Algonkian culture (Phelps 1980; 1983; Loftfield 1990; Bogdan and Weaver 1989; cf. Ubelaker 1974). Cemetery areas, containing clusters of individual and ossuary burials, have been recorded at the Broad Reach site (31CR218) (Mathis 1993c), on the central coast, and at the Baum site (31CK9) on the north coast (Phelps 1983).

Architecture

[North Carolina; house patterns; long houses, circular houses; Hanover, White Oak]

Direct and unequivocal evidence of Hanover-related architecture is also limited. However, Hanover pottery concentrations and possible Hanover-associated pit features were identified in and around oval and circular postmold patterns at the Broad Reach site (31CR218) (Mathis 1993a) and more recently at the Cape Island Site (31ON190), located on Topsail Island (Jones et al. 1997). However, a radiocarbon date of A.D. 1240^{c14} date attained from a postmold sample at Cape Island, suggests a Late Woodland time frame (Jones et al. 1997). Nevertheless, the patterns suggest structures ranging from about 4 to 6 meters in maximum diameter. Among the Late Woodland Siouan-speaking groups to the south and west, residential structures tend to be oval or rounded (e.g., Ward and Davis 1999).

White Oak architecture is, again, more or less consistent with Algonkian cultural patterns, and includes true longhouses and smaller rectangular structures. Longhouses have been documented at a number of sites on the central coast, and range from 4 to 6 meters in width and, in one example, over 18 meters in length (Loftfield 1979; Mathis 1993a; Loftfield and Jones 1995). In the largest instance, at Broad Reach (31CR218), an interior partition is evident (Mathis 1993a). No longhouse patterns have been identified south of Topsail Island.

White Oak Origins and the A.D. 536 Event

[Atlantic Seaboard, Eastern Woodlands, North Carolina; settlement, ceramics, mortuary, house forms; in situ, discontinuity, intrusion, Algonkian expansion, Algonkian retraction; Mockley, Colington Hanover, White Oak, Middle Woodland, Late Woodland]

The archaeological differences between the Hanover and White Oak assemblages, summarized above, are sufficient to suggest that latter is *not* an *in situ* developmental outgrowth of the former. Rather, the data suggest a marked discontinuity between the two and, furthermore, that White Oak was intrusive to the region. As noted above, Middle Woodland period precedents for shell-tempered ceramics, the Mockley series, and an intensive adaptation to estuarine resources, are found to the north, in the Chesapeake region, where they have been dated to as early as A.D. 200^{c14} (Egloff and Potter 1982; Potter 1993; cf. Byrd 1999). Mockley or Mockley-like ceramics have been identified in North Carolina, as far south as the central coast.

While a direct-line relationship between the Middle Woodland Mockley and Late Woodland Colington and White Oak assemblages has not been established with certainty, it has been proposed as evidence of the southern expansion of Algonkian culture (Potter 1993; Luckenbach et al. 1987; Herbert and Mathis 1996; Mathis 1995; David S. Phelps, personal communication). Osteological analyses of ossuary remains supports, albeit tentatively, the proposition that the expansion was not merely a gradual process of cultural interaction and exchange, but one involving the movement—migration—of people into the coastal region of North Carolina (e.g., Bogdan and Weaver 1989). Based on the distribution of shell-tempered ceramics, associated ossuary burials and longhouse architecture, the maximum extent of the expansion appears to have been about midway between the Neuse and Cape Fear rivers, near the southern end of Topsail Island.

If the early radiocarbon dates proposed for White Oak are more reliable than previously suggested, and the Mockley or Mockley-like ceramics are culturally related, then the initial effects of the Algonkian expansion may have reached the central coastal region by as early as A.D. 600^{c14}. However, the majority of the associated radiocarbon dates cluster between about A.D. 700 and A.D. 1400^{c14} (Eastman 1994a, 1994b; Herbert and Mathis 1996; Herbert 1999; Mathis 1999). Consequently, we can only be reasonably sure of an Algonkian presence by around A.D. 700^{c14}, and that the previous occupants of the area—the people who manufactured Hanover pottery—had ceded or abandoned the immediate area or incorporated with the new arrivals. As noted above, recent dates indicate that clay/grog-tempered pottery (Hanover or Hanover-like) continued to be manufactured in adjacent areas to the west and south until the 13th and 14th centuries, if not later. Based on the distribution of these later ceramics and their associated assemblages, a generalized subsistence strategy of hunting, gathering and fishing appears to have continued in much the same fashion as that of their so-called *Middle Woodland* predecessors. This stands in marked contrast to the intensive estuarine adaptation of the Algonkian "immigrants." It therefore appears likely that there was relatively little cultural assimilation, and that the displacement, at least in the estuarine zone of central North Carolina, was relatively rapid and complete.

The question is whether the Algonkian expansion could have been related, even distantly, to the impacts of the A.D. 536 event. The answer is an *un*equivocal *maybe*. Data presented in this volume and elsewhere document a period of up to 300 years of cooler and drier temperatures following, or at least temporally coincident to the event, with a concomitant decline in sea level (Brooks et al. 1989; Stahle et al. 1988; Tanner 1993; Walker et al. 1995). While the presumed cultural impacts of this could have been felt somewhat later along the southern Atlantic seaboard than to the north, it could have been a stimulus for the southern expansion, if not out-migration, of Algonkian populations from their northern homelands, possibly in response to locally significant pressures from their Iroquoian neighbors (Snow 1995). By at least A.D. 800^{c14}, significant cultural changes had already occurred or were underway throughout much of the eastern woodlands of North American, indicating a period of relatively widespread cultural flux and instability. In some instances, these changes appear to have involved population movements similar to that suggested for the outer coast of North Carolina (Anderson 1991; Snow 1995, 1996; Crawford and Smith 1996). Phelps (1983) has suggested a comparable date for the initial appearance of the *Cashie* phase in the northern inner coastal plain of North Carolina. The Cashie phase is related to the Iroquoian-speaking groups occupying the region prior to and until European settlement in the 17th and 18th centuries.

It has been suggested that the Middle to Late Woodland settlement changes along the central coast were an *in situ* adaptive response to micro- or macro-environmental changes, specifically changes in sea level and associated changes in ground moisture, vegetation, and shellfish availability (Claassen 1982:114; Reid and Simpson 1997; Loftfield 1976). Of particular note is Claassen's (1982) seminal *Mercenaria* study on the central coast which, among other notable contributions, identified periods of apparent stress in the local shellfish populations that can only be explained within the context of sea level fluctuations and related changes in salinity and water temperature. Recent analyses of the species composition and distribution of shell middens and shell-filled pit features on the North and South Carolina coast provide ample evidence for variations in the type, size and quantities of shellfish consumed, and in the patterns of deposition across the landscape (Loftfield 1984; Mathis 1993a, 1993b; Trinkley and Adams 1994; Gunn et al. 1995). In the case of the Hanover-White Oak pattern differences, our temporal controls are insufficient to make more than the most generalized and speculative comments. However, micro-variations in shellfish exploitation and deposition, some probably tied to idiosyncratic consumptive behavior, others to environmental conditions, are observed throughout the Middle and Late Woodland periods. And, while the environmental stress documented in the shellfish could be responsible for temporally specific alterations in site location and composition, the stark differences between virtually all aspects the local Middle and Late Woodland cultural assemblages are difficult to explain solely within the context of stress-related but *in situ* ecological adaptation. This is particularly evident when we consider the apparent persistence of otherwise Middle Woodland settlement and subsistence strategies, as well as ceramic manufacture, well into the Late Woodland in immediately adjacent territories.

It is conceivable that the environmental changes that occurred following the A.D. 536 event were sufficient to disrupt the subsistence strategies of the local Middle Woodland society, reducing the economic attractiveness of, and ultimately bringing about a general abandonment of, the Tidewater estuarine zone. The Algonkians, with their fully estuarine-oriented settlement-subsistence systems, simply expanded their territory to fill the vacant territory. It is of some consequence in this regard that the southernmost extent of the prehistoric Algonkian territory corresponds strikingly well with the southernmost extent of the open waters of the sounds and Tidewater.

Unfortunately, and not to belabor the point, we do suffer from a lack of temporal resolution, as well as the detailed environmental data required to fully address this issue. There is, nonetheless, an intriguing coincidence between the projected climatological degradation, sea level decline, and apparent movement of people across the landscape.

It is of comparable interest to note that there is also evidence that suggests that by the sixteenth century^{c14}, prior to the arrival of European settlers, the Algonkian territory had begun to contract. Although radiocarbon dates for Algonkian sites to the north range to as late as the early eighteenth century, the latest date thus far from the central region is A.D. 1460^{CAL} (Truesdell 1995). While this may be an artifact of our sample, we now have almost 70 radiocarbon dates from 16 sites with major White Oak components on the central coast. If the pattern suggested by these data proves valid, the apparent Algonkian contraction corresponds roughly with the middle to latter stages of the so-called "Little Ice Age," during which sea level could have fallen to a meter lower than present along the Atlantic Seaboard, a level comparable to that projected for the post-A.D. 536 event (Bradley and Jones 1993; Varekamp et al. 1992; Brooks et al. 1989; Walker et al. 1995). During those periods of low sea stand, vast areas of the present-day estuaries may have been opened to human occupation, but would also have significantly reduced the shellfish and fish habitat (tidal and intertidal). Since the sounds are much broader and deeper to the north of the Neuse River, and conceivably able to sustain greater shellfish and fish populations during the low stands, the southern Algonkian populations, with their intensive estuarine orientation, may have slowly migrated back to the north, where they were subsequently encountered by the early European explorers. In like manner, their initial appearance on the southern coast probably did not occur any earlier than about A.D. 700^{c14}, as suggested by the available radiocarbon dates, when sea level had returned to near-present levels following the post-A.D. 536 decline.

Rapid, pan-regional population movements and cultural change are theoretically accepted and historically documented throughout the world, some owing to political and economic factors of purely human definition, others to human responses to environmental pressures and stress. Drought, floods, and other climatologically generated events can and do result in temporary and sometimes permanent relocation, but generally on a fairly localized basis. Larger scale events, such occurred in A.D. 536^{cal}, could have been responsible for larger scale changes in the cultural landscape, like those observed along the Eastern Seaboard prior to about A.D. 800^{c14}. The data and models presented in this volume call attention to an important avenue of research that can and should ultimately lead to a far better understanding of coastal North Carolina prehistory; a prehistory, it seems, that could not be so completely removed or isolated from the rest of the world.

ACKNOWLEDGMENTS: An earlier version of this paper was presented at the 53rd Annual Meeting of the Southeastern Archaeological Conference, in Birmingham, Alabama, November 9, 1996. I wish to thank *Dr.* David G. Moore, who read it for me, without benefit of previewing either the paper or accompanying slides. Thanks are of course due Joel Gunn, not only for organizing the original A.D. 536 symposium, but for his untiring efforts in pulling the papers together for this volume. His comments and suggestions on the paper were most appreciated. I also thank Stephen R. Claggett, State Archaeologist and *Boss*, for allowing me to take the time for this and other non-bureaucratic and otherwise purely academic endeavors.

References

Anderson, David G. (1991) Examining Prehistoric Settlement Distribution in Eastern North America. *Archaeology of Eastern North America* 19:1-22.

Anderson, David G. (assembler and editor) (1996) *Indian Pottery of the Carolina: Observations from the March 1995 Ceramic Workshop at Hobcaw Barony*. Assembled and edited by David G. Anderson. Council of South Carolina Professional Archaeologists.

Bogdan, Georgieann, and David S. Weaver (1989) Report on the Human Skeletal Remains from the Flynt Site (31On305). Ms. of file, North Carolina Office of State Archaeology, Division of Archives and History, Raleigh.

Botwick, Brad and Ashley M. Neville (1998) Phase I and II Archaeological Investigations at Sites 31ON624, 31ON625 and 31ON664, Marine Corps Air Station New River, Onslow County, North Carolina. Ms. on file, Marine Air Corps Station New River, Jacksonville, NC, and the North Carolina Office of State Archaeology, Raleigh, NC.

Bradley R.S. and Phillip D. Jones (1993) "Little Ice Age" Summer Temperature Variations: Their Nature and Relevance to Recent Global Warming Trends. *The Holocene* 3,4:367-376.

Brooks, Mark J., Peter A. Stone, Donald J. Colquhoun and Janice Brown (1989) Sea Level Change, Estuarine Development and Temporal Variability in Woodland Period Subsistence-Settlement Patterning on the Lower Coastal Plain of South Carolina. In *Studies in South Carolina Archaeology: Papers in Honor of Dr. Robert L. Stephenson*, edited by G. Hanson and A. Goodyear, III, pp. 91-100. Occasional Papers of the South Carolina Institute of Archeology and Anthropology, The University of South Carolina, Columbia.

Claassen, Cheryl P. (1982) Shellfishing Patterns: An Analytical Study of Prehistoric Shell from North Carolina Coastal Middens. Ph.D Dissertation, Department of Anthropology, Harvard University.

Crawford, Gary W., and David G. Smith (1996) Migration in Prehistory: Princess Point and the Northern Iroquoian Case. American Antiquity 61(4):782-790.

Davis, Thomas W., and Kathleen M. Child (1996) Phase III Data Recovery at Site 31ON536 and Phase II Evaluation of the Prehistoric Component at Site 31ON534, Marine Corps Base Camp Lejeune, North Carolina. Ms. on file, Atlantic Division, Naval Facilities Engineering Commend, Norfolk, Virginia.

Daniel, I. Randolph (1999) Archaeological Excavations at Hammocks Beach West (31ON665): A Woodland Shell Midden on the North Carolina Coast. *Occasional Papers of the Phelps Archaeology Laboratory* No. 1, East Carolina University, Greenville, NC.

Eastman, Jane M. (1994a) The North Carolina Radiocarbon Date Study. *Southern Indian Studies* Vol. 42. Research Laboratories of Anthropology, University of North Carolina, Chapel Hill.

Eastman, Jane M. (1994b) The North Carolina Radiocarbon Date Study. *Southern Indian Studies* Vol. 43. Research Laboratories of Anthropology, University of North Carolina, Chapel Hill.

Egloff, Keith T. and Stephen R. Potter (1982) Indian Ceramics from Coastal Plain Virginia. *Archaeology of Eastern North America* 10:95-117.

Gunn, Joel D., Thomas G. Lilly, Cheryl Claassen, John Byrd and Andrea Brewer Shea (1995) Archaeological Data Recovery Investigations at Sites 38BU905 and 38BU921 Along the Hilton Head Cross Island Expressway, Beaufort County, South Carolina. Submitted to the South Carolina Department of Transportation by Garrow and Associates.

Haag, William G. (1958) *The Archaeology of Coastal North Carolina*. Louisiana State University Studies, Coastal Studies Series Number Two. Louisiana State University Press, Baton Rouge.

Handsman, Russell and Charles W. McNett (1974) The Middle Woodland in the Middle Atlantic: Chronology, Adaptation, and Contact. Paper presented at the 5th Middle Atlantic Archaeological Conference, Baltimore.

Herbert, Joseph M. (1999) Prehistoric Pottery Taxonomy and Sequence on the Southern Coast of North Carolina. *North Carolina Archaeology* Vol. 48:37-58.

Herbert, Joseph M., and Mark A. Mathis (1996) An Appraisal and Re-evaluation of the Prehistoric Pottery Sequence of Southern Coastal North Carolina. In

Indian Pottery of the Carolinas: Observations from the March 1995 Ceramic Workshop at Hobcaw Barony, assembled and edited by D. Anderson, pp. 136-189. Council of South Carolina Professional Archaeologists, Columbia, SC.

Jones, David C., Christopher T. Espenshade and Linda Kennedy (1997) *Archaeological Investigations at 31On190, Cape Island, Onslow County, North Carolina.* Prepared for the Island Development Group, Inc., Ringgold, Virginia, by Garrow & Associates, Inc.

Lawrence, David R. (1987) Appendix 3: Oyster Analysis: Stressed Oysters and Man-Made Structures at an Archaeological Site Near Sneads Ferry. In *Excavations at 31ON305, the Flynt Site at Sneads Ferry*, by Thomas C. Loftfield, Ms. on file at UNC-Wilmington.

Loftfield, Thomas C. (1976) *"A Briefe and True Report...." An Archaeological Interpretation of the Southern North Carolina Coast.* Ph.D. Dissertation, Department of Anthropology, University of North Carolina, Chapel Hill.

Loftfield, Thomas C. (1979) Excavations at 31 On 33, A Late Woodland Seasonal Village. Ms. on file, Department of Anthropology/Sociology, University of North Carolina, Wilmington.

Loftfield, Thomas C. (1984) Testing and Excavations at 31ON82. Ms. on file, University of North Carolina at Wilmington.

Loftfield, Thomas C. (1987) Excavations at 31ON305, the Flynt Site at Sneads Ferry, North Carolina. Ms. on file at UNC-Wilmington.

Loftfield, Thomas C. (1990) Ossuary Interments and Algonquian Expansion on the North Carolina Coast. *Southeastern Archaeology* 9: 116-123.

Loftfield, Thomas C., and David C. Jones (1995) Late Woodland Architecture on the Coast of North Carolina: Structural Meaning and Environmental Adaptation. *Southeastern Archaeology* 14:120-135.

Luckenbach, Alvin H., Wayne E. Clark and Richard S. Levy (1987) Rethinking Cultural Stability in Eastern North American Prehistory: Linguistic Evidence from Eastern Algonquian. *Journal of Middle Atlantic Archaeology* 3:1-33.

Marshall, Adam (1999) Interior Rim Impressions as an Indicator of Typological Relationships. *North Carolina Archaeology* Vol. 48:87-94.

Mathis, Mark A. (1993a) Broad Reach: The Truth About What We've Missed. In *Site Destruction in Georgia and the Carolina*, edited by D. Anderson and V. Horak. Readings in Archeological Resource Protection Series No. 2, Interagency Archeological Services Division, National Park Service, Atlanta.

Mathis, Mark A. (1993b) Beyond the Shell of Shell Middens. Paper presented at the Conference on Shell Midden Archaeology, Hobcaw Barony, Georgetown, South Carolina. October 22, 1993.

Mathis, Mark A. (1993c) Mortuary Processes at the Broad Reach Site. Paper presented at the 50th Annual Meeting of the Southeastern Archaeological Conference, Raleigh, NC. November 4, 1993.

Mathis, Mark A. (1995) The Carolina Algonkians: Comments from the Fringe (the Southern Frontier). Paper presented at the 27th Algonquian Conference, Chapel Hill, NC.

Mathis, Mark A. (1999) Oak Island: A Retiring Series. *North Carolina Archaeology* Vol. 48:18-36.

Mathis, Mark A. (n.d.) Field Notes: Cremation Excavation at 31Cr223, Carteret County, NC. On file, North Carolina Office of State Archaeology, Raleigh.

Phelps, David S. (1977) Recent Archaeological Research in Northeastern North Carolina. *Eastern States Archaeological Federation*, Bulletin 35, 36:19-20.

Phelps, Davis S. (1980) Carolina Algonquian Ossuaries. Paper presented at the 37th Annual Meeting of the Southeastern Archaeological Conference, New Orleans.

Phelps, David S. (1983) Archaeology of the North Carolina Coast and Coastal Plain: Problems and Hypotheses. In *The Prehistory of North Carolina, An Archaeological Symposium*, Division of Archives and History, North Carolina Department of Cultural Resources, Raleigh.

Potter, Stephen R. (1993) Commoners, Tribute and Chiefs: The Development of Algonquian Culture in the Potomac Valley. University Press of Virginia, Charlottesville.

Reid, William H. and Kay Simpson (1997) Phase II Investigations of Nine Prehistoric Sites and Phase I Survey of the P-028 Range Area, Greater Sandy Run Acquisition Area, Marine Corps Base, Camp Lejeune, Onslow County, North Carolina. Prepared for Marine Corps Base, Camp Lejeune, by the Cultural Resource Group, Louis Berger & Associates, Inc.

Robinson, Kenneth W. (1995) Archaeological Study, Replacement for Bridge No. 411 on Greene Street (SR 1531) over the Tar River, Greenville, Pitt County, North Carolina. Ms. on file, North Carolina Department of Transportation, Raleigh.

Robinson, Kenneth W. (1996) More on Migration in Prehistory: Accommodating New Evidence in the Northern Iroquoian Case. *American Antiquity* 61:791-796.

Scarry, John F. and C. Margaret Scarry (1997) Subsistence Remains from Prehistoric North Carolina

Archaeological Sites. North Carolina Archaeology World Wide Web site, http://www.arch.dcr.state.nc.us/susbsist/subsis.htm.

Shumate, M. Scott and Patti Evans-Shumate (2000) Archaeological Investigations at Long Point (31JN2) and Haywood Landing (31JN3) on the Croatan National Forest, Jones County, North Carolina. Ms. on file, U.S.D.A. National Forests in North Carolina, Asheville.

Snow, Dean R. (1996) More on Migration in Prehistory: Accommodating New Evidence in the Northern Iroquoian Case. *American Antiquity* 61:791-796.

Snow, Dean R., Kenneth W. Robinson (1995) Migration in Prehistory: The Northern Iroquoian Case. *American Antiquity* 60:59-79.

South, Stanley (1960) An Archaeological Survey of Southeastern Coastal North Carolina. Ms. on file, North Carolina Office of State Archaeology, Raleigh.

South, Stanley (1976) *An Archaeological Survey of Southeastern Coastal North Carolina*. The Institute of Archeology and Anthropology Notebook 8, The University of South Carolina, Columbia.

Stahle, David W., John G. Hehr and Malcolm K. Cleaveland (1988) North Carolina Climate Changes Reconstructed from Tree Rings: A.D. 372-1985. *Science* 240:1517-1519.

Tanner, William F. (1993) An 8000-Year Record of Sea Level Change from Grain-Size Parameters: Data from Beach Ridges in Denmark. *The Holocene* 3:220-231.

Trinkley, Michael (1989) An Archaeological Overview of the South Carolina Woodland Period: It's the Same Old Riddle. In *Studies in South Carolina Archaeology: Papers in Honor of Dr. Robert L. Stephenson*, edited by G. Hanson and A. Goodyear, III, pp. 73-89. Occasional Papers of the South Carolina Institute of Archeology and Anthropology, The University of South Carolina, Columbia.

Trinkley, Michael and Natalie Adams (1994) *Middle and Late Woodland Life at Old House Creek, Hilton Head Island, South Carolina*. Chicora Foundation Research Series 42, Columbia, South Carolina.

Truesdell, S. (1995) *Paleopathological and Paleoepidemiological Analysis of the Piggot Ossuary (31Cr14), Carteret County, North Carolina*. M.A. thesis, Wake Forest University, Winston-Salem.

Ubelaker, Douglas H. (1974) Reconstruction of Demographic Profiles from Ossuary Skeletal Samples: A Case Study from the Tidewater Potomac. Smithsonian Contributions to Anthropology No. 18.

Varekamp J. C., E. Thomas and O. van de Plassche (1992) Relative Sea Level Rise and Climate Change Over the Last 1500 Years. *Terra Nova* 4:293-304.

Walker, Karen J., Frank W. Stapor, Jr., and William H. Marquardt (1995) Archaeological Evidence for a 1750-1450 BP Higher-Than-Present Sea Level Along Florida's Gulf Coast. *Journal of Coastal Research Special Issue No. 17: Holocene Cycles: Climate, Sea Levels, and Sedimentation*, pp. 205-218.

Ward, H. Trawick and R.P. Stephen Davis, Jr. (1999) *Time Before History: The Archaeology of North Carolina*. University of North Carolina Press, Chapel Hill.

Wilde-Ramsing, Mark (1984) *Archaeological Survey and Testing on Prehistoric Shell Midden Sites in Niw Hanover County, North Carolina.* Master's thesis, Department of Anthropology, Catholic University of America, Washington, DC.

11. A COOLING EPISODE IN SOUTHWEST FLORIDA DURING THE SIXTH AND SEVENTH CENTURIES A.D.

Karen J. Walker

Florida Museum of Natural History, Gainesville, Florida

Coastal Southwest Florida and its Caloosahatchee IIA Period

[Coastal Southwest Florida; Stratification, Paleosalinity, Population Dynamics, Cultural Change, Episode; Calusa, Archaeology, Climate, Paleoenvironment; sixth/seventh Centuries]

Archaeologically, the sixth and seventh centuries in coastal Southwest Florida—a period known as "Caloosahatchee IIA"—are markedly different from the preceding and subsequent centuries. The differences are both environmental and cultural, and I have long pondered their co-occurrence with the documented A.D. 536^{cal} atmospheric event (Baillie 1995; Gill 1994:310-316; Strothers 1984) and the broader climatic cooling episode known in Europe as the Vandal Minimum (Bryson 1988; Gunn 1994:17, 19; Keigwin 1996:1506-1507). Whether or not a cause-and-effect relationship existed between what may have been a global event and a patterned set of subregional peculiarities—is highly debatable for any place and time (e.g., Kellogg 1988). But, as I have argued before (Walker 1992; Walker et al. 1994, 1995), South Florida—geologically, hydrologically, and ecologically— is, will be, and surely always has been sensitive to the impacts of rapid, human-scaled global climate change. Moreover, the nature of the region's archaeological deposits is such that the consequences of paleoenvironmental change, both subtle and catastrophic, are potentially preserved and therefore are potentially detectable.

So, are archaeological signatures for the A.D. 536^{cal} atmospheric event and/or the broader-scale Vandal Minimum detectable in Southwest Florida? I present some relevant archaeological patterns under four topics: 1) site stratification; 2) paleosalinity-gradient analysis; 3) faunal-population dynamics; and 4) cultural change. By "coastal Southwest Florida," for purposes of this paper, I mean the region roughly encompassing Charlotte Harbor, Pine Island Sound, San Carlos and Estero bays, and all associated coastal lands including the barrier islands (Figure 11.1). Chronological nomenclature for precolumbian South Florida departs from the rest of the precolumbian Southeastern U.S. (Cordell 1992; Marquardt 1992). At the end of the Late Archaic in Southwest Florida (about 500^{c14} B.C.), the Caloosahatchee Sequence (Figure 11.1) begins with Caloosahatchee I and continues through the historic-Indian period, Caloosahatchee V. I refer to several Caloosahatchee periods, primarily I and IIA, and occasionally IIB, III, and IV (Pineland contexts). The cultural shift from I to IIA occurs around A.D. 500^{cor}, generally correlating to the time of the A.D. 536^{cal} event and to the cultural shift from Middle to Late Woodland for the greater Southeastern U.S. Finally, following Gunn (1994:15-20) I use the word "episode" to refer to a concept of a flexible time period with uncertain boundaries and thus, amenable to refinement.

Wightman, Pineland, and Cash Mound Site Abandonment and Reoccupation

[Wightman, Pineland, Cash Mound; Stratigraphy, Geoarchaeology, Radiocarbon Dating; Shell Middens, Storm Deposits, Inundation, Human Abandonment; A.D. 50-800^{cor}, Caloosahatchee I and IIA]

The archaeological stratigraphies at three sites in southwest Florida (Figure 11.1) present a record of changing sea levels, varying storm frequencies and magnitudes, and human abandonment and reoccupation. The sites vary in exposure to the sea insuring that the observed phenomena are not purely of site specific significance.

The Wightman Site is on the bay side of Sanibel Island. Its stratigraphic sequence presents a lower, oyster-shell midden (marked "B" in Figure 11.2) dating to Caloosahatchee I during the fifth $century^{cor}$ A.D. (RC4 in Table 11.1). The midden includes shells of a few other bivalve and gastropod species, as well as pottery sherds and shell tools. This midden overlies a beach ridge with swash deposits (A) and is overlain by inundation strata in the form of an incipient, in situ oyster bar (C, along the top of Stratum B) and a shelly storm deposit (D). A second storm deposit (F) is also recorded. The two storm deposits, or "washovers," are separated by a thin midden (E) recording a short-lived human-reoccupation episode, probably also in the fifth century (RC2 in Table 11.1). These deposits are capped by a crown-conch-and-oyster shell midden (G) that dates to the seventh century within Caloosahatchee IIA period (RC1 in Table 11.1).

Wightman's strata are examined in much greater detail by Walker et al. (1994). In that article, I and my colleagues argue that the stratigraphic sequence portrays an episodic rise in sea level and increase in storm frequency during the fifth, and possibly fourth centuries. We further proposed that sea level during the seventh and possibly eighth centuries was at

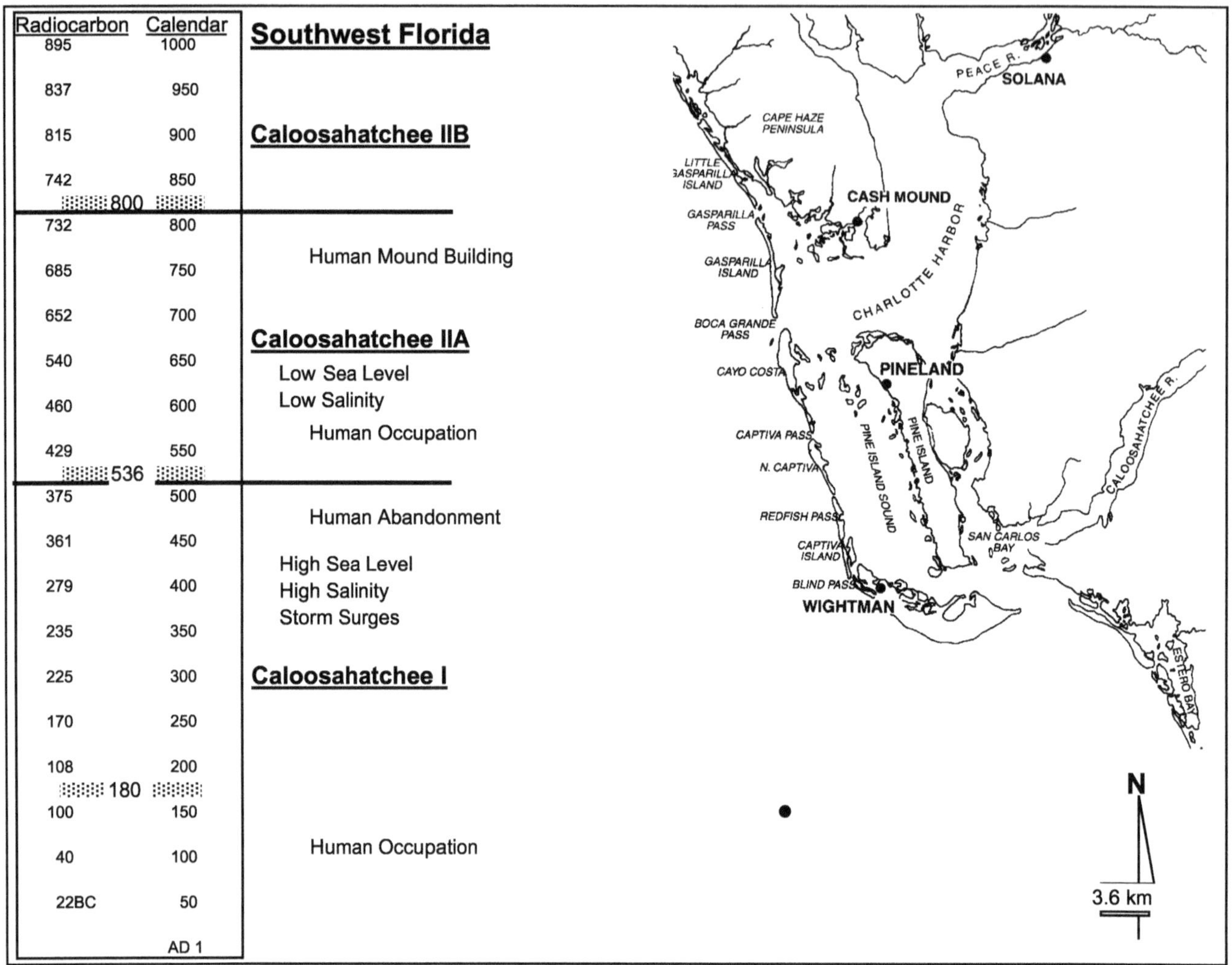

Figure 11.1. Map and Chronology of Southwest Florida. Map of coastal Southwest Florida showing locations of features cited, including the Wightman (8LL54), Pineland (8LL33), Cash Mound (8LL38), and Solana (8CH67) sites, and cultural chronology of Southwest Florida's Caloosahatchee I and II periods.

a lower elevation than that of the present, perhaps by 50 cm; it is during this time of low sea level that Wightman was reoccupied. Stapor and his colleagues' (1991) geological seriation of beach ridges on Sanibel Island independently supports our interpretations.

The Pineland Site on Pine Island is on the landward side of shallow Pine Island Sound (see Figure 11.1). It is in a more "protected" setting than Wightman. Here, both vertical and horizontal strata are relevant to our inquiry. In what today is pasture, a thin Caloosahatchee I shell midden (primarily oyster shell, labeled A in Figure 11.2) dates to A.D. 60-190^{cor} (RC4, Table 11.1), representing Pineland's first human occupation. The thin oyster-shell midden, also contains artifacts and charcoal and is interbedded within a non-cultural sand layer. The sand is overlain by a more substantial and slightly later Caloosahatchee-I oyster-shell midden (B) that dates to A.D. 190-340^{cor} (RC3, Table 11.1). Thick, sand-tempered plain pottery and charcoal also appear in the midden.

Midden B at the Pineland Site is covered by inundation and mangrove sediments (C). Interbedded within these is a thin (one-shell thick) deposit of sea-urchin remains and articulated surf-clam shells (D); these date to the late fourth or early fifth century (RC2, Table 11.1). Above these, and widely scattered within the sand, are fragments of sea-turtle bone and in at least one location, a section of porpoise vertebrae. I interpret the C and D strata to have been transported by non-cultural means. Whereas the sand that encases Stratum A and the lower sand in Stratum C may have been accreted slowly (e.g., in situ mangroves as the agent), Stratum D and the upper portion of C appear to record storm deposits. The Pineland Site area and all along the shoreline were abandoned during this time. Residents moved to "high" ground located in the interior pine flatwoods, as is evidenced by a feature known as the Low Mound (Walker et al. 1995), and later apparently abandoned Pineland altogether.

Like Wightman, when the Pineland Site dried during early Caloosahatchee IIA (RC1, Table 11.1), the site was resettled, as evidenced by an oyster-and-whelk shell midden (E). Although the RC1 date ranges from A.D. 400 to 540^{cor}, I think that this midden was not deposited until the sixth century. The reason is that three additional dates for this stratum/reoccupation tightly range between A.D. 530 and 660^{cor} (Beta-52551, -72997, -52708, Walker et al. 1995). This sixth-century midden is widespread across the Pineland Site indicating an extensive reoccupation. In one area two large shell mound complexes rise to 8 and 9 m above MSL; the bases of these mounds are expansive middens dating to the sixth and seventh centuries. These middens are today below MSL but they were not deposited in water. We estimate a sea-level at least 50 cm lower today's for the period of Pineland's Caloosahatchee IIA occupation (Walker et al. 1995). About two hundred years later, the swelling middens became more and more mounded, and less and less midden accumulated in the lowland areas so that by A.D. 900^{cor} (Caloosahatchee IIB), all midden was confined to the high, large mounds. We believe this is due to rising sea level.

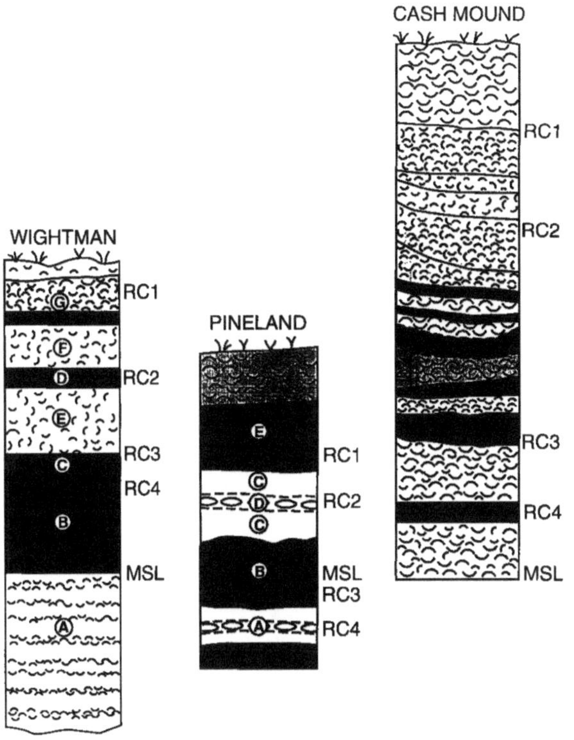

Figure 11.2. Stratification of Wightman (pre-mound sequence, as documented in Trenches 3 and 4, Walker et al. 1994), Pineland (pre-mound sequence as documented in Trench 11A, Walker et al. 1995), and Cash Mound (lower mound sequence, as documented in column-excavation A-1, Marquardt 1992; Marquardt 1985, field records). Semi-eliptical characters represent shells, both bivalves and gastropods. Size of characters indicate shell size relative to each other.

Cash Mound, in Turtle Bay (see Figure 11.1), is largely composed of oyster shells. An erosional cut into the mound complex along the water's edge was profiled and sampled for radiocarbon and archaeofaunal analysis (Marquardt 1992; Walker 1992). The profile records numerous strata of varying combinations and quantities of whole and crushed oyster shells, sand, fish bones, and charcoal (see Figure 11.2). Artifacts and other molluscan shells occur infrequently. Much of the deposit shown in Figure 11.2 ranges in calibrated radiocarbon dates, from about A.D. 50 to 370^{cor} (RC2, 3, 4, Table 11.1). The upper portion of the deposit dates to the seventh and/or eighth centuries (RC1, Table 11.1) with no representation of the intervening fifth and sixth centuries. Although no inundation deposit occurs here—surely due to the relatively high elevation of the Caloosahatchee I midden—we observe a time gap in occupation between Caloosahatchee I and IIA as at the other sites.

In sum, the Caloosahatchee I occupations at Wightman, Pineland, and Cash Mound—located in different parts of the coastal region—abruptly ended some time between A.D. 350^{cor} and 500^{cor}. The cause at Wightman and Pineland, and possibly Cash, appears be inundation and violent storms. All three sites were reoccupied in the sixth or seventh centuries by Caloosahatchee IIA people. This was during a time when sea level was lower, as noted in the stratigraphy at Wightman and Pineland, and presumably storm frequency and intensity were diminished.

Paleosalinity-Gradient Analyses of Sites and Periods

[Coastal Southwest Florida; Gradient Analyses; Estuarine Gradient, Estuarine Mollusks, Paleosalinity, Diversity; A.D. $50\text{-}800^{cor}$]

To examine paleoenvironmental change, I apply the ecological concept of an environmental gradient (Boesch 1977) to estuarine animal distribution and abundance over time. Analysis involves recognition of ecological zones ranging from fresh to salty oceanic water and the aquatic animals that inhabit these zones (Walker 1992; Walker et al. 1994). The zones and their associated fauna exist as a graded continuum at a regional scale. The gradient distribution of estuarine animals whose remains are common in southwest Florida's archaeological middens is one where invertebrate animals—primarily mollusks—are largely immobile compared to the vertebrate fauna which are mostly fishes (Walker 1992:275). Because of their immobility, the provide excellent indicators of rapidly changing environmental conditions.

In addition to a limited mobility, mollusks are generally restricted to narrow salinity ranges along the estuarine gradient (Wells 1961). Moreover, as you move along the gradient from fresh to salt water, overall faunal diversity increases (Wells 1961:252). These qualities are well illustrated by Eastern oyster (*Crassostrea virginica*) communities. Eastern oysters are found within a wide range along the estuarine gradient because the oysters have a wide-salinity tolerance (Butler 1954:479). But within that overall range, oyster bars can be classified as low-, mid-, or high-salinity bars. This is because the distribution and abundance of associate fauna— all those animals who live among and attached to the oysters—vary according to the salinity of the bar. Examples of molluscan associates are barnacles, mussels, crested oysters, ceriths, tulips, and drills (Butler 1954:486; Galtsoff 1964:435; Wells 1961:249-250). Bars at the mid- to low-salinity end of the gradient contain the healthiest oysters, the fewest predators (i.e., many of the associates), and the least diversity. Bars at the saltier end of the gradient contain the most stressed oysters, the most predators, and the highest diversity.

Thus, mollusks are excellent for paleoenvironmental gradient analyses because, for any given archaeological shell midden, the paleosalinity regime **and any change in that regime**, are reflected in its assemblage of oyster remains and their associates. This is because oysters that are collected by humans and transported in their shells, reflect their community of origin in that the gradient: salinity sensitive oyster associates incidentally enter the midden firmly attached to their host oyster shells. Also important is that the weight, perishability, and general abundance of oysters argue for collection at bars located near the human residence site (Walker 1992).

An unambiguous, single-species salinity marker found in middens is the crested oyster (*Ostreola equestris*, previously known as *Ostrea equestris*). It is an associate of the Eastern oyster; the crested oyster attaches itself to the Eastern oyster shells using them for a substrate to anchor themselves. South Florida's crested oyster is much smaller than the Eastern oyster, generally no longer than 20 mm. The smaller oyster has a well-defined range in the salinity gradient; it is absent in the low and medium salinity ranges; in the high-salinity range, crested oysters are abundant (Galtsoff 1964:12, 406; Galtsoff and Merrill 1962; Hofstetter 1959:1-2; Wells 1961:249, 253). The crested oyster's optimum salinity is 32 ppt and its lower limit is 28 or 29 ppt. These are all high-salinities in the estuarine system.

Applied to archaeological oyster middens, one can measure the paleosalinity of source bars by expressing a ratio of crested oysters to Eastern oysters (Table 11.2). Wightman's ratio for Caloosahatchee I is 1:6 implying a local high-salinity regime. In contrast, it's Caloosahatchee IIA oyster ratio is 0:346 implying a salinity too low for the presence of even a few crested oysters. Pineland's oyster ratios indicate that high salinities characterize periods I (1:4) and IIB (1:2) and a comparatively reduced salinity (1:18) characterizes IIA. At Cash Mound, a ratio (1:2) characterizes Caloosahatchee I, suggesting high salinity. In contrast, the seventh-century sample contains no crested oyster specimens despite the occurrence of at least 343 Eastern oysters (Walker 1992:283). Like at Wightman and Pineland, a lowered salinity is indicated for Caloosahatchee IIA at Cash Mound. All ratios from all sites support a retreat of the estuarine environment and lowering of sea level from Caloosahatchee I to IIA.

Wells (1961) measured the diversity of modern-day Eastern oyster bars in simple terms of Number of Taxa per bar in order to determine variation along the estuarine gradient. His work documents a continuum with high-diversity bars at the high-salinity end of the gradient and low-diversity bars at the low-salinity end of the gradient. Similar to the paleosalinity example above, this diversity concept can be applied to archaeological oyster middens to examine variation in the paleogradient, assuming that consistent oyster collection took place at local bars. Number of Taxa for Wightman's Caloosahatchee I midden (36 taxa) is significantly higher than its Caloosahatchee IIA midden (16 taxa), indicating a decreased salinity for the IIA (seventh-century) reoccupation. The Number of Taxa for the high-salinity washover is 46.

Shannon-Weaver Indices also were calculated for Wightman's strata (Walker et al. 1994:166); the results also support a decreased salinity. Numbers of Taxa at Pineland indicate that high diversity (average of 96 taxa), hence, high salinity is recorded for period-I samples. A lower diversity (average of 57 taxa), hence, lower salinity, is recorded for period-IIA samples dating to the sixth century. Interestingly, the diversity, hence salinity, increased during the IIB period (average 92 taxa), after A.D. 800^{cor} correlating with the Medieval Optimum, and decreased again during the III and IV periods (average 63 taxa) correlating with the Little Ice Age (Walker and Marquardt, in press). The three Caloosahatchee-I samples from Cash Mound average 39 taxa, lower than the 48 taxa of the one Caloosahatchee-IIA sample, and thus not fitting the expected pattern based on the oyster ratios cited above.

Overall, the three sites record a Caloosahatchee I to Caloosahatchee IIA decrease in salinity, suggesting a seaward shift in the regional estuarine gradient. Interpretation of this pattern, however, must include a consideration of temporal scales (seasonal, decadal, century) and the explanatory variables implied by each (Walker 1992:276-279). If each site is considered in isolation, the temporal scale of the midden samples cannot be determined with confidence. But when considered as a group and within the context of other lines of paleoenvironmental evidence, a long-term scale (fifty to a hundred years) is indicated (Stapor et al. 1991; Walker 1992; Walker et al. 1994, 1995).

Faunal Population Dynamics and Sea Level Change

[Coastal Southwest Florida; predator/prey relationship, population increase; *Melongena corona/Crassostrea virginica*, migratory ducks; A.D. 400 - 700^{cor}]

Shifts in several animal species apart from oysters also support ecologically significant changes in estuaries and sea level. The diachronic population dynamics of some invertebrates and vertebrates whose remains are common to southwest Florida's shell middens offer patterns of change focused on the Caloosahatchee IIA period. One example is the intertidal, common crown conch (*Melongena corona*). The scavenging crown conch feeds on many different living organisms and decaying flesh (Dalby 1989:709; Gunter and Menzel 1957:86-87; Hathaway and Woodburn 1961:53, 64; Menzel and Nichy 1958:136; Tabb and Manning 1961:577). Included in its diet is the Eastern oyster. Shellfish biologists agree that the present-day crown conch is not a "serious" predator to oysters. But these conclusions are biased toward a present-day focus on ***productive***, healthy oysters—in other words, they believe that crown conchs do little damage to healthy oysters, the focus of today's commercial industry.

Of great paleoenvironmental relevance, however, is that crown-conch abundance is usually associated with ***poorly producing*** oyster bars (Hathaway and Woodburn 1961:45, 60, 64). Conchs attack oysters that already are weakened by natural stresses such as summer high temperatures or prolonged exposure. Exposure of oysters can be caused by too crowded a bar or by a lowering of water level.

The Wightman Site has a stratum of crown conch shells that directly overlies the Caloosahatchee I inundation deposits (G, see Figure 11.2). The conchs outnumber the oysters almost 2:1, far exceeding the number normally found on a healthy oyster bar (Figure 11.3a). The conch shells were deposited when Wightman was reoccupied in the seventh century.

At Pineland, crown conch shells, dating to IIA or the sixth century, are very abundant compared to other time periods (Figure 11.3a). In one area of Pineland, the Brown's Mound Complex, the conch shells are interbedded in a massive deposit of lightning and pear whelk shells exhibiting essentially no sandy sediments or vertebrate faunal remains and very few artifacts. The deposit is contemporary with the crown conch suggest that the conchs were not alone in their population increase. Although whelks primarily feed on bivalves, they also include crown conchs in their diet when the opportunity presents itself. Thus, the crown conchs were surely enjoying a substantial boost in their food supply, while the whelks took advantage of the stressed oysters as well as the increased crown conch population. To identify the time of the crown-conch abundance, another dense deposit of conchs, in another area of Pineland—the Old Mound—was radiocarbon dated. Both deposits date (Beta-41299, -72997, Walker and Marquardt, in press) to the sixth century[cor].

Similar to Wightman and Pineland, Cash Mound's Caloosahatchee IIA sample shows an increase in crown conch over the Caloosahatchee I samples (Figure 11.3a). Again to identify the timing of the conch abundance, a sample of conch shells was taken from an area of Cash Mound where a dense deposit of the shells occurred. The shells were radiocarbon-dated with a resulting range of the fifth through sixth centuries (Beta-47886, Walker and Marquardt in press). A fourth site that has produced dense deposits of crown conch shells is the Solana Site (see Figure 11.1) at the mouth of the Peace River (Widmer 1986). There the shells radiocarbon-date to the fifth century (Walker et al. 1995; Widmer 1986). The abundance of crown conch was clearly widespread in the estuarian system.

Another example of inferred animal population change is recorded in Pineland sixth- and seventh-century strata. Winter middens of all time periods and from a variety of sites in the region contain bones of migratory ducks such as the lesser scaup (*Aythya affinis*). In a departure from this general pattern, 1/4"-screened winter-midden samples from Pineland's Caloosahatchee-IIA contexts produced an abundance of migratory-duck remains (Figure 11.3b).

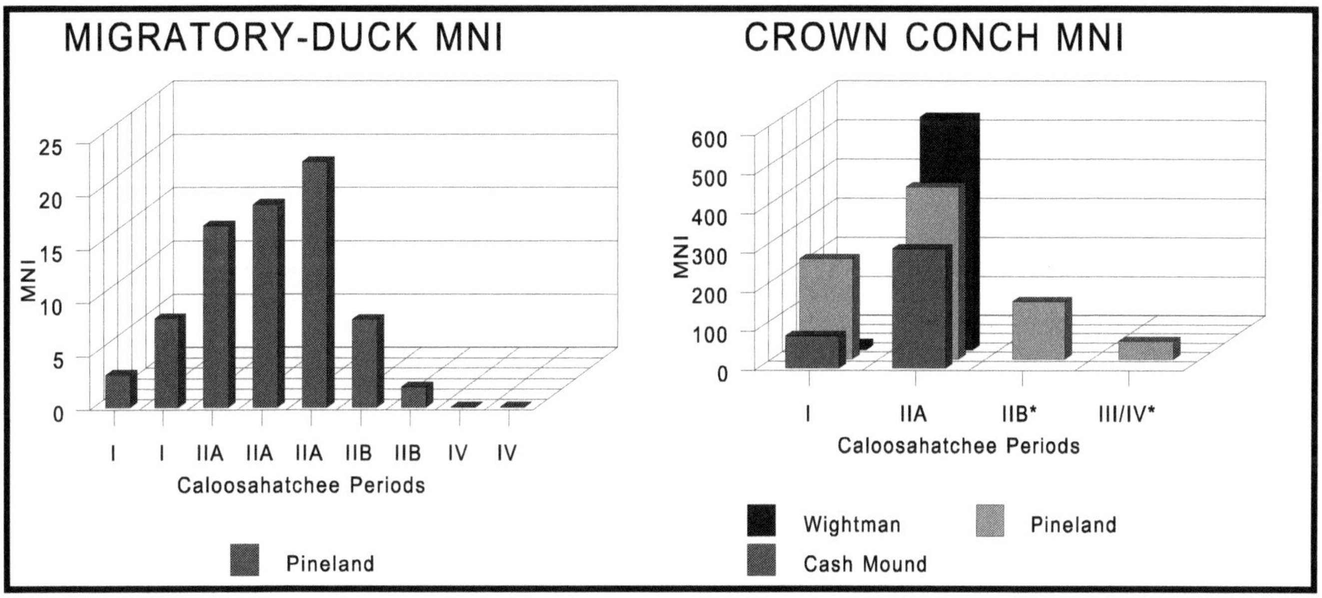

Figure 11.3a (left). Bar graphs of Crown conch *(Melongena corona)* MNI (1/4" screened) from Wightman, Pineland, and Cash Mound Caloosahatchee I and IIA (also IIB, III, IV from Pineland) contexts.
Figure 11.3b (right). Bar graph of migratory duck (Anatidae) MNI (1/4" screened) from Pineland's Calooshatchee contexts.

Numerous species of migratory birds winter in southwest Florida, and when winters are especially harsh in the Eastern U.S., wildlife refuges such as Ding Darling on Sanibel Island accommodate abnormally high numbers of ducks (Lou Hinds, personal communication, 1994). During "normal" winters, many of these ducks winter along the Eastern Shore in Maryland and in the Mississippi Delta region (Bellrose 1976); in cold winters, they fly farther south, to south Florida, seeking warmer climes.

What is the most parsimonious explanation for multiple-species population changes in the sixth century? The crown conch pattern cannot be explained by paleosalinity change as in the gradient examples above; the "active" salinity range of the conch is relatively broad, 20 to 32 ppt (Hathaway and Woodburn 1961:49). A purposeful change in collection strategy by residents seems unlikely unless the conchs were simply more available than usual (i.e., the population increased). An increase in the population of the oyster predating conch implies that Eastern oyster communities could have been stressed environmentally. What type of stress, then, can account for an increased food supply for conchs (i.e., weakened oysters), as well as the decreased salinities that are indicated by the crested oyster and other animals during the sixth and seventh centuries? A major hurricane is a possibility. But a hurricane would not explain what appears to be a long-term, regional episode of change (Walker 1992:285) nor why sites of the period have their "feet" wet today. Nor does it explain the sixth- and seventh-century change in the migratory duck population recorded at Pineland.

A comprehensive explanation is offered by a period of climatic cooling with a concomitant lowering of mean sea level. Oyster bars would gradually have become exposed more frequently (i.e., episodically) and for longer durations as waters receded. This would result in a steady supply of weakened oysters as sea level declined. If the magnitude of the sea-level decline were large enough, the crown conch's food supply would be abundant for some time and a significant seaward shift in the salinity gradient would occur, resulting in lowered salinities. My colleagues and I argue that our hypothesized sea-level rise of the fourth and fifth centuries (Walker et al. 1995) would have placed sea level above that of the present by perhaps a meter or even more. A drop from this position to a low of perhaps 50 cm below that of the present would result in a substantial gradient shift. Such a shift would affect not only the salinity regime of oyster communities, but would literally expose and weaken many, making them more vulnerable to crown conch and whelks. The radiocarbon dates (including the Solana one) for regional crown conch abundance range from the fifth to the seventh centuries. This suggests that the sea-level fall could have been episodic, as was the earlier sea level rise. Tanner's (1991; 1992) beach-ridge-based sea-level curve independently supports such a pattern (see Chapter 2).

Cultural Change during Inclining and Declining Sea Levels

[Coastal Southwest Florida; ; settlement, material culture, subsistence, climate, sea level; Caloosahatchee I and IIA]

As more and more coastal Southwest Florida sites are systematically studied and radiocarbon-dated, preliminary settlement patterns are emerging. Thus far, a gradual population increase and possibly sedentary (based on seasonality evidence from Pineland, Quitmyer, in press) coastal settlement characterize the long Caloosahatchee I period (500 B.C. to A.D. 500^{cor}). During the last 200 years (A.D. 300 to 500^{cor}) of Caloosahatchee I, occupation of the coast of Southwest Florida was diminished; the coast was possibly even abandoned in favor of interior South Florida, a region dominated by the Belle Glade peoples. I suggest that this movement inland did take place and was in great part a response to an episode of storminess and higher-than-present sea-level— an episode that correlates with the fourth- and fifth-century sea level transgressions of Tanner's (1991; 1992) Gulf of Mexico sea-level curve; this is shown by the jagged episodic curve (see Chapter 2, this volume), and also with the "Wulfert High" of Stapor and his colleagues' Southwest Florida sea-level curve (1991). At a broader scale, the entire Caloosahatchee I period correlates with the globally known warm and stormy Roman Optimum (Denton and Karlén 1973; Gunn 1994). The culmination of Southwest Florida's stormy sea-level rise correlates with the end of the Roman Optimum.

Around A.D. 500^{cor}, the beginning of Southwest Florida's Caloosahatchee IIA (Cordell, in press), people again populated the coastal region, this time bringing a plethora of changes and innovations: greater numbers of people, the beginnings of ceremonial mound building, the addition of a new pottery technology (Belle Glade culture), a decreased diversity of shell tools, an expanded fishing technology, and possibly a greater availability and use of lithic resources. In addition, there were dramatic increases in dietary reliance on whelks, conchs, and migratory ducks, and a reduced reliance on oysters. One of the most dramatic changes seen in this reoccupation is that some sites were established seaward of today's shoreline, documenting a sea level lower than that of today by as much as 50 cm (Walker et al. 1994, 1995). These mounded middens were the beginnings of what became massive shell mounds. I think that drought in interior South Florida and an end to the storminess may have been push and pull factors in the coastal reoccupation. Caloosahatchee IIA correlates with the sixth- through eighth-century regressions of Tanner's sea-level curve (1991; 1992; Chapter 2) and with Stapor and his colleagues' (1991) "Buck Key Low." Southwest Florida's IIA period also correlates with the globally known cool and dry Vandal Minimum, including the A.D. 536^{cal} event (Bryson 1988).

Toward the end of IIA, about A.D. 800^{cor}, mound building in Southwest Florida intensified, possibly a response in part to the beginning of another sea-level rise—this time correlating with the warm, global Medieval Maximum. This time, instead of moving inland away from rising waters, coastal residents literally moved upward away from rising waters. Both Tanner (1991) and Stapor et al. (1991) record Gulf of Mexico transgressions for the ninth through eleventh centuries. Tanner (1992) also independently documents a Danish transgression for the same period.

The Medieval Maximum and Roman Optimum transgressions contrast with the intervening Vandal Minimum regression, thus serving to closely define the temporal parameters of the sixth- and seventh-century regression, and presumably also the cool-climate episode. All three sea-level and climate episodes independently support the archaeological patterns that I have presented in this chapter. Cultural response in South Florida to climatic shifts and/or sea-level fluctuations may not follow patterns exhibited by other southeastern US pre-Columbian peoples to the north. In climatic schemes (Oliver and Hidore 1984), South Florida is subtropical and a warming episode accompanied by an increase in storm frequency and intensity (Galli 1989) is not a welcome change. Conversely, a cool and calm period might be welcomed as long as drought does not become debilitating. For South Florida's precolumbian residents, indeed the terms "Optimum" and "Minimum" may seem to be used backwards, reflecting the bias of northern Europe where the climatic episodes are best documented and were first recognized (Gunn 1994:17).

Archaeological Signatures for a Climatic Cooling?

[Southwest Florida; radiocarbon dating; sea-level fall/climatic cooling, cultural change; A.D. 536, Caloosahatchee I and IIA, Middle and Late Woodland]

So, **are** archaeological signatures for the A.D. 536 event detectable in Southwest Florida? The temporal limitations of radiocarbon dating, of course do not allow pin-pointing the A.D. 536 event, but what I can say is that an extended decline in sea level with its implied cooling episode explains much about the environmental and cultural changes that characterize Southwest Florida's Caloosahatchee I to IIA transition, a time generally correlating with the greater Southeast's shift from Middle to Late Woodland periods. It is also clear that multiple lines of evidence, as well as more refined dating techniques, narrow the likelihood of detecting event-time scale impacts on humans and other species.

ACKNOWLEDGMENTS: The Florida Department of State, the National Endowment for the Humanities, the Maple Hill Foundation, the Ruth and Vernon Taylor Foundation, the National Science Foundation, and the Calusa Constituency funded various components of the field research and analyses that led to this article. I am grateful to my colleague William Marquardt, principal investigator of those grants, and the Florida Museum of Natural History for their support. Kelly Barber prepared the Figure 1 map and Figure 2. Thanks, also, to Joel Gunn for organizing the 1996 SEAC A.D. 536 Symposium and coordinating the production of this book with impressive skill.

References

Baillie, M. (1995) Patrick, Comets and Christianity. *Emania* 13:69-78.

Bellrose, F. (1976) *Ducks, Geese & Swans of North America.* Stackpole Books, Harrisburg, PA.

Boesch, D. (1977) A New Look at the Zonation of Benthos along the Estuarine Gradient. In *Ecology of Marine Benthos*, edited by B. Coull, pp. 245-266. University of South Carolina Press, Columbia.

Bryson, R. (1988) Civilization and Rapid Climatic Change. *Environmental Conservation* 15:7-15.

Butler, P. (1954) Summary of Our Knowledge of the Oyster in the Gulf of Mexico. In *Gulf of Mexico: Its Origin, Waters, and Marine Life*, pp. 479-489. Fishery Bulletin 89, U. S. Fish and Wildlife Service, vol. 55. Washington, D. C.

Cordell, A. (1992) Technological Investigation of Pottery Variability in Southwest Florida. In *Culture and Environment in the Domain of the Calusa*, edited by W. Marquardt, pp. 105-189. Institute of Archaeology and Paleoenvironmental Studies, Monograph 1. University of Florida, Gainesville.

Cordell, A. (in press) Technological Investigation of Pottery Variability at the Pineland Site Complex. In *The Archaeology of Pineland: A Coastal Southwest Florida Site Complex, A.D. 100-1600*, edited by K. Walker and W. Marquardt. Institute of Archaeology and Paleoenvironmental Studies, Monograph 4. University of Florida, Gainesville.

Dalby Jr., J. (1989) Predation of Ascidians by *Melongena corona* (Neogastropoda: Melongenidae) in the Northern Gulf of Mexico. *Bulletin of Marine Science* 45:708-712.

Denton, G. and W. Karlén (1973) Holocene Climatic Variations—Their Pattern and Possible Cause. *Quaternary Research* 3:155-205.

Galli, G. (1989) Is Holocene Storm-Generated Stratification in Florida Bay a Reflection of Solar Storm Cycles? *Palaeogeography, Palaeoclimatology, Palaeoecology* 76:169-185.

Galtsoff, P. (1964) The American Oyster *Crassostrea virginica* Gmelin. *U.S. Fish and Wildlife Service Fishery Bulletin* 64:1-480. Washington, D. C.

Galtsoff, P. and A. Merrill (1962) Notes on Shell Morphology, Growth, and Distribution of *Ostrea equestris* Say. *Bulletin of Marine Science of the Gulf and Caribbean* 12:234-244.

Gill, R. (1994) *The Great Maya Droughts.* Unpublished Ph.D. dissertation, Department of Anthropology, University of Texas, Austin.

Gunn, J. (1994) Introduction: A Perspective from the Humanities-Science Boundary. *Human Ecology* 21:1-23, Special Issue.

Gunter, G. and R. Menzel (1957) The Crown Conch, *Melongena corona*, as a Predator Upon the Virginia Oyster. *The Nautilus* 70:84-87.

Hathaway, R. and K. Woodburn (1961) Studies on the Crown Conch *Melongena corona* Gmelin. *Bulletin of Marine Science of the Gulf and Caribbean* 11:45-65.

Hofstetter, R. (1959) *The Texas Oyster Fishery.* Bulletin No. 40. Texas Game and Fish Commission, Austin.

Keigwin, L. (1996) The Little Ice Age and Medieval Warm Period in the Sargasso Sea. *Science* 274:1504-1508.

Kellogg, D. (1988) Problems in the Use of Sea-Level Data for Archaeological Reconstructions. In *Holocene Human Ecology in Northeastern North America*, edited by G. Nicholas, pp. 81-104. Plenum Press, New York.

Marquardt, W. (1992) Recent Archaeological and Paleoenvironmental Investigations in Southwest Florida. In *Culture and Environment in the Domain of the Calusa*, edited by W. Marquardt, pp. 9-57. University of Florida, Institute of Archaeology and Paleoenvironmental Studies, Monograph 1. University of Florida, Gainesville.

Menzel, R. and F. Nichy (1958) Studies of the Distribution and Feeding Habits of Some Oyster Predators in Alligator Harbor, Florida. *Bulletin of Marine Science of the Gulf and Caribbean* 8:125-145.

Oliver, J. and J. Hidore (1984) *Climatology: An Introduction.* Charles E. Merrill Publishing Company, Columbus.

Quitmyer, I. (in press) Precolumbian Harvest of Estuarine Resources and Archaeological Site Seasonality at the Pineland Site Complex, Charlotte Harbor, Florida. In *The Archaeology of Pineland: A Coastal Southwest Florida Site Complex, A.D. 100-1600*, edited by K. Walker and W. Marquardt. Institute of Archaeology and Paleoenvironmental Studies, Monograph 4. University of Florida, Gainesville.

Stapor, F. Jr., T. Mathews, and F. Lindfors (1991) Barrier-Island Progradation and Holocene Sea-Level History in Southwest Florida. *Journal of Coastal Research* 7:815-838.

Strothers, R. (1984) Mystery cloud of AD 536. *Nature* 307:344-345.

Tabb, D. and R. Manning (1961) A Checklist of the Flora and Fauna of Northern Florida Bay and Adjacent Brackish Waters of the Florida Mainland Collected During the Period July, 1957 Through September, 1960. *Bulletin of Marine Science of the Gulf and Caribbean* 11(4):552-649.

Tanner, W. (1991) The "Gulf of Mexico" Late Holocene Sea Level Curve and River Delta History. *Gulf Coast Association of Geological Societies, Transactions* 41:583-589.

Tanner, W. (1992) Late Holocene Sea-Level Changes from Grain-Size Data: Evidence from the Gulf of Mexico. *The Holocene* ?:249-254.

Walker, K. (1992) Zooarchaeological Perspectives on Charlotte Harbor Maritime Adaptation. In *Culture and Environment in the Domain of the Calusa*, edited by W. Marquardt, pp. 265-366. Institute of Archaeology and Paleoenvironmental Studies, Monograph 1. University of Florida, Gainesville.

Walker, K. and W. Marquardt (in press) Excavations at the Pineland Site Complex. In *The Archaeology of Pineland: A Coastal Southwest Florida Site Complex, A.D. 100-1600*, edited by K. Walker and W. Marquardt. Institute of Archaeology and Paleoenvironmental Studies, Monograph 4. University of Florida, Gainesville.

Walker, K., F. Stapor, Jr., and W. Marquardt (1994) Episodic Sea Levels and Human Occupation at Southwest Florida's Wightman Site. *The Florida Anthropologist* 47:161- 179.

Walker, K., F. Stapor, Jr., and W. Marquardt (1995) Archaeological Evidence for a 1750-1450 BP Higher-than-Present Sea Level Along Florida's Gulf Coast. *Journal of Coastal Research Special Issue No. 17: Holocene Cyclic Pulses and Sedimentation*, pp. 205-218.

Wells, H. (1961) The Fauna of Oyster Beds, with Special Reference to the Salinity Factor. *Ecological Monographs* 31:239-266.

Widmer R. (1986) *Prehistoric Estuarine Adaptation at the Solana Site, Charlotte County, Florida.* Florida Division of Archives, History and Records Management, Bureau of Archaeological Research, Tallahassee, Florida.

Table 11.1. Radiocarbon Dates[a,b] for the Wightman, Pineland, and Cash Mound Stratigraphic Sequences Presented in Figure 11.2.

RC Number	Wightman 8LL54	Pineland 8LL33	Cash Mound 8LL38
RC1	A.D. 650-750^{cor} Beta-47887 M. corona	A.D. 400-540^{cor} Beta-60770 M. campechiensis	A.D. 660-780^{cor} Beta-16281 C. virginica
RC2	A.D. 290-420^{cor} Beta-71513 M. campechiensis	A.D. 380-510^{cor} Beta-60769 S. solidissima	A.D. 50-250^{cor} Beta-16280 C. virginica
RC3	A.D. 450-590^{cor} Beta-49170 B. sinistrum	A.D. 190-340^{cor} Beta-60768 M. campechiensis	A.D. 100-280^{cor} Beta-16279 C. virginica
RC4	A.D. 410-570^{cor} Beta-56533 B. sinistrum	A.D. 60-190^{cor} Beta-72994 B. sinistrum	A.D. 220-370^{cor} Beta-16278 C. virginica

[a] These radiocarbon dates are both ^{13}C-adjusted and calibrated using the Pretoria Calibration Procedure, which Beta Analytic, Inc. began using in 1994.

[b] All of these radiocarbon dates have been published previously using an older calibration method (Marquardt 1992; Walker et al. 1994 1995), resulting in slightly different ranges than those presented here.

Table 11.2. Crested Oyster (*Ostreola equestris*) to Eastern Oyster (*Crassostrea virginica*) Ratios for the Wightman, Pineland, and Cash Mound Sites Comparing Late Caloosahatchee I (A.D. 100 to 500^{cor}) Midden Samples to Caloosahatchee IIA (A.D. 500 to 800^{cor}) Midden Samples.

SITE	CALOOSA-HATCHEE I	CALOOSA-HATCHEE IIA	CALOOSA-HATCHEE IIB
Wightman	1:6	0:346	
Pineland	1:4	1:18	1:2
Cash Mound	1:2	0:343	

VI. CHANGES IN SOCIETY: ADJUSTING TO ELEVATED ENVIRONMENTS

Mountainous terrains have a great potential for revealing the interactions of cultures and climates. Elevated and exposed to the upper winds, they can be the canaries of world physiographies. Something that has interested me since spending a summer excavating at the Alamo Mission in San Antonio, Texas, is the relatively great number of people from western North Carolina and eastern Tennessee in the Alamo battle of 1836. It occurred near the end of two decades of global cold precipitated by the Mount Tambora eruption of 1815, the largest volcanic eruption in the Holocene. In subsequent years refugees from the highland South flooded first into the U.S.-Mexico no-man's land along the western Louisiana border and then into Texas. In this pair of chapters, Jerry Lilly and Paul Webb, and Ruth Wetmore and her colleagues explore the effects of global cooling on the Appalachian Summit of western North Carolina and the ridge and valley province of northwestern Georgia. Their studies reveal very different responses by peoples only a few miles apart, but perhaps vastly separated by the irregular character of mountain environments. Because of the exposed character of these environments, they offer an opportunity to examine changes in social organization. As I will discuss later, there are interesting parallels across the Atlantic Ocean in Europe.

The Etowah Valley of northwestern Georgia possesses one of the most spectacular arrays of Native American monumental architecture in North America. Huge, then-colorful mounds at the Etowah archaeological site attest to the organizational prowess, taste, and significant populations of a complex society, recently suggested by a suite of studies to be state organizations (Barker and Pauketat 1992). In the middle Etowah Valley of northern Georgia, Jerry Lilly and Paul Webb have excavated a large multicomponent site (Cartersville, Woodstock, and Cherokee phases) in north Georgia. Features from the site transcend the sixth century and the time of the generally recognized "Late Woodland Problem," a period of presumed collapse or rapid change of proportions and character only here being substantially defined and debated (see Anderson chapter). Before the sixth century, the Etowah area was populous and prominently represented in sites. During the sixth century, population diminished and/or dispersion, and a there was a transition to the Late Woodland, which is nearly archaeologically invisible.

Lilly and Webb believe there is a potential correlation between the climatic events following A.D. 536^{cal} and the period of substantial culture change following the Middle Woodland period Cartersville phase. They examine past studies of the valley and reflect on their recent research at the Hickory Log archaeological site northwest of Atlanta, Georgia. This includes a re-evaluation of the regional chronology based on calibration of radiocarbon dates. Ceramics decorated with check stamped designs, which are the principal diagnostic artifacts of the Cartersville phase, are essentially gone from the region by about A.D. 400-550^{c14}, and are supplanted first by simple stamped decorations, and then by a variety of subregion-specific designs. The subregionalization of ceramics designs probably reflects a breakdown in the former widespread regional social organization into independent provinces. Domesticated food remains present in many Middle Woodland pit features attest to the important role that horticulture played in the Cartersville diet. It could also indicate that the Cartersville society had difficulty adapting to rapidly changing environmental conditions. The ultimate disappearance of Cartersville, and a subsequent 200 year period of subregionalization of ceramic styles, could at least in part be a consequence of the A.D. 536 event and subsequent environmental instability.

Ruth Wetmore, Ken Robinson, and David More examine the A.D. 536^{cal} issue in nearby western North Carolina, though at much higher altitudes. At European contact, the fertile valley floors of the Appalachian Mountains were inhabited by sophisticated horticultural mound building people, and had been for centuries. The Middle and Late Woodland periods of the mountains of western North Carolina span from approximately A.D. 200 to 1000^{cor}. Radiocarbon dates have been obtained from a number of Woodland sites for this period with the greatest concentration occurring between A.D. 550 and 700^{cor}. This correlates with the A.D. 536^{cal} aftermath. Wetmore and her co-authors use research on cold air drainage in the southern Appalachians to inform their findings from a microclimate perspective. Their data are primarily from 1912-1916. They review ceramic designs and settlement patterns from Woodland period phases, including the Connestee phase, to ascertain whether substantial cultural changes or continuity characterized this approximately 800-year period. Their analyses indicate the persistence of Connestee ceramics throughout the period with minimal stylistic change (surface treatment or form). The presence of sherds with complicated-stamping from other Woodland ceramic types imported from other regions, but found in Connestee sites, and the selective adaptation of new technologies in the mountains, indicate multiple inter-regional contacts through the period. The authors explore settlement pattern and subsistence mode data from excavated contexts to uncover the influences of climate on mountain cultural adaptations. The apparent underlying cultural continuity through the period of study suggests that influences of the A.D. 536^{cal} phenomenon will appear in the frequency and density of mountain people and their cultural residues, rather than in population movements like those discussed by Mathis for the Coastal Plain (see Mathis chapter), and perhaps by Woodall for the Piedmont (see Woodall chapter).

Taken together, the two chapters appear to suggest that important changes in social organization followed A.D. 536^{cal} in the ridge and valley or northern Georgia, but minor changes in settlement pattern and technology followed in the mountains. However, as discussed in the following section, when viewed in another perspective, that of frequencies rather than qualitative cultural changes, there appear to be internal adjustments at some social and density levels that require attention.

Parallel Events: Synchronization Across the Atlantic

One of the pressing issues for comparing affects of climate change across oceans is understanding the synchronization of climates on both sides, in this case across the Atlantic Ocean between North America and Europe, and between historic and prehistoric cultures. Without precise knowledge of the climate processes that synchronize European and American climate, the understanding of climate change effects becomes fogged. Lamb (1977) believes that there could be an approximately 100 year lag time between, for example, cooling in North American and Europe. This is reasonable if one assumes that the oceans cool more slowly than the atmospheric climate. The northern hemisphere westerlies carry, in a major part, continental air into the Southeastern United States from Canada and east Eurasia, especially in winter. These same air streams, however, cross the Atlantic Ocean before becoming the weather and climate of Europe. If the oceans are warm, this greatly warms and moistens areas of Europe that would otherwise be only marginally habitable (see Broecker [1995] for a discussion). However, it has recently been discovered that roving pools of warm and cold water in the Atlantic also influence down airstream climate at decadal time scales.

The trends in cultural change discussed by Wetmore and her colleagues could bear on the synchronization issue if one compares frequency domain indicators in the southeastern United States with those of Europe. The Appalachian Summit cultural trends can be detected in frequencies of radiocarbon dates. One of their number, Ken Robinson, has a new date on a Connestee Phase site. It is A.D. 680^{c14}, corrected (710^{cor}). They now have a cluster of ten dates on the Connestee Phase from A.D. 550 to 750^{cor}. This period is after the collapse of the Hopewell Interaction Sphere in the Midwest and its equivalent Cartersville culture in the Southeast as discussed by Lilly and Webb, and after A.D. 536^{cal}. In Europe, the parallel period to this date cluster is an era of good times for the Carolingian state supported in part by an extensive trade network around the North Sea. If higher frequencies of radiocarbon datable features in Appalachian archaeological sites can be taken to represent an era of increased population and well-being, the two corresponding cultural trends could well be indicative of a period of stable and benevolent climate on both sides of the Atlantic some decades after the A.D. 536^{cal} event. This period seems to fall between the immediate negative impact of the A.D. 536 event itself, and thus a break with an earlier cluster of Connestee dates, and the nadir of subsequent cooling recorded in Europe as a series of sever winters between A.D. 760 and 840^{cal} identified by Lamb (1977).

The termination of the Appalachian Connestee date cluster is also within the range of the nadir of the A.D. 536^{cal} aftermath sea level fall implying globally cold conditions (see Tanner chapter). Cultural phenomena in Europe that correspond to the sea level nadir and the cold winters include Charlemagne's appearance in Rome on Christmas day of A.D. 800^{cal}, the great Viking onslaught of A.D. 829^{cal}, probably precipitated by an exceptional conditions, since the Nile River froze that year, and, as one might anticipate, the collapse of the Carolingian trade network on the North Sea in the face of the Viking assault. All of this suggests approximately parallel cultural changes on both sides of the Atlantic; the precise timing of these parallel events needs to be given careful attention. Clearly there is great potential for studying the processes of climate and cultural change through the various linkages offered by parallel events (see Gunn [1994] for a discussion of parallel events methodology).

Global Change Assessment. Especially, the comparison of European and Appalachian cultural sequences during the A.D. 536 event and its aftermath suggests to me an interesting parsing of affects of climate change. Events appear to precipitate a different suite of cultural phenomena than episodes or trends of climate. Bad years and bad decades of extreme character precipitate migration and related manifestations. Mathis' and Jones's chapters are also helpful in this regard. Episodes, on the other hand, result in a much more divers, regionally tailored, traditionally flavored responses as discussed by Lilly and Webb.

In the cases of the ridge and valley and mountain populations discussed in this concept set, it is worth noting that during the milder climates preceding A.D. 536^{cal}, ridge and valley populations joined the world wide tendency to aggregate small societies into large societies. Apparently in the mountains of western North Carolina, cultures maintained their traditional social organization of small divided societies conditioned by a geography of mountain valleys. The evidence indicates that following A.D. 536^{cal}, people of the ridge and valley converted to a social organization strategy more like that of the mountain cultures. This suggests that following A.D. 536^{cal} microenvironments defined by the valleys of the ridge and valley province presumed a greater role than before, enough to require special attention by local social organizers to their individual valley's characteristics. Is this a tendency encouraged by colder and less stable global climates? Can it be a simple function of higher altitude climates descending into lower elevations as the Earth system cools?

References

Barker, Alex W., and Timothy R. Pauketat, editors (1992) *Lords of the Southeast: Social Inequality and the Native Elites of Southeastern North America.* Archaeological Papers of the American Anthropological Association, Number 3, American Anthropological Association, Washington D.C.

Broecker, Wallace S. (1995) Chaotic Climate. *Scientific American* 273:62-69.

Gunn, Joel (1994) Global Climate and Regional Biocultural Diversity. *In Historical Ecology*, edited by Carole L. Crumley, pp. 67-97. School of American Research, Santa Fe.

Lamb, H. H. (1977) Climate: Present, Past and Future: Climatic History and the Future. Vol 2, Methuen and Co. Ltd., London.

12. THE A.D. 536 EVENT AND THE MIDDLE TO LATE WOODLAND TRANSITION IN NORTHERN GEORGIA

Thomas G. Lilly Jr. and Paul A. Webb

TRC Garrow Associates, Inc., Chapel Hill, North Carolina 27514Historic Events and Prehistoric Times

[Southeastern U.S., northern Georgia, Hickory Log site; microcosmic setting, holism/reductionism, archaeology, alternative models; cultural change, temporal resolution, social organization; Middle Woodland, Late Woodland, A.D. 400-900^{c14}]

In northern Georgia (Figure 12.1), the A.D. 536 event would have coincided with an elusive transition between the Middle and Late Woodland periods (Table 12 1). Whether or not this transition was in some way related to a significant climatic change is the primary focus of this chapter. Recent research at the Hickory Log site (9CK9), in the Etowah Valley northwest of Atlanta provides data that are posed here as a microcosmic setting of human activity during this transition.

Whatever its precise nature, the A.D. 536^{cal}, event coincided with, perhaps even caused, a pronounced change in world climate. Sea level (Tanner 1992, 1993) and tree ring (Stahle et al. 1988) data suggest that the period between about A.D. 550 and 850^{cal} was marked by a cold interval. Average temperatures during this climatic downturn may have been as much as 3° F lower than during the preceding period, and likely accompanied other changes in rainfall patterns, length of growing seasons, and related variables (Baillie 1995).

Assuming that such changes were in fact global, we thought it worthwhile to seek their traces in the culture history of the southeastern United States. However, we try not be reductionist in outlook seeking to explain all aspects of human behavior in terms of a year of clouds or decades of cold. We view human cultures as complex systems, in which climate changes are one of a myriad of factors that might affect individuals and groups as they interact with their social and physical environments. Investigation of the possible role of climatic events is, in a holistic context, only one avenue through which we can hope to gain a broad understanding of human behavior and the processes of cultural change.

In assessing the possible cultural effects of such change, we first consider the issue of chronology. Today, in a way that is novel to our usual sense of prehistoric, a specific date is being suggested to an archaeological tradition comfortable with rounding time periods and phases to the nearest century. Understanding human responses to past environmental changes is hampered in prehistory by both the uncertainty of timing and the rate of environmental change, as well as by a lack of resolution in the archaeological record as it currently structured. However, this case is different. We have excellent control over one aspect of the problem—we know that something happened in A.D. 536^{cal}(Baillie 1995). Despite our best efforts, we cannot at present date prehistoric cultural manifestations with comparable precision. For the purpose of our argument, we will focus on cultural changes between about A.D. 400 and 650^{C14}, and will also consider cultural changes dating as late as A.D. 900^{C14}.

What sort of changes are we looking for? Answering this question requires consideration of the likely environmental effects of the event as well as the nature of the Woodland period societies. The direct physical results of a dry fog in a non-coastal locale would probably be limited to what has been described by historic chroniclers— failure of some crops, especially those grown near the limits of their range, suppressed productivity of some wild plants, and potential related effects on animals. Any of these impacts, of course, could have cascading effects on human populations. The results would vary greatly depending on many aspects of a society's relationship to its biological and physical environment, and could include the breadth of the subsistence niche, the extent of the society's reliance upon agriculture, their food storage technology, their demographic structure, and foreign relations. In short, while it is easy to imagine that a single bad agricultural year could have ever increasing economic and social repercussions throughout a society, we might likewise imagine a circumstance in which such a change was easily buffered, resulting in few if any long-term repercussions. It is our talk to sort though these alternative scenarios and find which is most plausible and supportable in the light of our archaeological data on the relevant time period.

Like other contemporaneous societies in the Southeast, the Woodland period occupants of northern Georgia are best described as tribal societies, small-scale and non-hierarchical. As a general rule, they show long-term population aggregation, made possible either by the multi-seasonal availability of predictable resources or by storing surpluses of domesticated and/or wild foods. Ritual activities crosscut family ties to integrate groups, and power and prestige are earned rather than inherited. Individual groups remain neighbors through kinship ties and exchange networks, which perhaps serve to buffer local shortages in food and other resources (Gregg 1991). Such societies are

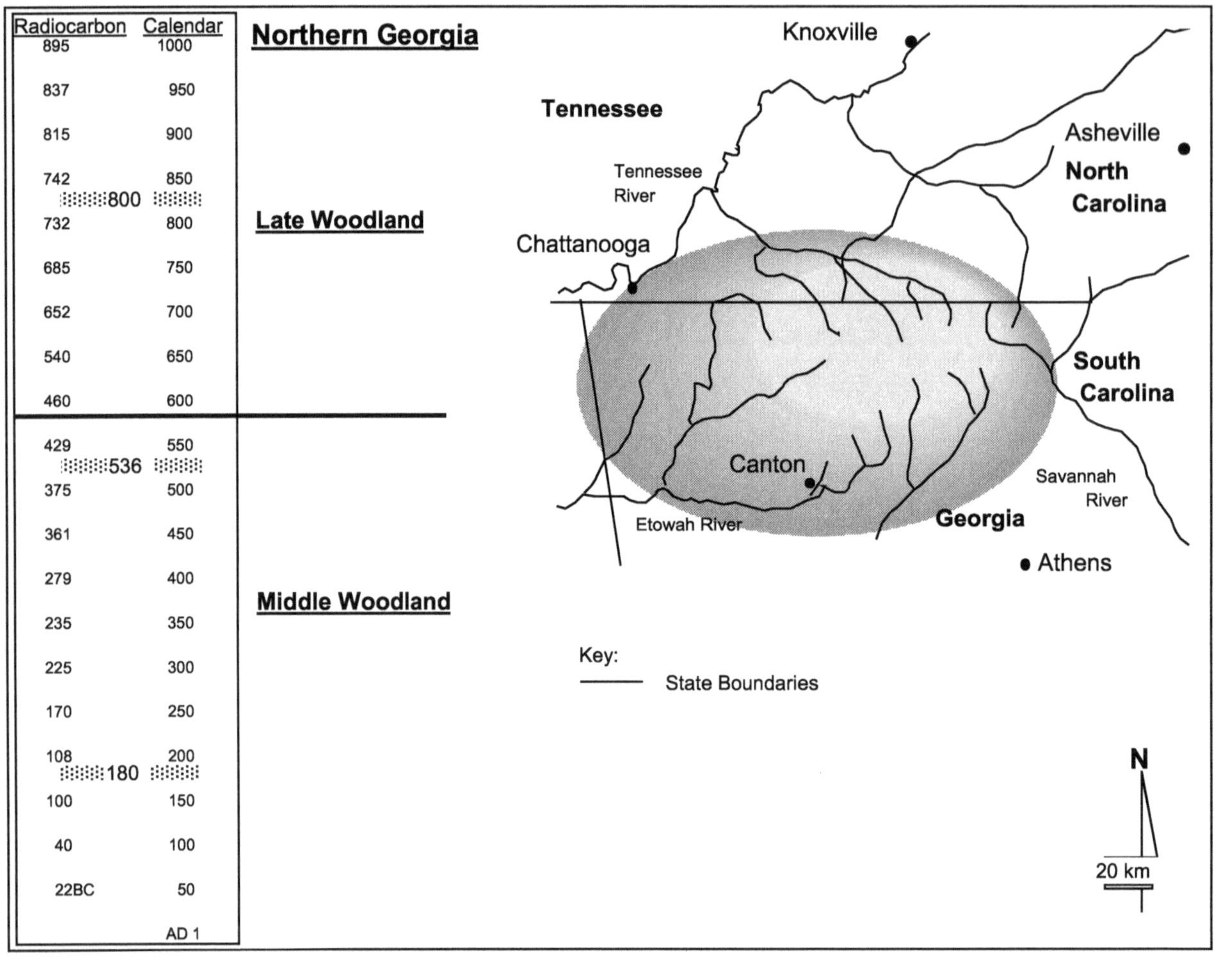

Figure 12.1. Map and Chronology of Upland Georgia.

"organizationally fluid," frequently being characterized by population movement, shifting alliances, and "a continually changing social environment." (Milner et al. 1991:258). In short, the societies for whom we are seeking evidence of change were, by definition, designed to be flexible. This could both diminish our ability to observe cultural change through archaeological research and make detectable changes all the more credible.

The Evidence from Northern Georgia

[Northern Georgia, Hickory Log site; ceramic motif changes, Woodland chronology; social integration, check stamp, stone hoes, cultivation, may grass, sunflower, squash, nuts, simple stamp, ritual, territory, stone mounds, transition, Swift Creek complicated stamp, population redistribution, Late Woodland problem; Early Woodland, Cartersville, 300 B.C.-A.D. 300^{c14}, Late Cartersville, Late Woodland]

Northern Georgia (see Figure 12.1) is loosely defined as north of the Piedmont Fall line (Garrow 1975). The Etowah Valley, which includes parts of the Piedmont and Ridge and Valley physiographic provinces, has been the source of data collected over fifty years and is our region of focus. Our study of cultural change in this region will draw on a number of previous synthetic works (e.g., Anderson 1985; Caldwell 1957, 1958; Garrow 1975; Rudolph 1991; Williams 1994; Wood and Bowen 1995), as well as new information from our Hickory Log site.

Like many Woodland sequences in the eastern U. S., the cultural sequeoce in northern Georgia (see Figure 12.1) is largely based on changes in ceramic motifs. The outline of the Etowah Valley sequence was first contrived in the 1940s and 1950s. It was based on Wauchope's (1966) extensive Depression-era survey of northern Georgia and Caldwell's (1957, 1958) and other's work in the Allatoona Reservoir area. More recent research has enriched understanding of the sequence considerably, adding data on subsistence practices and other factors of human life. Nonetheless, period, phase, and social boundaries during this period are still based largely on ceramics.

The Middle Woodland period in the Etowah Valley is traditionally considered to begin about 300 to 200^{C14} B.C. with the appearance of Cartersville ceramics alongside the preceding Dunlap, and perhaps Mossy Oak, ceramic types. The earliest Cartersville ceramics are check-stamped, which appear to have supplanted fabric-impressed designs in the region by the first century B.C. Cartersville simple stamped ceramics appear by about 200^{c14} B.C. and both types co-occur until about A.D. 300^{c14}. Vessel forms are primarily conoidal-based jars, frequently with podal supports; small bowl forms also occur. Cartersville ceramics are found over a broad expanse of the Piedmont and adjacent ridge and valley, and related types such as Deptford are found across neighboring regions. The region-wide similarity in ceramic motifs during this period can be interpreted as reflecting considerable social integration, as can the occasional presence of Hopewell Interaction Sphere-related items (Jefferies 1976). The Hopewell Interaction Sphere was a vast trading network centering in the upper Ohio Valley and encompassing items of trade from most of temperate North America (Brose and Greber 1979); it was contemporary with the Cartersville Period.

Cartersville sites are abundant in the Etowah Valley, and are concentrated along river terraces and similar environments. An intensive early Cartersville occupation at Hickory Log appears to date between about 100 B.C. and A.D. 300^{C14}, and is marked by the presence of numerous circular structures and abundant pit features. Cartersville check-stamped ceramics are the dominant type, comprising a majority of the Middle Woodland assemblage, with slight admixture of simple stamped and complicated stamped materials.

The presence of stone hoes at many Cartersville sites, including Hickory Log, suggests that horticulture was becoming increasingly common during this period. This is confirmed by the widespread appearance of maygrass in storage pits, which is present in about 25 percent of the Cartersville flotation samples at Hickory Log, and may have been especially important as a late spring/early summer food. Lesser amounts of such cultivated plants as sumpweed and sunflower also occur at this and other sites, and it is increasingly evident that maize, as well as squash, was also present in at least small quantities. Evidence of domesticated corn (*Zea maize*) was recovered at both the Cane Island site (9PM209, Wood 1981) and at the Rush site (9FL164); at the latter, Wood and Ledbetter (1990) recovered a single cupule that returned an uncorrected date of A.D. 660±100^{C14}.

A decline in subsurface storage facilities from the earlier Kellogg Phase has been interpreted as suggesting a lessened emphasis on mast crops during Cartersville, but the ubiquity of these items in samples from Hickory Log indicates that, at least at this site, nuts continued to be used in quantity.

Our clear understanding of the Cartersville manifestation dissipates as we approach the elusive terminus of the Middle Woodland period. Simple stamping, as much a temporal conundrum as nondescript plain wares, increasingly supplants check stamping in the Cartersville assemblage, and the latter is traditionally said to be largely if not totally gone by about A.D. 400^{C14}. Simple stamped vessels with podal supports continue to be made, although the supports appear to be more and more rudimentary. Complicated stamped ceramics with Swift Creek designs appear in small quantities within Cartersville assemblages as early as A.D. 400^{C14}, but do not appear in northern Georgia in the same elevated frequency as they do in the southern part of the state. The motifs from Hickory Log include B-Complex. Simple stamping appears to continue until at least A.D. 500^{C14}, but few if any clearly *sixth through eighth century* simple stamped sites have been recognized.

At Hickory Log, the latter part of Cartersville is evident in an almost pure simple stamped component, which is located in a different portion of the site from the earlier check-stamped occupation. Vessel forms associated with the simple stamped component include large, everted rim, globular jars that are more similar to later forms following Cartersville than the preceding Early Cartersville conical vessels. This component is also associated with a cluster of 17 rock-filled burials, some of which contain up to three individuals and are associated with bar gorgets, mica, and triangular and expanding-stemmed projectile points. The rock filled burials were probably once stone mounds, and indicate both increased mortuary ceremonialism and to visually mark territory.

The post-Cartersville transition to the Late Woodland world has traditionally been seen as a major disjunction in an otherwise orderly Woodland ceramic and cultural sequence. This discontinuity was graphically illustrated by Caldwell in his dissertation, where he postulated major changes in regional influences between about A.D. 500 and 900^{C14}. As Caldwell noted, an era of ceramic homogeneity is replaced by one of increased stylistic diversity in which plain, simple stamped, and various complicated stamped types appear both individually and in combination across the region; it creates a chronological visionary's delight with the ceramic twister's base pinpointed near the A.D. 536^{cal} date. This diversity thwarts most archaeological methods that estimate population sizes and probably contributes to the apparent lack of visibility of Late Woodland sites in the region, which has been aptly termed the "Late Woodland Problem" by Teresa Rudolph (1991). Rudolph also notes that this lack of visibility could result from an increasing shift towards occupation of floodplains (see Wetmore et al., this volume), resulting in site burial and thus invisibility. Another contributing factor appears to be population movements resulting in certain regions, such as portions of the Etowah Valley, being abandoned. Other evidence, including site file data compiled by Mark Williams (1994), supports the notion of a redistribution of populations rather than a region-wide decline in Late Woodland sites or population.

Like the Etowah Valley in general, Hickory Log witnesses a major fall-off in occupation during the Late Woodland. Such early Late Woodland ceramic types as Swift Creek and Napier, which are seen as the primary motif markers of this period, are represented by only a few artifacts and features. The available data seem to confirm patterns noted elsewhere, including an increasing frequencies of bowls in vessel assemblages and a continuing use of cultigens such as squash.

Order is not restored to the Late Woodland sequence in this region until sometime after A.D. 800^{C14}, with the appearance of Woodstock complicated stamped ceramics. These ceramics show some motif design affinities with earlier Napier, but are dominated other vessel forms such as large, everted rimmed jars. The Woodstock occupation of Hickory Log contains a large palisaded enclosure, similar in many ways to that found by Caldwell at the type site Woodstock Fort. Both maygrass and maize are common in flotation samples, and it seems clear that cultigens played a major role in the Woodstock diet.

Etowah Valley, Hickory Log, and the Late Woodland Problem

[Etowah Valley, Hickory Log; radiocarbon date calibration; calibration and traditional cultural chronologies, global correlations, historical correlations; Middle-Late Woodland transition]

Our discoveries at Hickory Log reflect the general pattern of the Etowah River watershed, though with great clarity. Intensive occupation during the early Middle Woodland correlates with a regional pattern of higher population density and stability, at least as indicated by the continuity in ceramic design motifs. This is followed by an apparent population decline during the late Middle Woodland, and then a period of near abandonment of both the site and drainage during the early Late Woodland. This change seemingly parrots a decline in interregional exchange and ceremonial activity, as evident in diminished construction of both earthen and rock mounds. Not until the end of the Late Woodland are there indications of increased population and regional order, apparently against a backdrop of conflict as signaled by palisaded villages.

For the purposes of this study, the timing of the apparent Middle to Late Woodland transition is vital. Our initial pass through the Hickory Log data suggested to us that a decline in ceramic homogeneity and a population shift were occurring by at least A.D. 400^{C14}, well in advance of A.D. 536^{cal}. This picture changed, however, when the effects of calibration of radiocarbon dates to calendar dates were taken into account. During the period between A.D. 300 and 600^{C14}, calibration adds up to 140 years to radiocarbon dates. Thus, an uncalibrated date of A.D. 385^{C14}, near the traditional, radiocarbon-based, end of Cartersville, becomes the hypothesized A.D. 536^{cal} cultural event.

We have not yet established a data base of calibrated Woodland dates from our region like that of Wetmore et al., and thus have not attempted a systematic reappraisal of the effects of calibration on the traditional Woodland sequence. However, while the overall duration of the Cartersville phenomenon does not appear to change, it appears obvious that the "end of Cartersville," which is traditionally placed sometime around A.D. 400 to 500^{C14}, must be pushed forward about 100 years in calibrated time. If the Middle-Late Woodland transition is to be examined in the broader context of historical events and global environment, minimally dates have to be calibrated. Other temporal refinements can follow.

We entered this study feeling that there was no apparent correlation between A.D. 536^{cal} and the timing of cultural change, indirectly siding with Caldwell when he said in the 1950s that we may never know what happened at the end of Cartersville. However, we now have at least a temporal correlation between the atmospheric event and the decline of the previously stable Cartersville cultural phenomenon. Similarly, the subsequent period of apparent colder climate seems to correspond with a period of increased ceramic localization and population movement/ dispersion throughout this region. Like many of the other studies recounted in this book, notable cultural change at A.D. 536^{cal} appears to have been presaged by lesser shifts of cultural emphasis (the Late Cartersville Phase) by a couple of centuries (see Houston, this volume).

As we noted at the beginning of this chapter, correlation is only the first step in the search for culture-climate relations. It is now clear that, though we still have miles to go, we have to consider global climate factors as one of the influences in significant cultural changes. As part of this consideration, those of us working primarily in cultural resource management need to actively integrate climatic studies in our evaluation of site significance, and not relegate them to a short section at the end of our environmental overview. We also wish to suggest increased consideration of calibration and calendrical dates in the construction of our cultural sequences. Through tree ring and sea level studies, we are now gathering increasingly precise data on past climates, while oxidized carbon ratio dates and other non-radiocarbon based techniques are providing us with calendrical results that are not expressed in radiocarbon years. If we are to interpret accurately our archaeological data in light of these data, we must alter our approach. Perhaps it is time for a concerted effort to construct a systematic, calibrated chronology of the Woodland and Mississippian Southeast in which we look anew at received wisdom of the timing of events in prehistory. Such a recalibration of our approach will provide us with a more precise backdrop in which to investigate both environmental and cultural change.

References

Anderson, David G. (1985) Middle Woodland Societies on the Lower South Atlantic Slope: A View from Georgia and South Carolina. *Early Georgia* 13 (1 & 2):29-66.

Baillie, M. G. L. (1995) Patrick, Comets, and Christianity. *Emania* 13:69-78.

Brose, David S., and N'omi Greber (1979) *Hopewell Archaeology: The Chillicothe Conference.* Kent State University Press, Kent, Ohio.

Caldwell, Joseph R.1957) *Survey and Excavations in the Allatoona Reservoir, Northern Georgia.* Unpublished Manuscript on file at the Laboratory of Archeology, University of Georgia, Athens.

Caldwell, Joseph R. (1958) *Trend and Tradition in the Prehistory of the Eastern United States.* Memoirs of the American Anthropological Association 88.

Garrow, Patrick H. (1975) The Woodland Period North of the Fall Line. *Early Georgia* 3(1):17-26.

Gregg, Susan A., (Editor) (1991*Between Bands and States.* Center for Archaeological Investigations, Southern Illinois University at Carbondale Occasional Paper No.9.

Jefferies, Richard W. (1976) The Tunacunnhee Site: Evidence of Hopewell Interaction in Northwest Georgia. *Anthropological Papers of the University of Georgia* Number 1. Department of Anthropology, Athens.

Milner, George R., Virginia G. Smith, and Eve Anderson (1991) Conflict, Mortality, and Community Health in an Illinois Oneota Population. In *Between Bands and States,* edited by Susan A. Gregg. Center for Archaeological Investigations, Southern Illinois

University at Carbondale Occasional Paper No.9:245-264.

Rudolph, Teresa P. (1991) The Late Woodland "Problem" in North Georgia. In *Stability, Transformation, and Variation: the Late Woodland Southeast,* Plenum Press, New York, pp. 259-283.

Tanner, William F. (1992) Late Holocene Sea-Level Changes from Grain-Size Data: Evidence from the Gulf of Mexico. *The Holocene* 2:164-170.

Tanner, William F. (1993) An 8000-Year Record of Sea-Level Change from Grain-Size Parameters: Data from Beach Ridges in Denmark. *The Holocene* 3:220-231.

Wauchope, Robert (1966) *Archaeological Survey of Northern Georgia.* Memoir 21, Society for American Archaeology.

Williams, Mark (1994) Archaeological Site Distributions in Georgia: 1994. *Early Georgia* 22(1):35-76.

Wood, William Dean (1981) *An Analysis of Two Early Woodland Households from the Cane Island Site, 9PM209.* Wallace Reservoir Project Contribution Number 4, Department of Anthropology, University of Georgia.

Wood, William Dean, and William R. Bowen (1995)Woodland period Archaeology of Northern Georgia. *Georgia Archaeological Research Design Paper No.* 9, University of Georgia, Laboratory of Archaeology Series, Report Number 33.

Wood, William Dean, and R. Jerald Ledbetter (1990) *Rush: An Early Woodland Site in Northwest Georgia.* Prepared for the Georgia Department of Transportation. Southeastern Archaeological Services, Inc., Athens, Georgia.

Table 1. The Northern Georgia Woodland Sequence (1,000 B.C.-A.D. 1000^{C14})

DATE RANGE		PERIOD	PHASE
A.D. 750	A.D. 1000	LATE WOODLAND	Woodstock
A.D. 500	A.D. 750		Late Swift Creek/Napier
300 B.C.	A.D. 500	MIDDLE WOODLAND	Cartersville/Early Swift Creek
1,000 B.C.	300 B.C.	EARLY WOODLAND	Dunlap

13. WOODLAND ADAPTATIONS IN THE APPALACHIAN SUMMIT OF WESTERN NORTH CAROLINA: EXPLORING THE INFLUENCE OF CLIMATE CHANGE

Ruth Y. Wetmore, Kenneth W. Robinson, and David G. Moore

Appalachian Mountain Environment and Paleoclimate

[Appalachians, Appalachian Summit; climate change, cultural area, rainfall, microenvironments, physiography, thermal belts; Pleistocene-Holocene vegetation sequence, tree rings; Holocene, A.D. 570-950^{cal}]

The A.D. 536 event, documented in climatic and historic records in numerous locations around the world, initiated a cooling trend that coincided in many places with cultural changes. To identify and evaluate the event's influence on Appalachian Summit cultures, we consider regional climatic conditions first. After posing alternative cultural strategies for dealing with abrupt climate change, we examine the regional archaeological record for evidence of correlation between cultural changes, and A.D. 536^{cal} and its aftermath. Pertinent data include regional radiocarbon dates, Woodland period subsistence practices, and settlement patterns from two western North Carolina surveys. This combination of information on when people lived in the mountains, where they lived, and what they ate provides a basis for assessing both their probable and observable adjustments to sudden environmental change and colder global climate and its local expression.

The Appalachian Mountain region of western North Carolina is bounded on the east by the Blue Ridge Mountains, and on the west by the Unaka and Great Smoky Mountains (Figure 13.1). This 6,000-square mile area contains the highest peaks in the Appalachian Mountain system (Stuckey 1965:19-20). Kroeber (1939:95) referred to the region as the "Appalachian Summit." Elevations in western North Carolina range from 1,000 feet above mean sea level (amsl) along the southern and eastern edges of the Blue Ridge to a high of 6,684 feet amsl on Mount Mitchell. Valley floors in the region are generally between 1,000 to 3,000 feet amsl. Given this variation, the Southern Appalachians possess a range of climates and environments as diverse as the habitats found from Georgia to Newfoundland, and encompass most forest environments that occur in the eastern half of the continent (Sutton and Sutton 1986).

Altitude is the single most important source of climatic variability in the region, and has an important effect on rainfall (Carney 1960:2). The annual rainfall averages more than 80 inches in some areas: the highest in the eastern United States. In winter, the mountain ranges moderate the weather by forming a partial barrier to cold waves moving southeast from the Midwest, either by deflecting them south, or decreasing their intensity during ascent and passage on the western slopes. Mean temperatures in the Southern Appalachians range from 36-42° F. in January and 68-78° F. in July, providing a frost-free season of 150-190 days (Lee 1955:9-12).

Numerous mountain microenvironments are fostered by diverse combinations of landform, elevation, aspect, slope, vegetation, and precipitation. Thermal belts are an important microenvironment generating feature of high relief terrain. First described in the Appalachians by Silas McDowell in the 1880s (Chickering 1884), thermal belts are small mountain slope areas with considerably higher minimum temperatures than adjacent areas. Caused by cold air drainage to lower valleys and an inversion of warm and cool air, thermal belts in the Appalachian Mountains are centered at an average of 1150 feet (350 meters) above valley floors (Cox 1923). Temperatures on valley floors can fall from 15 to 30° F. below those in a thermal belt up slope (Ward 1925:130). Other studies show that growing seasons are one to two weeks longer in thermal belts and that snow melt is earlier (Barry 1992:218).

Climatic studies indicate that during the Pleistocene (Ice Age) era, the Appalachian Mountains were boreal forests and perhaps tundra at higher elevations (Whitehead 1973). Increased rainfall and warming during the late Wisconsin glacial period caused deciduous (hardwood) trees to replace conifer forests as the dominant overstory species between 14,500 and 10,500^{c14} B.C. (Delcourt and Delcourt 1985:19). Essentially modern climate and forest conditions were established by 4,000^{c14} years ago. For most of the Holocene (11,000^{c14} B.C.-present), temperate forests of oak, chestnut and hickory have predominated in the Southern Appalachian Mountains (Shelford 1963:18). At Flat Laurel Gap near

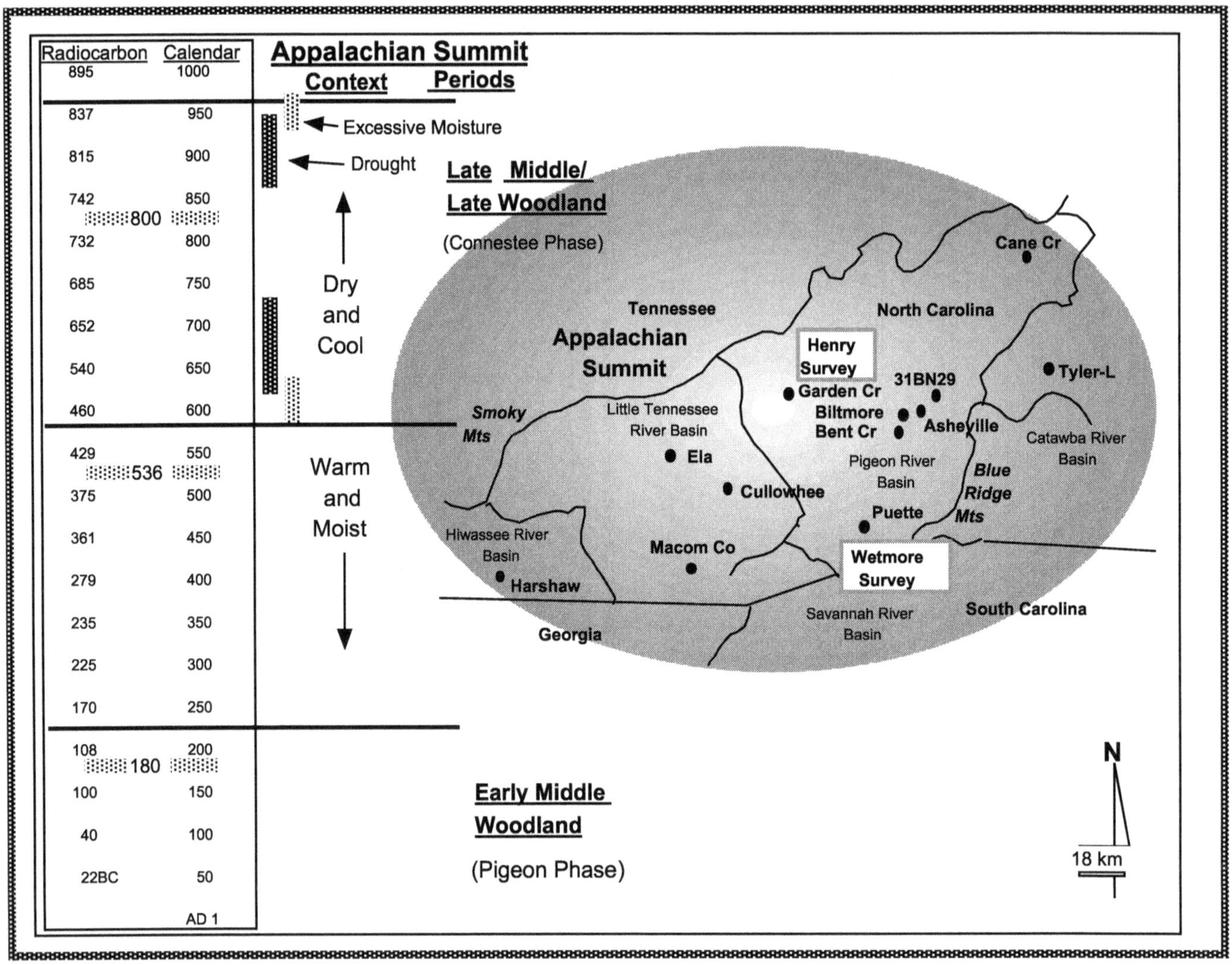

Figure 1. Map and Chronology of the Appalacian Summit.

Mount Pisgah, Shafer (1984:72) cites the late Holocene cooling trend in the southern Appalachians as a probable factor in the persistence of relict boreal plants to the present.

More recent studies provide fine scale climatic information for the Woodland period, which in the Appalachian Summit dates from around 700 B.C.-A.D. 1000^{c14} (Keel 1976:17-19). A global study based on glacial advance and retreat cycles indicates warm global climate conditions between 500^{c14} B.C. and A.D. 600^{c14}, followed by colder weather between A.D. 600-900^{c14} (Denton and Karlen 1973). Due to its sensitivity to the major climate-determining factors, sea level also is used as a measure of global temperatures (Tanner, Chapter 2).

Prehistoric North Carolina climates have been reconstructed from baldcypress tree rings to before and after A.D. 536^{cal}. Because tree ring width is a measure of spring/early summer rainfall, it has a high degree of correlation with corresponding historic data (Stahle et al. 1988:1518). The tree ring data indicate a series of statistically significant alternating dry and wet periods on several time scales, with a general pattern of paired 30-year cycles of wet and dry conditions. Extended wet periods (see Figure 13.1) were observed between A.D. 570-630^{cal} and between A.D. 950-1040^{cal}, with longer periods of drought between A.D. 630-730^{cal} and between A.D. 890-950^{cal} (Stahle et al. 1988:1518). On the basis of these climatic studies, the southern Appalachians would have been characterized by warm and wet weather from about 500 B.C.-A.D. 600^{cal}, followed by generally colder and drier conditions from about A.D. 600-900^{cal}.

Biological and Cultural Responses to Climatic Change

[Appalachians; ecology, tree rings; plant adaptations, cultural adaptations, drought and cooling, ecotonal shifts; Holocene, Woodland period, A.D. 372-700^{cal}]

The presence of many diverse microenvironments would probably reduce the impact of short-term changes in temperature on humans, although long-term changes in the environment could require greater adaptive flexibility. Climatological records indicate that the mountain growing season of about 150-days is half as long as that of the North Carolina coastal plain (Clay et al. 1975:100). Sustained cooler weather could reduce the plant resources available to humans and consequently preferred game animals. It is difficult to determine whether prolonged episodes of colder weather would have caused significant change in plant communities. Deciduous nut trees—hickory, chestnut, walnut, and acorns—which supported mountain human and game animal populations during the Woodland Period, require some winter cold for best crop development, although water shortages during the growing season adversely affect nut productivity as well (Yearbook of Agriculture 1941:415).

Plant responses to climatic change range from adaptation to migration/colonization or extinction. While changes in vegetation can lag behind climatic changes (Pacala and Hurtt 1993:59), some species exhibit rapid adaptation to changes in moisture, temperature, and even pesticides (Geber and Dawson 1993:181). The great temporal range of numerous plant species testifies to their ability to survive drastic climatic change if the locations are favorable. When Pleistocene ice sheets expanded, North American forest species migrated down the Appalachian Mountain chain and persisted in refugia along the Gulf Coast and in Mexico, while their European counterparts became extinct because routes to appropriate habitats in other localities were blocked (Geber & Dawson 1993:182).

Present-day mountain vegetation is adapted to generally high volumes of rainfall. The effects of the periodic droughts noted above during the Woodland Period are difficult to assess, since the magnitude of change and seasonal distribution from present-day annual rainfall is unknown and the paleoclimatic data remain rare. Stahle et al.'s (1988) coastal plain tree ring data for the years A.D. 372-700^{cal} are based on small samples of two to ten rings. The few pollen series in the mountains are not sufficiently resolved to show if precipitation decreased during or after the A.D. 536^{cal} event (Delcourt and Delcourt 1985; Shafer 1984). Furthermore, one researcher warns that "still life [e.g., pollen analysis] reconstructions fail to do justice to the dynamic processes active in any environmental setting" (Hammett 1986:1).

We will assume that the pollen and tree ring studies described above accurately reflect prehistoric conditions in the Appalachian Summit, and consider the effects that long-term drought (Stahle et al. 1988) and cooling (Denton and Karlen 1973) would have on the region's tribal societies. We can suggest that the availability of diverse microenvironments would reduce the impact of short-term temperature and rainfall changes. Vertical movement would figure into seasonal collecting rounds, as specific wild plant foods mature weeks apart in floodplain and upland settings. MacArthur (1972) calculated that each 1° C. increase in temperature corresponds to about 100 kilometers of latitude or 150 meters of elevation.

Long-term changes in the environment could require more decisive cultural responses. If drought or sustained cooler weather reduced the available plant resources, it would probably be reflected in diminished size or number of settlements, or shifts to more favorable locations. Schneider and Lander (1984:372-373) suggest the following archaeologically visible cultural adaptations to reduce vulnerability to environmental stress:

1. Diversification rather than specialization, i.e., reliance on a few plants or animals;

2. Storage of food;

3. Conversion of surplus food into valuables to be stored and traded for food as needed;

4. Establishment of social relationships that provide access to food resources of other regions.

We will evaluate these four cultural adaptations to environmental stress against pertinent aspects of Appalachian Summit Woodland cultures during the study period.

Appalachian Summit Woodland Cultural Patterns

The cultural factors we examine include subsistence practices, settlement patterns, and evidence of inter-regional contacts. Survey and excavation data, ceramics, and radiocarbon dates are used to compare diachronic climatic and cultural changes. Precise correlations were not anticipated, but the intersection of cultural patterns with environmental patterns can provide landscape traces indicating directions for further research. While the entire Woodland era is considered in general (Table 13.1), we emphasize the Middle/Late Woodland Connestee Phase during which the A.D. 536^{cal} event occurred.

Subsistence Practices

[Appalachians, Sandymush Creek, Newfound Creek; surface survey, landform/terrain analysis, ethnobotany, subsistence analysis; nuts, fruits, small grains; Early Woodland, Swannanoa, Middle Woodland, Pigeon, Connestee, Late Woodland]

Current information indicates that Woodland subsistence practices encompassed a wide range of habitats. Woodland period regional overviewers have inferred "a broad adaptation to local resources" from the presence of Early Woodland Swannanoa and Early Middle Woodland Pigeon phase occupations in all microenvironments (Bass 1977:81; Keel 1976:231; Purrington 1983:132, 136). This pattern of utilizing diverse upland and lowland habitats continued throughout the subsequent Late Middle Woodland Connestee phase.

Available ethnobotanical evidence reveals systematic use of wild plant foods throughout the Woodland Period. The percentage of nutshells from Swannanoa features at the Warren Wilson site (31BN29, see Figure 13.1) was highest for acorn, followed by hickory and black walnut (Simpkins 1984; Yarnell and Black 1985:100). The same three species are present in Connestee features at the Ela (31SW5), Macon County Industrial Park (31MA185) and Harshaw Bottom (31CE41) sites. At Harshaw Bottom, hickory nutshell predominates, although acorn:hickory ratio calculations indicate that acorn was probably more important economically (Gremillion 1989:3,11). In addition to these three species, chestnut shell was present in F.108-A which provided a radiocarbon date of A.D. 654^{cor}. The analysis of fifteen pollen samples from the Bent Creek site (31BN335) revealed oak and chestnut pollen in fourteen and hickory pollen in twelve samples. The presence of *Ericaea* (huckleberry or blueberry) and *Rubus* (raspberry, blackberry or dewberry) was also identified in Middle Woodland contexts at this site, the later from unit 3 which produced a radiocarbon date of A.D. 746^{cor} (Fish 1990:8). Although not from a dated context, acorn, hickory nut, walnut, butternut, hazelnut, blackberry and polygonum seeds were recovered from at the Biltmore site (31BN174) (Oliver 1988:1-8).

In addition to wild plant foods, there is additional evidence for use of several domesticated small grain and seed crops during the Connestee Phase. Two small grain crops, little barley and maygrass, and species usable as greens, were present in Connestee feature fill at the Harshaw Bottom site with a date of A.D. 654^{cor} (Gremillion 1989:4). Chenopodium and other fruiting plants were recovered at the Tyler-Loughridge site (31MC139) (Robinson 1994). At the Bent Creek site (31BN335), other species considered economic indicators because of their quantity or aggregation included *Typha* (cattail), *Tubiflorae* (sunflower family), and *Graminae* (grass family) (Fish 1990:4-6).

Less faunal data are available, due in part to poor bone preservation in generally acidic soils. Examination of Connestee feature contents at the Cane Creek (31ML3), Ela (31SW5), and Macon County Industrial Park (31MA185) sites revealed the presence of deer, black bear, small mammal, turtle, and bird bones. Fish hook blanks, fish vertebrae beads and fresh water shell beads at the Cane Creek site (31ML3) suggest additional dietary resources (Keel and Egloff 1984:36-38). Remains of white-tailed deer, black bear, opossum, box turtle and turkey were identified at the Biltmore site (31BN174) (Holm n.d.:9), but contexts of the finds are not known.

Settlement Patterns: Time Trends

[Appalachian Summit, North Carolina, Buncombe County, Madison County, Transylvania County; graphing, time series; decreasing site frequencies; Early Woodland, Middle Woodland, Late Woodland]

Both the location and frequencies of Woodland sites change over the course of the Woodland period. To refine our understanding of pattern changes, we examined the results of two archaeological surface surveys near the center of our study area. The Henry Survey includes the Sandymush Creek and Newfound Creek watersheds (see Figure 13.1), a 31,812-hectare area in Buncombe and Madison counties (Henry 1996:195). The creeks are tributaries of the French Broad River. Conducted over a nine year period, a few sites were revisited as many as ten times. The Wetmore Survey (1993) of Transylvania County (see Figure 13.1) provides comparative data to the Henry survey. It includes the headwaters of the French Broad and Savannah Rivers. The floodplain of the French Broad River near the Henry survey is about 1,800 feet amsl, while the Savannah River floodplain in the Wetmore survey area is about 2,100 feet amsl.

In general, Early Woodland sites in the region are found in all microenvironments. By Middle Woodland times, preference emerges for site locations on or near floodplains (Purrington 1983:136). Henry (1996) found 403 Woodland site components from four periods. (A site component is the occupation of one site for one period of time; thus a site that was occupied during both the Early and Middle Woodland periods has two components.) Site component frequencies decline over time in the Henry Survey area (Figure 13.2A). The 192 Early Woodland, Swannanoa phase components constitute 48 percent of the occupations, followed by 92 Pigeon components (24 percent), 76 Connestee components (19 percent), and finally drop to 36 Late Woodland components (9 percent) (Henry 1993: Table 1). Late Woodland occupations were identified by temporally diagnostic projectile points, not ceramics (Henry 1996:158). Most regional studies do not define a Late Woodland period (Bass 1966; Purrington 1983; Wetmore 1993).

Wetmore's analysis of settlement patterns was based on state site forms, and data from a recent archaeological survey conducted for the Transylvania County Historic Preservation Commission (Wetmore 1993). Of the 194 Woodland components in this study, 41 (21 percent) belonged to the early Swannanoa phase, 67 (34 percent) to the Pigeon phase, and 67 (34 percent) to the Connestee phase, a result somewhat the opposite of the Henry survey with regard to frequency trends. This comparison of the results of the two surveys could suggest a preference for more southerly and south facing contexts during the Connestee phase; this should be given focused attention in the future.

Further support for this assertion is found by comparing the Henry and Wetmore study results with to those in other areas. The Henry survey trend toward fewer occupations through time is similar to findings in the upper Watauga Valley (Purrington 1983:115-116). On the other hand, Transylvania County's increase in site frequency compares favorably to settlement pattern data from the Lower Little Tennessee River Valley (Davis et al. 1982:192). The Henry study area and the Lower Little Tennessee River valleys are relatively less elevated than Transylvania County and the upper Wataga Valley, so the importance of elevation needs to be given study.

Settlement Patterns: Terrain Trends

[Appalachian Summit, Tennessee; graphing, time series; landforms, stream ranks, exposure; Early Woodland, Middle Woodland, Late Woodland]

Three terrain characteristics of sites reveal something of subregional/survey area changes in settlement preferences: landform, stream rank, and site exposure (slope face). In the Henry survey, more than 60 percent of all Woodland components were found on lower toe slope landforms (Figure 13.2B), defined as the first rise from the floodplain to the adjacent upland (Henry 1996:149). Through time, the frequency of components on upland slope, bench, and ridge locations declined, while those on floodplains increased. Using approximately the same landform categories in Transylvania County, more than half of the Swannanoa and Pigeon components and 73 percent of the Connestee components were situated on first terraces or floodplains (Wetmore 1993). The tendency for Connestee to move to the floodplain is clearly visible in Wetmore's results.

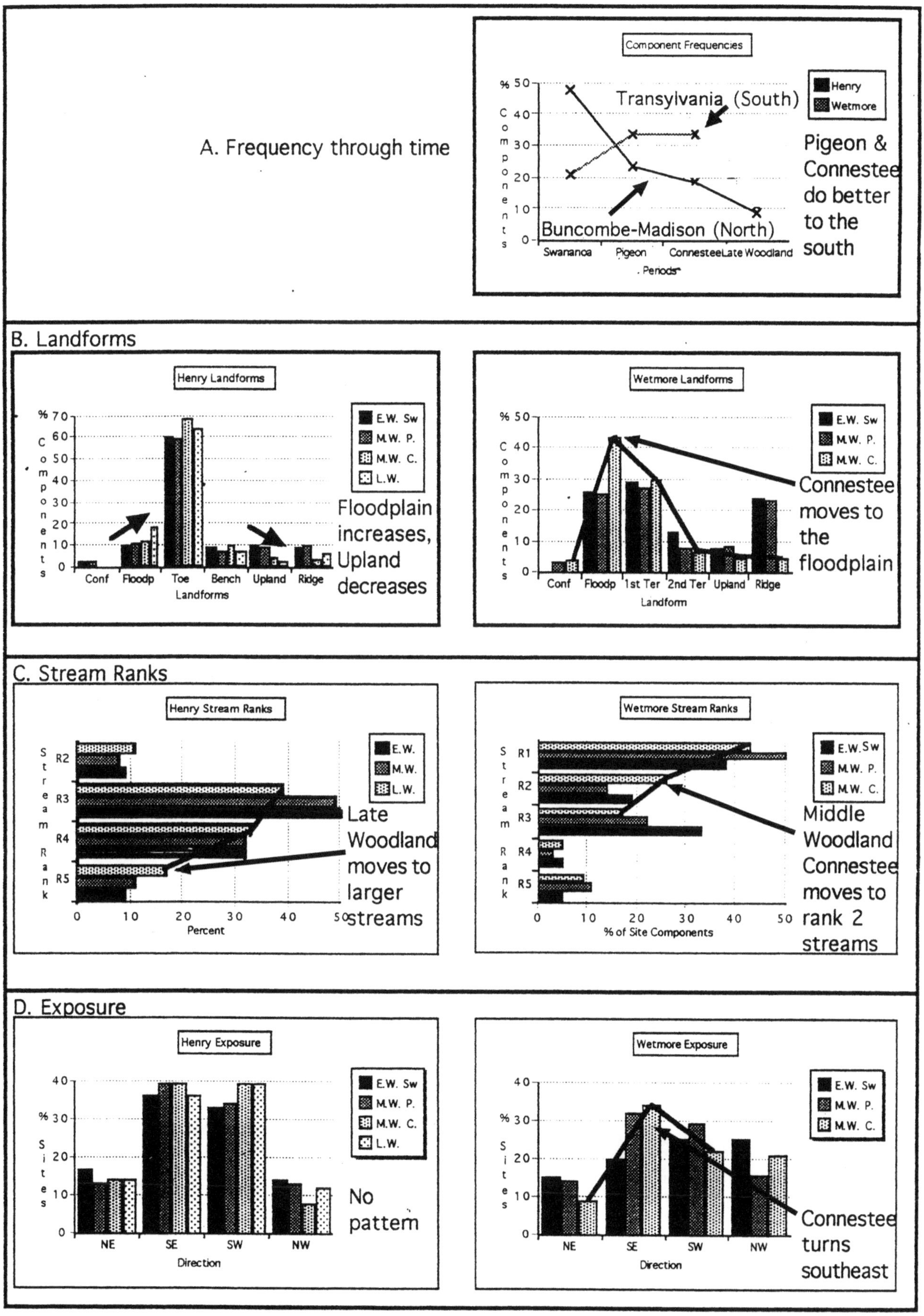

Figure 2. Site Frequency (A), Landforms (B), Stream Rank (C), and Exposure (D) through the Woodland Periods.

Stream size (rank 1 smallest to rank 5 largest) was analyzed although the two studies can not be compared directly; streams in the Henry survey were classified by the stream's rank at its mouth (Henry 1996:147), while Wetmore reported stream rank at the site. The Henry Survey finds Early and Middle Woodland patterns to be similar (see Figure 13.2C) while the Late Woodland sites are clearly located on larger and lower streams. Wetmore's study finds a shifting pattern in which Early Woodland sites on rank 1 and 3 streams converge on rank 2 streams during the Middle Woodland. This could indicate a focus on more reliable spring heads.

Site exposure (Figure 13.2D) in the Henry survey favored southeast and southwest orientations, while southeast facing slopes dominate the Wetmore study. The lack of change in the Henry survey results could be a result of the relatively restricted watersheds in which the survey was conducted. Wetmore's study, however, shows a change from a relatively indifferent site orientation during the Early Woodland to a preference for southeast orientation during the Middle Woodland Connestee phase. Southeast facing camp sites are generally interpreted as a preference for locations exposed to first warmth of the day.

The results of the terrain analyses minimally suggest a repatterning of settlement during the Middle and Late Woodland toward warmer southeast facing slopes and toward floodplains or lower terrace/slope locations. In Transylvania County they were more focused on reliable upland water courses at rank 2 streams. There appears to be a region-wide tendency toward more southerly circumstances. The standard interpretation of these shifts would be that Middle Woodland people were adjusting their habits to accommodate cooler and drier conditions. A shift toward occupation of higher elevations in Transylvania County needs to be examined as it would appear to contradict the pattern. This could provide an entrée to explorations of the thermal belts question. Obviously, a great deal of further work is needed to explore these tendencies including statistical analysis of the data examined graphically here. In order to define meaningful settlement patterns, additional factors need to be examined, such as site size, site function, and duration of occupation. Standardization of definitions and methodology is needed to make region-wide comparisons meaningful. This would also facilitate the use of data in studies that focus on watershed or drainages, which are more likely to reflect Woodland settlement patterns than surveys bounded by modern governmental boundaries.

Connestee Radiocarbon Dates

[Appalachian Summit; radiocarbon dating, calibration; Connestee ceramics, Ice House Bottom; Middle Woodland, Connestee, A.D. 300-900^{cor}, Late Woodland, Mississippian]

Radiocarbon dates obtained from western North Carolina research projects during the last two decades now clearly define the chronological range of the Connestee Middle/Late Woodland phase, which spans the A.D. 536^{cal} event (Figure 13.3). The 17 dates are calibrated to calendar dates (using version 3.0.3c of the CALIB program [Stuiver and Reimer 1993a, 1993b], and are shown with a 2-sigma error range). Most of the dates were obtained from archaeological contexts associated with ceramics identifiable to one of the Woodland phases.

The earliest two dates (A.D. 306^{cor} and A.D. 360^{cor}) are from pit features containing Connestee ceramics at the Tyler-Loughridge site just east of the Blue Ridge escarpment (Robinson 1994). These dates support Keel's (1976:19) original proposal of Connestee phase inception near a Connestee date of A.D. 397^{cor} from the Icehouse Bottom site (40MR23) in Tennessee (Chapman 1973; Eastman 1994b:88).

The largest cluster of dates extends from A.D. 544 to 746^{cor}, a range of about 140 years; note the steps in Figure 13.3 before and after these dates. They are followed by another cluster of four ninth century A.D. dates from three sites (A.D. 840-892^{cor}). The 11 dates in the large cluster represent eight widely separated sites in the region. The contexts of these dates indicate that Connestee ceramics were used from A.D. 300 to 900^{cor}. Connestee ceramics from geographically separated areas and different age contexts are virtually identical in general appearance, and preliminary analyses indicate no major differences by site in relative percentages of surface treatments. This range supports Keel's (1976:19) thesis that the use of Connestee ceramics continued from the Middle Woodland into an unnamed Late Woodland period that extended to A.D. 1000^{cor}.

Other radiocarbon dates help define the Woodland Period terminus and the inception of the Mississippian Period. Few archaeological correlates have been defined for this transition, in this or other regions of the Southeast (Purrington 1983; Schroedl et al. 1990; Rudolph 1991). Currently, A.D. 1028^{cor} from the Tyler-Loughridge site (31MC139) is the earliest Pisgah Phase (Mississippian Period) date yet obtained in western North Carolina (Robinson 1994). Other dates from the transition inspire less confidence. A date of A.D. 1005^{cor} from Bent Creek (31BN335) was associated with mixed ceramics that included Woodland sherds and unidentified specimens (Baker and Hall 1990:46). A postmold at the Macon County Industrial Park site produced the date of A.D. 1020^{cor}, but was not associated with any cultural materials (Western Carolina University Archaeological Laboratory Site Records; Eastman 1994a:53). A date of A.D. 1162^{cor}, was obtained from contexts containing Mississippian Napier-like ceramics at the Cullowhee Valley site (Moore 1994), but contamination from the prevalent Late Prehistoric material at the site is suspected.

Associations of Connestee with Other Regional Ceramic Types

[Appalachian Summit; ceramic typology; Connestee, Candy Creek, Napier, Swift Creek; Middle Woodland, Connestee]

Further evidence for long usage of Connestee ceramics is provided by associated ceramic types from other regions. These ceramic types from other regions span several periods and thus verify that Connestee ceramics persisted while ceramic designs in other regions came and went. The Middle Woodland limestone-tempered Candy Creek ceramic type of eastern Tennessee dating between A.D. 200-600[cal] (Lewis and Kneberg 1946; Schroedl et al 1990:178) is the earliest type found in direct association with Connestee ceramics. In the Appalachian Summit (see Figure 13.1), Candy Creek sherds were recovered from Connestee excavation units and features at Harshaw Bottom (Robinson 1989) and from Connestee features at the Ela site (31SW5) (Wetmore 1990). Connestee and Candy Creek ceramics also are found together at several southeastern mound sites that participated in the Hopewell Interaction Sphere. These include Garden Creek site (31HW2) in North Carolina (Keel 1976), Icehouse Bottom (31MR23) in Tennessee (Chapman 1973, Chapman and Keel 1979:118), the Tunacunnhee site (9DD25) in northwest Georgia (Jefferies 1976), and the Walling site in Alabama (Walthall 1985:245). Thus, the exchanges that mixed sherds from various adjoining regions were reciprocal; the makers of Connestee ceramics were not isolated in the mountains.

Following the collapse of the Hopewell Interaction Sphere, complicated stamped Swift Creek ceramics (Wauchope 1966) from Georgia reveal contacts between that region and the Appalachian Summit (see Lilly and Webb, this volume). A radiocarbon date of A.D. 654[cor] was obtained from a feature at the Harshaw Bottom site containing both Connestee and Swift Creek sherds (Robinson 1989). Connestee and Swift Creek sherds were associated in features at the Ela site (Wetmore 1990:157), and are present at the Garden Creek (Keel 1976:116), Macon County Industrial Park, and Puette sites (Holden 1966:93; Southerlin and Espenshade 1991).

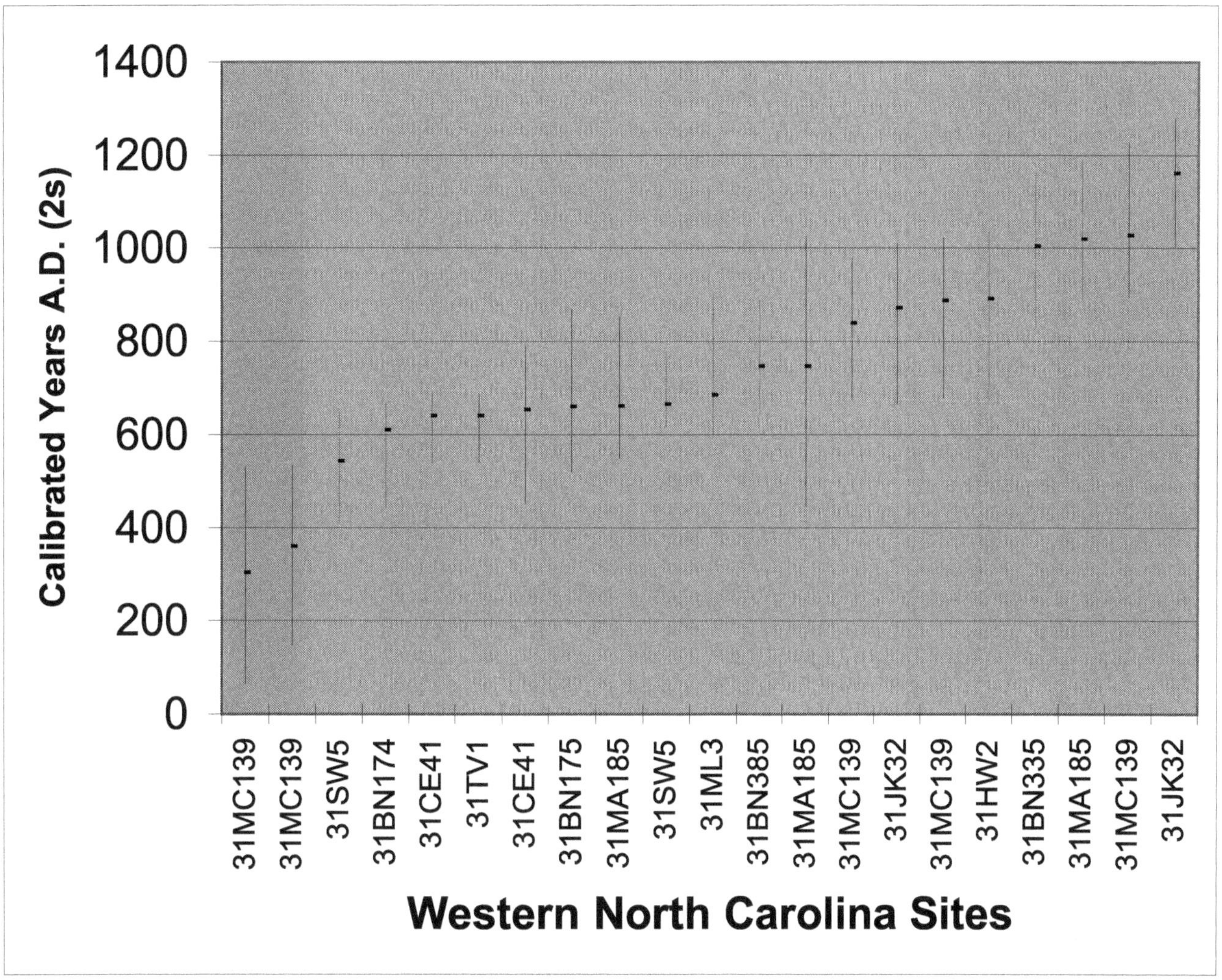

Figure 3. Western North Carolina Connestee Radiocarbon Dates.

Napier ceramics (Wauchope 1966:57-60), somewhat later than Swift Creek but also from Georgia, are reported with two dates obtained from Cullowhee Valley site features of A.D. 872[cor] and A.D. 1162[cor] (Moore 1994). The earlier date places Napier ceramics with most regional sequences, but the latter one is believed to postdate the Woodland Period. Napier sherds were present at Garden Creek (Keel 1976:116) and Napier-like sherds were associated with Connestee sherds in two excavation units at Harshaw Botton (Robinson 1989).

Complicated stamped sherds documented at other Woodland sites in western North Carolina could be variations of the

Swift Creek type, or represent local variations of a Late Woodland complicated stamped type as described by Keel (1976:48) and Dickens (1976:12, 184, 198) who suggest a derivation from the Napier and Woodstock stamped types of northern Georgia. For example, a radiocarbon date of A.D. 600^{cor} was obtained from a feature at 31BN175 containing several rectilinear complicated stamped sherds identified as Connestee (Hall and Baker 1993:13). It is also possible that some complicated stamped sherds belong to later ceramic types such as Woodstock or Lamar.

Connestee external associations with the Candy Creek, Swift Creek, and Napier types thus support the contention that Connestee endured for six hundred years, from about A.D. 300 to 900^{cor}.

Evidence of Climate-Related Cultural Change

[Appalachian Summit; subsistence; Hopewell Interaction Sphere; long distance trade; Middle Woodland]

Having examined the expected climatic effects of the A.D. 536^{cal} event and regional Woodland cultural patterns, we return to the proposed cultural responses to climatic change enumerated earlier, to see if any cultural shifts correspond to this climatic event.

(1) Subsistence throughout the Woodland period in the Appalachian Summit was based on utilization of diverse ecological microenvironments (Bass 1977:181; Keel 1976:231; Purrington 1983:136). Connestee features providing ethnobotanical information show reliance on hunting and a broad variety of wild plant foods. Woodland features with radiocarbon dates before and after A.D. 536^{cal} do not show changes in food remains, or increased diversification in subsistence activities. While most of the subsistence data comes from larger sites located on or above major stream floodplains, there is no evidence of intensified horticulture or greater use of aquatic resources.

(2) With regard to food storage, Connestee ceramics show no change in vessel form or paste composition throughout this cultural phase. If storage became more important after the A.D. 536^{cal} event, either changes in ceramic form/composition or evidence of increased storage from excavated features might be expected to support this functional adjustment. At present, neither can be demonstrated.

(3) If surplus food was converted into durable valuables stockpiled for time of need, one would expect increased frequency of raw material, tool, or ceramic caches following the A.D. 536^{cal} event. Long distance trade with other regions existed during Connestee times as discussed below, but whether subsistence and status goods were received for Appalachian materials has not been determined.

(4) The establishment of social relationships with other areas is well documented for at least two periods during the Connestee Phase. Participation of Appalachian Summit populations in the Hopewell Interaction Sphere prior to the A.D. 536^{cal} event is documented by the presence of Ohio Hopewell-type sherds, copper, and prismatic blades of Flint Ridge, Ohio chalcedony at Garden Creek (Keel 1976:118-120, 136-40). A single prismatic blade of this chalcedony was recovered also at 31BN174 (Moore 1994) and Connestee ceramics are found at Hopewell interaction sites in other physiographic provinces.

Centuries later, between A.D. 500 and 750^{cor} long distance trade to the west and southwest, is inferred from the presence of Swift Creek and Napier sherds at Appalachian Summit Connestee sites. The date of A.D. 654^{cor} from the Harshaw Bottom site could reflect the operation of social relationships following the A.D. 536^{cal} event, but additional dated contexts containing Swift Creek sherds are needed to determine if this interaction existed prior to the A.D. 536^{cal} event.

The distribution of dated Woodland sites shows that 65 percent of them fall between A.D. 544^{cor} and A.D. 746^{cor}. Intervals of a century or more appear between A.D. 360^{cor} and 544^{cor}, A.D. 746^{cor} and 840^{cor}, and the last Connestee date of A.D. 892^{cor} and first Mississippian date of A.D. 1005^{cor} (Figure 13.3). The continuity of radiocarbon dates throughout the Appalachian Summit postdating the A.D. 536^{cal} event suggests little or no disruption that could be linked to drastically changed climatic conditions. The wide geographic distribution of dated sites between A.D. 550 and 750^{cor} cuts across the ranges of warm to cool and wet to dry cycles documented in glacial and tree ring patterns and may indicate that mountain microenvironments buffered the effects of extended climatic shifts.

The impact of long-term climatic cooling would undoubtedly be more serious for populations dependent on cultivated crops than for groups with diversified hunting/gathering strategies. It is possible that the cooler and drier conditions of the early seventh century might have delayed the introduction of maize into the Appalachian Summit, despite its presence in eastern Tennessee around A.D. 200^{c14} (Chapman and Crites 1987:352).

The long duration of Connestee ceramics throughout the Appalachian Summit and adjacent areas implies considerable stability. This does not mean that Woodland cultures were static. There is abundant ceramic evidence of interactions with other regions; Dickens (1980) argued that Appalachian Summit participation with the Hopewell Interaction Sphere encouraged the adoption of new traits that could be seen in Connestee ceramic variability. New technologies appearing in the region included the bow and arrow (Blitz 1988; Cobb and Nassaney 1995:208), which generally coincided with a shift from formal standardized core to expedient core technologies (Parry and Kelly 1987:288-289). Two other innovations found elsewhere in the Southeast at this time—maize agriculture and shell-tempered pottery—do not appear in the Appalachian Summit until the Mississippian Period.

To summarize, the series of Appalachian Summit radiocarbon dates shows a concentration of Connestee dates between A.D. 550 and 700^{cor}, with use of Connestee ceramics extending several centuries before and after this range. Dated Woodland complicated-stamped ceramics in Connestee contexts affirm the duration of this ceramic tradition and place it firmly in a regional context. Two studies of Woodland site distributions show quite different settlement patterns, and specific areas for further study were suggested.

The combination of climatic and cultural studies (Gunn and Adams 1981; Bryson 1994) opens a promising new research area. The contributions of archaeology to this interdisciplinary endeavor are obvious. While we are better able to address the question of the A.D. 536[cal] event on Woodland cultures than even ten years ago, this preliminary consideration has identified no cultural shifts that coincide or appear related to this event. Understanding climatic/cultural interactions will require more data and more types of data. Some can be supplied by existing methods: additional radiocarbon dates, subsistence data, and detailed regional studies. Pollen and charcoal analyses similar to those conducted in the lower Little Tennessee River (Cridelbaugh 1984) will help clarify the specific roles of climatic conditions and human impacts on the landscape. The challenge is to develop new tools to meet the scope of questions we are now asking.

ACKNOWLEDGMENTS: The authors thank Gary Henry for permission to use selected data from his survey of the Newfound and Sandymush Creek watersheds in settlement pattern comparisons. Linda G. Hall provided ethnobotanical data and radiocarbon dates from her work at 31BN175 and 31BN335. Mr. and Mrs. Ed Wilson and Mr. Steve Wilson permitted testing at the Cane Creek site (31ML3) to obtain a radiocarbon for this important site.

References

Baker, C. Michael and Linda G. Hall (1990) *The Bent Creek Site: A Woodland Traditional Settlement within the French Broad River Basin.* Report on file at Office of State Archaeology, Raleigh, North Carolina.

Barry, Roger G. (1992) *Mountain Weather and Climate.* 2nd edition. Routledge, New York.

Bass, Quentin. R., II (1977) *Prehistoric Settlement and Subsistence Patterns in the Great Smoky Mountains.* NPS Contract No. CX500050211 between the National Park Service and the University of Tennessee.

Blitz, J. H. (1988) Adoption of the Bow in Prehistoric North America. *North American Archaeologist* 9:123-145.

Bryson, Reid A. (1994) On Integrating Climatic Change and Culture Studies. *Human Ecology* 22:115-128.

Carney, Charles B. (1960) *Climates of the States: North Carolina.* U.S. Department of Commerce Weather Bureau Climatology of the United States No. 60-31. Government Printing Office, Washington, D.C.

Chapman, Jefferson (1973) *The Icehouse Bottom Site, 40MR23.* Department of Anthropology, University of Tennessee, Report of Investigations No. 13.

Chapman, Jefferson, and Gary D. Crites (1987) Evidence for Early Maize (Zea Mays) from the Icehouse Bottom Site, Tennessee. *American Antiquity* 52:352-354.

Chapman, Jefferson, and Bennie C. Keel (1979) Candy Creek-Connestee Components in Eastern Tennessee and Western North Carolina and Their Relationship with Adena-Hopewell. In *Hopewell Archaeology: The Chillicothe Conference*, pp. 157-161, edited by D. Brose and N. Greber, Kent State University Press, Kent, Ohio.

Chickering, J. W., Jr. (1884) Thermal belts. *American Meteorology Journal* 1:213-218.

Clay, James W., Douglas M. Orr, Jr., and Alfred W. Stuart (1975) *North Carolina Atlas: Portrait of a Changing Southern State.* University of North Carolina Press, Chapel Hill.

Cobb, Charles R., and Michael S. Nassaney (1995) Interaction and Integration in the Late Woodland Southeast. In *Native American Interactions: Multiscalar Analyses and Interpretations in the Eastern Woodlands*, edited by M. Nassaney and K. Sassaman, pp. 225-226, University of Tennessee Press, Knoxville.

Collins, Susan M. (1977) *A Prehistoric Community at the Macon County Industrial Park.* North Carolina Archaeological Council Publications in Archaeology 2, Division of Archives and History, Raleigh, North Carolina.

Cox, H. J. (1923) Thermal Belts and Fruit Growing in North Carolina. *Monthly Weather Review*, Supplement, 19. Government Printing Office, Washington, D.C.

Cridelbaugh, P. A. (1984) American Indian and Euro-American Impact on Holocene Vegetation in the Lower Little Tennessee River Valley, East Tennessee. Ph.D. dissertation, University of Tennessee, Knoxville.

Davis, R. P. Stephen, Jr., Larry R. Kimball, and William W. Baden (editors) (1982) *An Archaeological Survey and Assessment of Aboriginal Settlement within the Lower Little Tennessee River Valley.* Report submitted to the Tennessee Valley Authority, Norris, Tennessee.

Delcourt, Hazel R., and Paul A. Delcourt (1985) Quaternary Palynology and Vegetational History of the Southeastern United States. In *Pollen Records of Late-Quaternary North American Sediments*, edited by V. M. Bryant, Jr., and R. G. Holloway, pp. 1-37, American Association of Stratigraphic Palynologists Foundation.

Denton, George, and Wibjorn Karlen (1973) Holocene Climatic Variations - Their Pattern and Possible Cause. *Quaternary Research* 3:155-205.

Dickens, Roy S. (1976) Cherokee Prehistory: The Pisgah Phase in the Appalachian Summit Region. University of Tennessee Press, Knoxville.

Dickens, Roy S. (1980) Ceramic Diversity as an Indicator of Cultural Dynamics in the Woodland Period. *Tennessee Archaeologist* 5:34-45.

Eastman, Jane M. (1994a) The North Carolina Radiocarbon Date Study (Part I). *Southern Indian Studies* Vol. 42.

Eastman, Jane M. (1994b) The North Carolina Radiocarbon Date Study (Part II). *Southern Indian Studies* Vol. 43.

Fish, Suzanne K. (1990) Results of the Bent Creek Site Pollen Analysis. Appendix C in *The Bent Creek Site: A Woodland Traditional Settlement within the French Broad River Basin*, by C. M. Baker and L. G. Hall.

Geber, Monica A., and Todd E. Dawson (1993) Evolutionary Responses of Plants to Global Change. Chapter 11 In *Biotic Interactions and Global Change*, edited by P. Kareiva, J. Kingsolver, and R. Huey, Sinnauer Associates Inc., Publications. Sunderland, Massachusetts.

Gremillion, Kristen J. (1989) Plant Remains from 31Ce41, Cherokee County, North Carolina. In Archaeological Excavations within the Alternate Pipeline Corridor Passing Through the Harshaw Bottom Site (31Ce41), Cherokee County, North Carolina by K. W. Robinson.

Gunn, Joel, and Richard E. W. Adams (1981) Climatic Change, Culture, and Civilization in North America. *World Archaeology* 13:87-100.

Hall, Linda G., and C. Michael Baker (1993) *Data Recovery at 31BN175: The Biltmore Estate, Buncombe County, North Carolina*. ER 93-7128. Report on file at Office of State Archaeology, Raleigh.

Hammett, J. E. (1986) Ethnohistory of Aboriginal Landscapes and Land Use in the Southeastern United States. M.A. thesis, University of North Carolina at Chapel Hill.

Henry, V. Gary (1996) A Longterm Site Survey of Sandymush and Newfound Creeks, Buncombe and Madison Counties, North Carolina. In *Upland Archeology in the East*, Symposium No. 6. Special Publication 38, pp. 145-186, Archeological Society of Virginia.

Holden, Patricia Padgett (1966) *An Archaeological Survey of Transylvania County, North Carolina*. Master's thesis, Department of Anthropology, University of North Carolina, Chapel Hill.

Holm, Mary Ann (n.d.) *Identification of Faunal Remains, 31BN174*. Report on file, Western Office, North Carolina Department of Archives and History, Raleigh.

Jefferies, Richard W. (1976) *The Tunacunnhee Site: Evidence of Hopewell Interaction in Northwest Georgia*. University of Georgia Anthropological Papers No. 1. Athens, Georgia.

Keel, Bennie (1976) *Cherokee Archaeology: A Study of the Appalachian Summit*. University of Tennessee Press, Knoxville.

Keel, Bennie, and Brian J. Egloff (1984) The Cane Creek Site, Mitchell County, North Carolina. *Southern Indian Studies* XXXIII 33:3-44.

Kroeber, A. L. (1939) *Cultural and Natural Areas of North America*. University of California Publications in American Archaeology and Ethnology 38. University of California Press, Berkeley.

Lee, William D. (1955) *The Soils of North Carolina: Their Formation, Identification and Use*. North Carolina Agricultural Experiment Station Technical Bulletin 115. North Carolina State College, Raleigh.

Lewis, T. M. N. and Madeline Kneberg (1946) *Hiwassee Island, an Archaeological Account of Four Tennessee Indian Peoples*. University of Tennessee, Knoxville.

MacArthur, R. H. (1972) *Geographical Ecology*. Harper and Row, New York.

Moore, David G. (1984) Biltmore Site (31BN174).

Moore, David G. (1994) Cullowhee School Site (31JK32).

Oliver, Billy (1988) *Ethnobotanical Analysis of Samples from 31BN174*. Report on file, Office of State Archaeology, Raleigh.

Pacala, Stephen W., and George W. Hurtt (1993) Terrestrial Vegetation and Climate Change: Integrating Models and Experiments. In *Biotic Interactions and Global Change*, edited by P. Kareiva, J. Kingsolver, and R. Huey, pp. 57-74, Sinauer Associates Inc. Publishers, Sunderland, Massachusetts.

Parry, William J., and Robert L. Kelly (1987) Expedient Core Technology and Sedentism. In *The Organization of Core Technology*, edited by J. Johnson and C. Morrow, pp. 285-304, Westview Press, Boulder, Colorado.

Purrington, Burton L. (1983) Ancient Mountaineers: an Overview of the Prehistoric Archaeology of North Carolina's Western Mountain Region. In *The Prehistory of North Carolina: an Archaeological Symposium*, edited by M. Mathis and J. Crow, North Carolina Diviskon of Archives and History, Raleigh, North Carolina.

Robinson, Kenneth W. (1989) Archaeological Excavations within the Alternate Pipeline Corridor Passing through the Harshaw Bottom Site (31CE41), Cherokee County, North Carolina. Clearinghouse Project Number CH-87-C-0000-0882, report on file, Office of State Archaeology, Raleigh, North Carolina.

Robinson, Kenneth W. (1994) *Archaeological Survey in McDowell County, North Carolina: 1988-90*. Warren Wilson College, Swannanoa, North Carolina.

Robinson, Kenneth W., David G. Moore, and Ruth Y. Wetmore (1996) Advances in Understanding Woodland Chronology and Settlement in the Appalachian Summit Region of Western North Carolina. Paper given at Conference on Integrating Appalachian Highlands Archaeology, New York State Museum, Albany.

Rudolph, Teresa P. (1991) The Late Woodland "Problem" in North Georgia. In *Stability, Transformation and Variation: The Late Woodland Southeast*, edited by M. Nassaney and C. Cobb, pp. 259-283, Plenum Press, New York.

Schneider, Stephen H., and Randi Lander (1984) *The Coevolution of Climate and Life*. Sierra Club Books, San Francisco.

Schroedl, Gerald F., C. C. Boyd, Jr., and R. P. Stephen Davis, Jr. (1990) Explaining Mississippian Origins in East Tennessee. In *The Mississippian Emergence*, edited by B. Smith, pp. 175-196, Plenum Press, New York.

Shafer, David S. (1984) Late-Quaternary Paleoecologic, Geomorphic, and Paleoclimatic History of Flat Laurel Gap, Blue Ridge Mountains, North Carolina. Master's of Science thesis, University of Tennessee, Knoxville.

Shelford, Victor E. (1963) *The Ecology of North America*. University of Illinois Press, Urbana.

Simpkins, Daniel J. (1984) *A Diachronic Ethnobotanic Study of the Warren Wilson Site (31Bn29), North Carolina*. Master's thesis, Department of Anthropology, University of North Carolina, Chapel Hill.

Southerlin, B. G., and Christopher T. Espenshade (1991) An Archeological Survey of the Brevard 44KV Transmission Line Corridor, With Limited Testing at the Puette Site (31TV1), Transylvania County, North Carolina. Report submitted to Duke Power Company, Charlotte, North Carolina.

Stahle, D. W., M. K. Cleaveland, and J. G. Hehr (1988) North Carolina Climate Changes Reconstructed from

Tree Rings: A. D. 372 to 1985. *Science* 240:1517-1519.

Stuckey, Jasper L. (1965) *North Carolina: Its Geology and Mineral Resources*. North Carolina Department of Conservation and Development, Raleigh.

Stuiver, M. and P. J. Reimer (1993a) Extended C14 Database and Revised CALIB Radiocarbon Calibration Program. *Radiocarbon* 35:215-230.

Stuiver, M., and P. J. Reimer (1993b) CALIB User's Guide Rev. 3.0.3. *Quaternary Research Center*, University of Washington, Seattle.

Sutton, Ann, and Myron Sutton (1986) *Eastern Forests*. Alfred A. Knopf, New York.

Walthall, John A. (1985) Early Hopewellian Ceremonial Encampments in the South Appalachian Highlands. In *Structure and Process in Southeastern Archaeology*, edited by R. Dickens, Jr., and H. T. Ward, pp. 243-262, University of Alabama Press, Tuscalusa.

Ward, Robert Decourcy (1925) *The Climate of the United States*. Ginn & Company, Boston.

Wauchope, Robert. (1966) *Archaeological Survey of Northern Georgia*. Memoirs of the Society for American Anthropology 21, Salt Lake City.

Wetmore, Ruth Y. (1990) The Ela Site (31SW5): Archaeological Data Recovery of Connestee and Qualla Phase Occupations at the East Elementary School Site, Swain County, North Carolina. Draft report on file, Western Carolina University Archaeology Laboratory, Cullowhee.

Wetmore, Ruth Y. (1993) *An Archaeological Survey of Transylvania County, North Carolina*. Report on file, Office of State Archaeology, Raleigh.

Wetmore, Ruth Y., K. W. Robinson, and David G. Moore (1996) Woodland Adaptations in the Appalachian Summit of Western North Carolina: Exploring the Influence of Climatic Change. Paper presented at Southeastern Archaeological Conference, Birmingham, Alabama.

Whitehead, Donald R. (1973) Late-Wisconsin Vegetation Changes in Unglaciated North America. *Quaternary Research* 3:621-631.

Yarnell, Richard A., and M. Jean Black (1985) Temporal trends indicated by a survey of Archaic and Woodland plant food remains from Southeastern North America. *Southeastern Archaeology* 4:93-106.

Yearbook of Agriculture (1941) *Climate and Man*. U.S. Department of Agriculture, Washington, D.C.

VII. CHANGES OF SEASONS: SLIPPAGE OF SEASONAL PATTERNS

This couplet of discussions focus on the middle elevations of the Atlantic Slope, the Appalachian Mountain/Blue Ridge Front piedmont. J. Ned Woodall works through a confluence of global climate, vegetation, and human interactions that could have transpired had there been a number of years of globally colder than usual spring weather following the A.D. 536 event. The sequence of impacts on human food collecting habits would have produced changes in settlement patterns on the landscape much like that observed, however tenuously, in the archaeological record for the upper Yadkin River region in North Carolina. From this model of global-local linkages between resources and decision making, Woodall develops a research agenda for future seasons of archaeological survey and excavation. Meanwhile, two hundred miles south along the Blue Ridge Front in the Savannah River valley, archaeological attention has been lavished on the Savannah River Plant, a sprawling area dedicated to the United State's nuclear power industry since World War II. Thanks to this surfeit of data, David Anderson is able to push the envelope of climate-culture linking to the greatest depth of detail anywhere in the southeastern United States. Anderson reviews the tendencies of cultural changes along the length of the Savannah River valley and in nearby regions. His chapter provides a synthesis of the foregoing chapters on the southeastern United States as well.

Woodall avers that, although there are limited numbers of well-dated sites, the period between about A.D. 550 and 750^{c14} appears to contain a significant change in Native American culture in the northwest North Carolina piedmont. Interriverine areas are largely abandoned in favor of floodplain settlements, sites become much larger and more sedentary, and ceramics change from small, relatively thin-walled bowls of Yadkin ceramics, to large, thick-walled storage jars of the Uwharrie type. The presence of Yadkin pottery in floodplain sites has encouraged a gradualist explanation of these changes. That is to say that people moved to the floodplain because of an ongoing process of cultural change from nomadic wild food gatherers to settled horticulturists. If instead this change was intertwined with a sudden severe climatic change, ceramics likely would not respond as quickly as settlement patterns, and age estimation of sites using gross ceramic categories for dating would obscure the sudden shift in lifeways. In the Yadkin floodplain, sites yield the remains of domesticated plants that do not appear on upland locales. These include characteristic maize and other plants of the eastern United States horticultural complex: they had been cultivated for at least 2,000 years in adjacent regions. Large storage pits and a variety of fish also are found in the apparently later sites. The A.D. 536 event could have stimulated the change by creating a set of "push-pull" forces. Slipping frosts to later in the spring would dramatically affect the production of mast crops such as acorns, killing white oak flowers and preventing fruiting that same year, and destroying red oak fruiting the following year. The archaeologically demonstrable importance of acorns to the upland settlements would be the push. The pull would be a change in the rainfall from a winter to summer pattern, making horticultural more productive in the drought-sensitive sandy soils of the Yadkin River floodplain. The resulting change would tether communities to the floodplain habitat, and induce a population growth rate that would inhibit a return to the prior, more mobile pattern, even with the return of warmer climate in the Medieval Maximum (A.D. 950-1250^{cal}). Since other regions in the Southeast had long since make this settlement and demographic transition, was the magnitude of the A.D. 536 event a critical factor in finally forcing a Woodland adjustment in the Yadkin River Valley?

David Anderson's contribution serves two purposes at the same time. First, he focuses on the Savannah River watershed of the southern Appalachian piedmont, and thus his paper is paired with that of Ned Woodall for a packaged evaluation of climate and culture processes at the extremes of the Appalachian foothills. At the same time, drawing on his breadth of experience in the southern Atlantic Slope region, Anderson provides a synthesis of the cultural change perspectives in the southern Appalachian summit, Carolina/Georgia piedmont, and coastal plain. Many of these changes were supported by evidence in previous chapters on the Southeast, and with others he takes exception. The study of the Woodland is so variant from usual archaeological phenomena that its investigation requires open minds and open discussion, for which Anderson provides a model.

While time resolution presents a problem to identifying event–based processes in prehistory of the region, Anderson has developed perspectives from working with annual tree rings during the historical and Mississippian periods that promises bridges to investigations of the deeper reaches of prehistory. He reviews several methods that will help refine prehistoric time resolution making it more comparable with historical and event–driven cultural processes.

Anderson surveys the interval between roughly A.D. 400 and 600^{c14} in Georgia and South Carolina, which marks the transition between two major temporal sub periods, the Middle Woodland, which is dated from roughly 300 B.C. to A.D. 500^{c14}, and the Late Woodland A.D. 500 to 1000^{c14}. In some parts of the region, dramatic assemblage change is indicated, as for example in portions of the coastal plain where Deptford and Swift Creek ceramics are replaced by

Wilmington and related cord and fabric impressed pottery designs. In other areas, in contrast, appreciable cultural continuity is indicated, as for example in portions of the mountains, where Connestee assemblages occur during much of both the Middle and Late Woodland. While the temporal resolution of events occurring within these archaeological cultures is currently coarse-grained, a number of promising methods exist that should permit much greater precision in dating in the years to come. These include dendrochronology; dendrochronologically-derived high precision radiocarbon dating, or wiggle matching; archaeomagnetic dating; and the development of fine-grained ceramic sequences based on the occurrence of individual vessel decorative and design motifs. The possible consequences of climate perturbation, such as is inferred to have occurred following A.D. 536^{cal}, on local archaeological cultures are explored, and a number of scenarios advanced to explain the patterning that is observed in the local archaeological record. Population migration and replacement, as well as settlement and organizational change, appear to be occurring during this interval. How such changes were triggered is currently unknown, but the paleoclimate data offer a possible contributing influence.

Shifting Seasonal Precipitation and Cultural Change

In an earlier presentation, Woodall found that cultural changes during the Little Ice Age (A.D. $1250\text{-}1920^{cal}$) in the Yadkin Valley conformed to a modeled shift in conditions implied by colder global climate shifting precipitation into late spring and summer. In his chapter, he raises the question of whether that or similar seasonal slippages in rainfall pattern would have been caused by global cooling after the A.D. 536 event and its aftermath. The changes that seem to be more gradual in their inception, such as the Medieval Maximum or the Little Ice Age Minimum, were attended by shifting ocean currents and jet stream patterns. Would an A.D. 536 event that suddenly clouded the sky also shift the jet stream and affect precipitation?

Woodall believes that there could be significant differences between abrupt events and long term shifts that can be exploited. It would not be surprising to find that the coastal plain was less affected because the nearby warm waters of the Gulf Stream flowing up the coast of North America. Perhaps the Gulf Stream did not have time to cool significantly in the 15 years following the A.D. 536 event. At the same time, it would not be a surprise to find that the Midwest was devastated because of its exposure to the continental interior climate. Furthermore, there could be strong contrasts during an event and more homogeneity between the regions over the course of a trend. Since we are dealing with both the A.D. 536 event and the subsequent 200-300 year trend, the study interval could offer the best of both worlds to study differences between the two time scales. These questions provide a theoretical position/model for these two chapters, and indeed an important proposition for the book. How the Yadkin River valley and the Atlantic Slope piedmont figure into the event side of the equation generally could be very informative since it is at the maritime/continental interface.

Though again not directly in the era of A.D. 536^{cal}, Anderson's research has shown that at historical annual scales as during the Spanish colonial period on coastal Georgia and South Carolina, and during the Mississippian period of the early second millennium, runs of dry and wet years appear to influence the storage capacity of critical domestic crops. On occasion, affects of these runs of years, such as the Spanish abandonment of the South Carolina coast, can be identified. Though still in critical stages of study, these episodes of human and global climate interaction suggest the tenor of global change impacts and point the way to understanding grander scale disturbances such as A.D. 536^{cal}.

Global Change. These chapters squeeze between the mountain and coastal chapters. They present mechanisms for the influence of global climate change on cultures, and I think thereby open the door for further research that is targeted on the global climate change issue. David Anderson has offered a Piedmont Savannah River impact model in *American Antiquity*. Ned Woodall brings into play a model of global-local linkages currently under development and being tested by new excavations in the northwestern piedmont of North Carolina.

Anderson's work parallels Woodall's at opposite ends of the Atlantic Slope Appalachian piedmont. This provides a limited but informative variation in both latitude and exposure to maritime and continental environmental domains. A possibility of buffering from extreme events by oceans at the lower elevation and latitude end could be extremely important. This is brought into the light in Anderson's discussions of the lower Savannah River cultural changes. The lower Savannah River is a lush estuarine environment that could well have buffered inhabitants from climate impacts, but it appears that those inhabitants were disturbed by the movements of others more severely disrupted societies from the north, movements documented in Mathis's chapter.

Tree ring climate indicators inform both Woodall's and Anderson's thinking, exposing the environmental side of the equation at annual time resolution and placing unusual new demands on the less resolved archaeological data. Anderson discusses some of the issues in increasing archaeological time resolution raise by the questions posed in this book. He suggests that new methods of resolving time scales offer great potential for meeting the schedules of greater time resolution demanded by event sensitive archaeology.

14. THE NORTHWEST CAROLINA PIEDMONT: POSSIBLE EFFECTS OF THE A.D. 536 EVENT

J. Ned Woodall

Wake Forest University, Winston-Salem, North Carolina 27109

Cultural Change in the Northern Piedmont

[Piedmont, North Carolina, Yadkin River; ceramic analysis; cultural change; early, middle and Late Woodland, A.D. 0-1000^{c14}]

Although this chapter discusses the potential impact of the A.D. 536 event in the North Carolina piedmont (Figure 14.1), I acknowledge that present understanding of the prehistory of the subregion does not provide any convincing demonstration of such an event or of its consequences. Instead, we can observe that, between A.D. 0 and 1000^{c14}, there occurred a substantial change in settlement patterns and subsistence strategies of the prehistoric inhabitants. This may be due to a sudden catastrophic or, as is commonly assumed but seldom stated explicitly, it may be a result of incremental changes in the human ecology of the region. Although a form of "gradualism" is, I believe, the paradigm in vogue in the study of southeastern prehistory generally and North Carolina specifically, it is interesting to contemplate the possibility of an alternative scenario: could the observed changes have been the result of emergency decision-making by people 1,400 years ago?

An understanding of the period in question involves interpretation of the local ceramic sequence that includes three major wares. The earliest is the Yadkin series, which commonly is viewed as belonging to the Early Woodland (see Figure 14.1). It is followed by the Uwharrie series that marks the Middle Woodland, and finally the Dan River series is Late Woodland and continues until historic contact with Native Americans. There is no Mississippian period in the northwest piedmont as it is commonly understood. The terms Early, Middle and Late Woodland are defined, formally and temporally, in terms of local cultural change and have no necessary relationship to the traditional Mississippian cultures of the Ohio Valley and their chronology. Although this could be confusing to some, it is based on recognizing that the northwest piedmont did not participate in, nor was affected by, the rather spectacular achievements of Adena and Hopewell moundbuilders over the Appalachian Mountains to the west; rather, it presents a less dramatic but still fascinating record of prehistoric change.

Yadkin Ceramics and Sites

[North Carolina Piedmont; ceramic analysis.; upland sites; Early Woodland, A. D. 1^{c14}]

The earliest ceramic assemblages contain the Yadkin ceramic types, or series, along with a close typological relative, the rare and enigmatic Badin ware (Coe 1964). At present, in all of North and South Carolina there are only five dated Yadkin sites and only four of them are in the piedmont. Available dates fall between 400 B.C. and A.D. 1^{c14} (Eastman 1994:27; Webb and Leigh 1995:29). The situation is hardly better with the Uwharrie series, which follows Yadkin: there are nine dates with the earliest five falling between A.D. 671 and A.D. 1159^{c14} (Eastman 1994:27). After A.D. 1200^{c14} there is a relative abundance of dated sites, almost all belonging to the well-known Dan River series. The relative age of Yadkin, Uwharrie and Dan River pottery is not in dispute, but the dearth of absolute dates make it impossible to determine the timing of cultural changes and replacements. This is an important issue, however, because the shift from Yadkin to Uwharrie pottery also marks a distinctive change in settlement pattern and subsistence strategies. It is that change that could be a consequence of the of the A.D. 536 event. In summary, *something* happened between A.D. 1 and (approximately) 700^{c14} to effect this change, and it could be instructive to consider a catastrophe model.

Early Woodland or Yadkin series sites tend to occur in the interriverine uplands of the piedmont. These uplands are low, rounded hills dissected by numerous springs and small streams that feed into the major rivers. They supported a mixed oak-hickory forest on heavy clay soils, and the long growing season of almost 200 days and high 114 cm of precipitation per year provided a rich bounty of wild foods. The sites are very small, and perhaps because of the erosion of the uplands following the Euro-American occupation, few have any subsurface remains. An exception is the E. Davis Site in Forsyth County, located adjacent to a spring head. The site covers less than 1,000 m^2, situated near a large outcrop of bedrock from which issued the spring, flowing in a steep little valley. This may account for the fact that the site had never been plowed, making it unique for the area. Vessel fragments found only a few centimeters below the

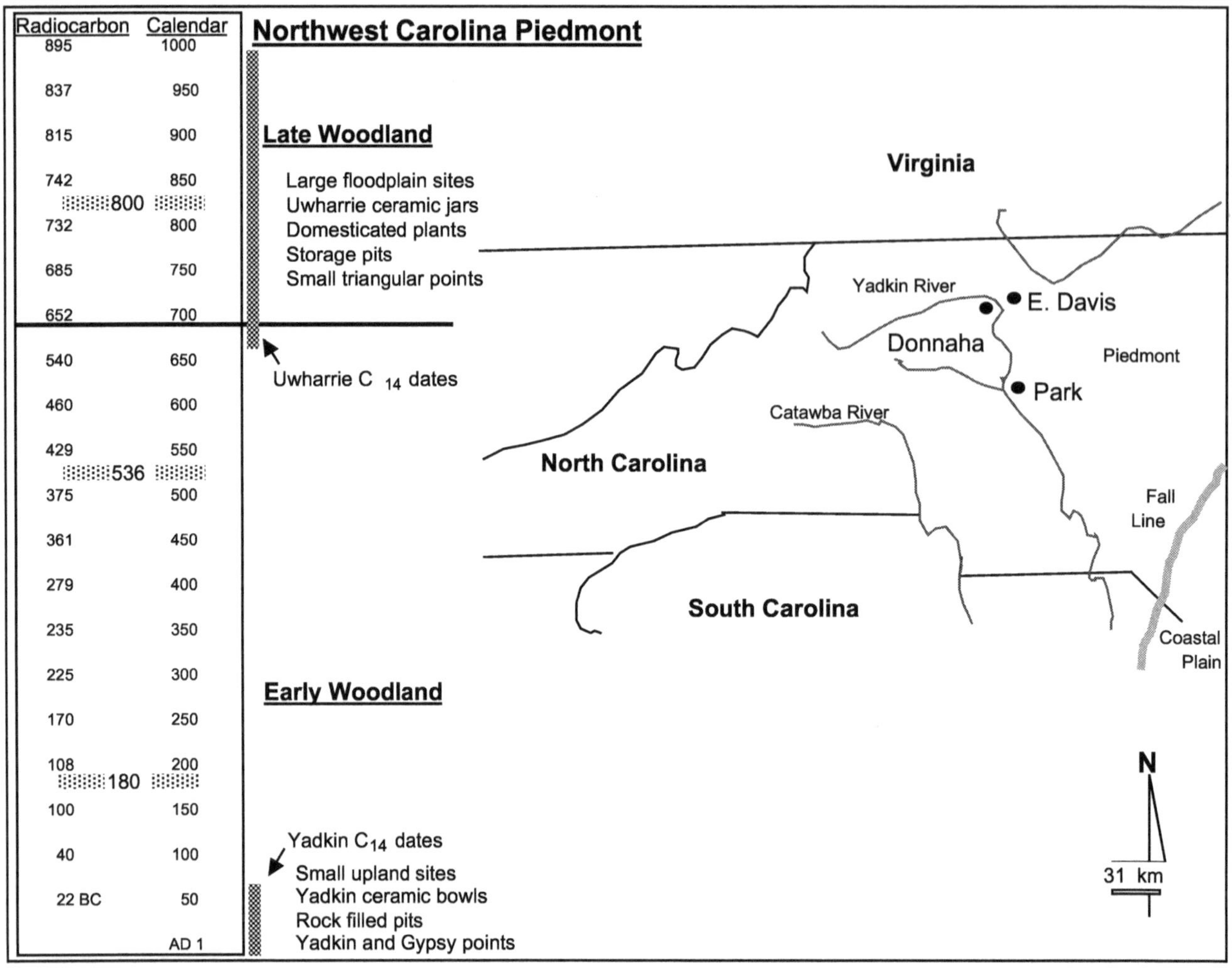

Figure 14.1. Map and Chronology of Piedmont North Carolina Yadkin Watershed.

present surface could be refitted to form a large segment of a typical Yadkin series fabric-impressed bowl. These bowls, found at other Early Woodland sites, are small open vessels that could be nested if necessary for easy transport. The site also contained several small (ca. 30 cm diameter) rock-filled pits of ambiguous function along with projectile points of the Yadkin and Gypsy types. These points, nine in all, were made of various metavolcanic materials from at least eight sources. The debitage included both metavolcanic flakes (not locally available) and the local milky quartz. There was no bone from this, almost certainly due to the highly acidic soils. Plant remains include charred acorns and hickory nut shell. No tropical or indigenous domesticated plants have been found associated with any Early Woodland site in the northwest piedmont.

A very similar site located a few kilometers away contained another small rock-filled pit along with fabric and check-stamped Early Woodland vessel fragments. At least one of these vessels was a straight-sided, thick jar of modest size (about 25 cm orifice diameter). Other than the contents of the single feature, aboriginal refuse was very scarce (Idol, Trage and Kirchen 1996). The Yadkin upland sites appear to be short-term residential locations, quite possibly camps that are part of a seasonal round settlement system. In that way they are similar to the late Archaic sites of the region.

Uwharrie Ceramics and Sites

[Geography; Methods, Second method, etc.; Topics; Chronology]

The upland Early Woodland sites strongly contrast with the Middle Woodland sites of the region, i.e., sites with a ceramic assemblage dominated by the Uwharrie series wares. These sites often are very large, invariably situated on the floodplains of major streams, and contain a distinctive artifact and ecofact inventory. Typical is the Parker Site, on the Yadkin River (see Figure 14.1). The site covers over 1 ha, with a midden about 50 cm thick, numerous post holes, trash-filled storage pits and human burials; two radiocarbon dates suggest an occupation between A.D. 650 and 1000[c14] (Newkirk 1978). Although wild foods were abundantly represented, also found was maize (including one charred cob fragment) and the seeds of disturbed-habitat plants such as knotweed (*Polygonum* sp.) and prickly poppy (*Argemone* sp.; Crawford 1978). In the lower levels of the Donnaha Site, the Middle Woodland component yielded domesticated sumpweed (*Iva annua*) and curcurbits as well as maize (Mikell 1987). The Uwharrie ceramic assemblage typical of the Middle Woodland in the area is dominated by large conoidal based jars, tempered with coarsely crushed quartz, and either cord or net-impressed on the exterior. The vessel size and the attributes of the paste and interior surface indicate they were designed for storage—likely for foodstuffs—rather than cooking (Woodall 1990:81-82). Certainly the size of the vessels, their fragility and the typical constriction at the neck would make easy transport difficult. Although a detailed analysis of the stone tools from the Parker Site has not been done, a casual inspection of the artifacts and debitage shows that the metavolcanic flakes used are much reduced in variability when compared to the E. Davis inventory, even though the Parker assemblage is many times larger. In addition, the projectile points from Parker show less variability as well, being composed almost altogether of small triangular forms (Newkirk 1978).

Middle Woodland sites such as Parker and Donnaha are not common in the 70 km surveyed section of the Yadkin River valley. Keeping in mind that overbank deposition could have reduced their visibility, known sites occur about every 5 to 10 km. When present, however, the sites are large, unlike the Late Woodland occupations (i.e. post-A.D. 1200^{c14}) that occur abundantly and represent both tiny hamlets as well as major settlements (Barnette 1978).

Modeling Climate, Vegetation, and Culture Change

[North Carolina Piedmont; regression, river discharge; seasonal slippage;1958-present, Vandal Minimum, Medieval Maximum, A.D. 900-1450^{cal}]

The question I hope to address is, "Can the observed changes be explained in terms of the A.D. 536 event"? As I noted at the beginning of this chapter, there is no straightforward means to answer the question using the artifacts currently available. There is, however, evidence bearing on the relationship between global climate and the density of riverine settlements in the northwest piedmont, and that relationship allows construction of a model that can inform future investigations. At present there is no clear indication of exactly what, if any, changes to the climate were triggered by the event. For present purposes, I assume that the most likely consequence was the depression of mean annual temperatures, and if so, this probably affected the amount and timing of the annual precipitation pattern. The relationship between temperature and rainfall can be seen in Figure 14.2, which charts fluctuations in the discharge of the South Yadkin River over a period from 1958 to 1990. The South Yadkin was chosen for this study because it is undammed; it heads in the Brushy Mountains, which divide the Yadkin and South Yadkin drainages, and empties into the Yadkin River near Salisbury, North Carolina (see Figure 14.1). It would appear to represent a good surrogate for the Yadkin proper, and its variation in rate of flow must approximate that of the Yadkin itself. The period between 1958 and 1990 contains some globally cold years such as 1976 (which approximated the Little Ice Age) and the globally hot years of the 1980's, as well as years of more moderate temperatures (Woodall and Gunn 1993). The Figure 14.3 data are smoothed to reduce but not eliminate the effects of several major floods.

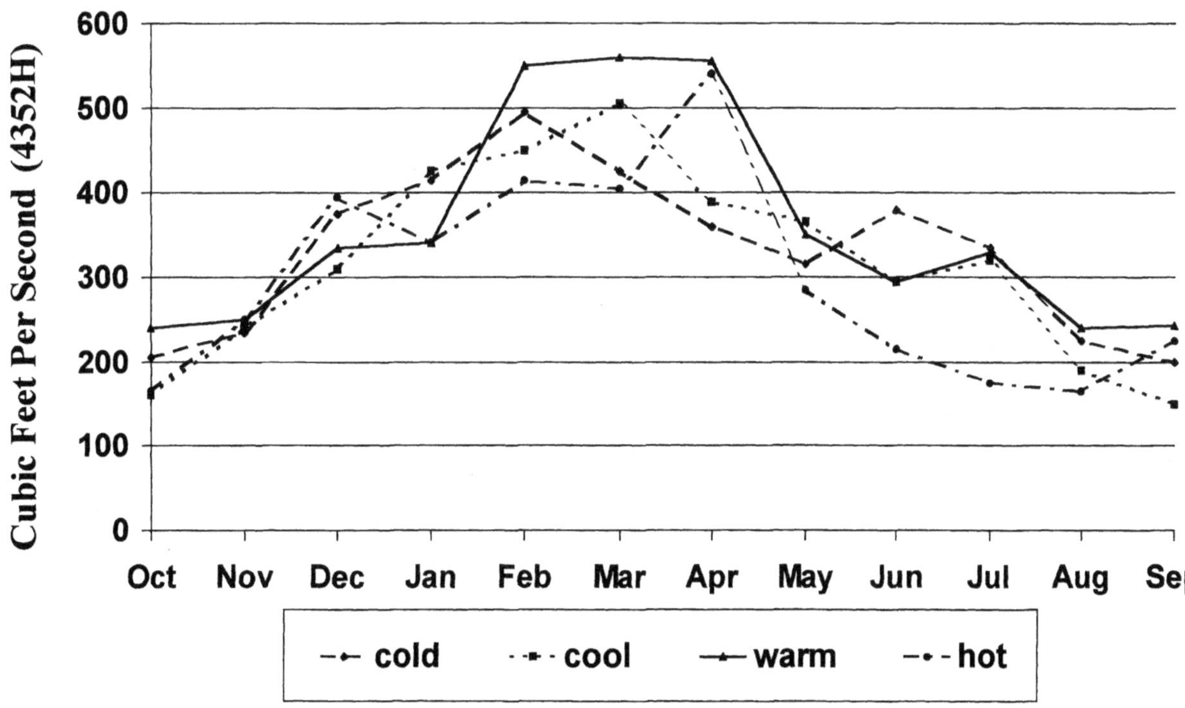

Figure 14.2. Fluctuations in South Yadkin River Discharge between 1958-1990.

Of particular interest are the months of May and June, late spring and early summer. Those months are critical to the growth of tropical and native cultigens. Years of cool or cold global annual temperatures provide more moisture during those months. The rainfall is especially important in the

Yadkin's floodplain, where archeological sites are found adjacent to the main stream on soils of loose, friable sand or sandy loam. Those soils are highly permeable and thus sensitive to drought; no moisture is provided by the water table because the river is a mature, well-entrenched stream. Absent late spring-early summer rainfall and almost any crop will fail, as they did in the ruinous summer of 1986.

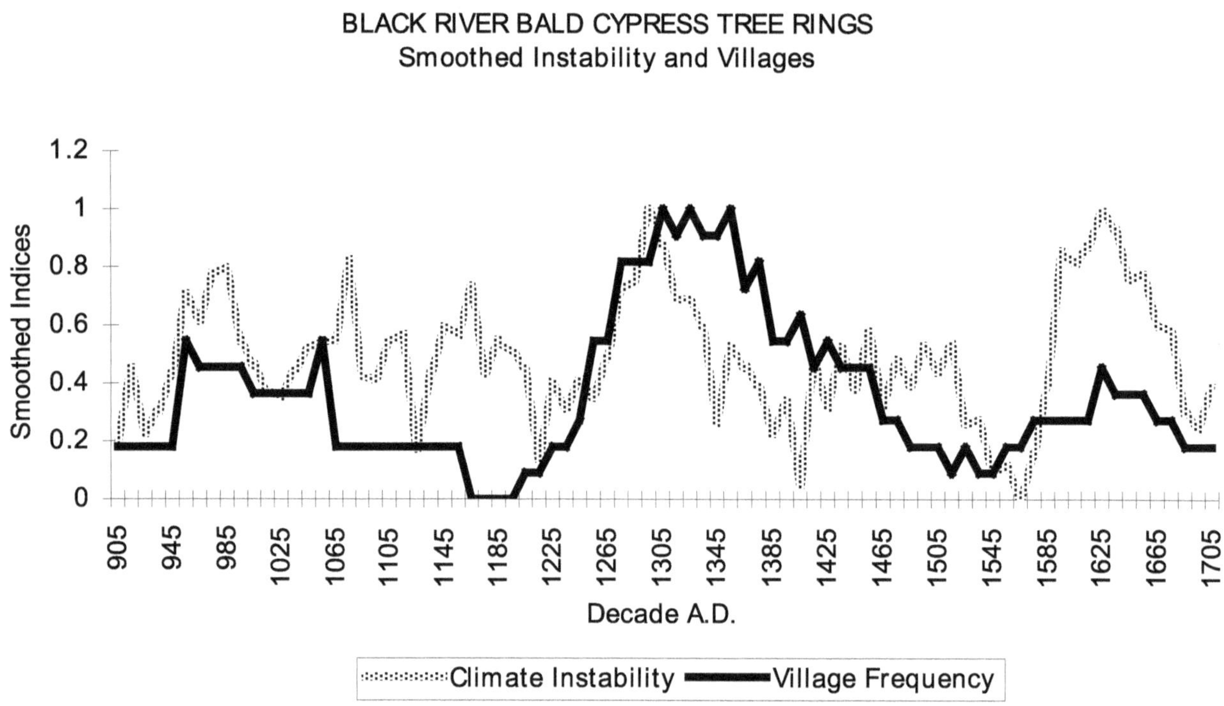

Figure 14.3. Decadal Instability in Baldcypress Treerings and Frequency of Villages.

Another aspect of food production, and an opposite effect of changes in temperatures, can be observed in the uplands. While cold years tend to increase rainfall in the early summer, the cold can also depress the production of mast in the area woodlands. Oak trees flower in the spring and, following pollination, produce their acorn crop either that fall (for white oaks) or the fall of the next year (red oaks). Although there is a natural fluctuation in mast for red oaks (a 2-year cycle) and white oaks (a 3-year cycle), there is a strong tendency for this fluctuation to be species-specific (i.e., not specific to individual trees). A good mast year for one white oak is probably a good year for most white oaks. For both species, however, mast production is strongly affected by mean daily maximum spring temperature. The later the date of spring frost during spring when flowering occurs, the lower the viable acorn production (Sork, Bramble and Sexton 1993). For this century the average last frost in the northwest North Carolina piedmont is April 11. If that date were shifted as little as 10 days toward the end of April, it would have a disastrous effect on acorn production, killing the flowers of white and red oaks and preventing fruiting that fall (for white oaks) and the next fall (for red oaks). These trees do not have an ability to acclimate of course, so successive unusually cold springs would affect populations of white-tailed deer, turkey, squirrels, and of course human populations either as dependent on those animals or as direct consumers of acorns. If the A.D. 536 event led to even a decade of notably colder temperatures, settlement and subsistence patterns may have been influenced by the "push" of a depleted mast crop and the "pull" of a more reliable and productive riverine habitat emphasizing cultivation of storable seed crops.

It is beyond question that the people of the Early Woodland were tracking the production of the oaks and accommodating shortfalls in acorn production with various risk-reduction strategies. In Mithen's (1990) apt terms, these were thoughtful foragers. They certainly used the Yadkin floodplain, probably seasonally, and were acquainted with its products (Webb and Leigh 1995). And they probably were acquainted with domesticated plants, which had been cultivated in neighboring Tennessee for at least 2,000 years. Apart from the modern archaeologist's wariness of prime movers generally, and saltation events specifically, there is no reason the change from early to Middle Woodland could not have been accomplished in a decade. Also, Tanner's sea level curve (see Chapter XX) suggests that the conditions imposed by A.D. 536^{cal} probably developed as a deepening trend over the subsequent 300 years.

Another indication of the effect of climate on settlement pattern is provided by a perceived relationship between climatic instability and site frequency in the Yadkin's floodplain. Using archeological data from the Late Woodland, a seriation of ceramic assemblages from sites in the Yadkin valley—anchored by several radiocarbon dates—was performed for 31 sites (Woodall 1990). Using the Black River bald cypress tree ring data compiled by Stahle et al.

(1988), the year-to-year changes registered in the tree rings were collapsed to decades and the variability within decades used to create an index of climatic instability (Woodall and Gunn 1993). Figure 14.3 compares climatic instability with Late Woodland site frequency, and suggests a positive relationship. It is not clear whether this reflects an increased population, increased rate of site relocation, or a food production strategy that involved situating personnel in various locations, perhaps as a risk-reduction strategy to accommodate uncertain rainfall. We do know that the three centuries following the A.D. 536 event experienced dramatic swings in precipitation and temperature before the stability of the Medieval Maximum, and if the Late Woodland pattern was pursued, then increased frequency of riverine habitat usage can be expected.

To summarize the model, I posit that the A.D. 536 event produced later frosts in the spring that reduced the mast crop of the uplands and strongly affected dietary staples of the people, namely acorns, deer and turkey. At the same time riverine floodplains were made more attractive by improved growing conditions, i.e., more late spring-early summer rain, and the result was the formation of sedentary communities more reliant on stored food resources, including tropical and native cultigens. Those tethered populations along the rivers then experienced a common consequence of reduced mobility, namely a higher rate of population growth, making it difficult to return to the mobile pattern of the Early Woodland and initiating further growth of settlements that remained bound to the rivers until historic contact. A more systematic procurement of lithic materials was initiated to provision these concentrated populations, replacing the previous pattern that drew from a variety of quarries probably imbedded in the catchment of several different bands linked by kin ties. Ceramic vessels become larger, responding with changes in form to meet storage needs in vermin-infested villages.

Proposed Research Agenda for the Early-Middle Woodland Transition

[North Carolina Piedmont; inference, hypothesis; radiometric dating, excavation, lithic and ceramic pattern recognition; early and Middle Woodland]

This model is largely inductive of course, and suffers all the liabilities of that approach. It can still be asked whether any of these change processes actually occurred, and for this we must depend on the future testing of derived hypotheses. Most obvious is the need for additional dated Early Woodland sites, along with more dates on the Uwharrie components along the rivers. Deep testing along the rivers is necessary to confirm the absence of sedentary Early Woodland sites there, and much more information is required for secure knowledge of the Early Woodland subsistence pattern, including the predicted absence of storage vessels. Most interesting are the changes in social systems that must have accompanied a shift from small mobile bands to larger villages.

The present state of knowledge is limited regarding Woodland social organization in the northwest piedmont. One observation that may prove instructive, however, contrasts the varied pottery of the Early Woodland with the rather monotonous Uwharrie wares. The role of ceramic style in the creation and maintenance of social networks is a complicated issue, but it seems probable that the homogeneity of Uwharrie ceramics is informing on a social process that is quite different from that of the Early Woodland. Ceramic styles, perhaps reinforced by other lines of evidence such as patterning of dwellings in villages and work habits revealed through lithic tools may provide future avenues of investigation into these important questions.

The prehistory knowledge base in the northwest piedmont of North Carolina has been accumulated through more than a decade of research in the upper and middle Yadkin watershed (Woodall 1990). This prehistory clearly poses possibilities for linking global climate change impacts to specific human-environment interactions in the subregion. That climate changes affected the prehistory of the region cannot be doubted. For example, Anderson et al. (1995) have demonstrated the correspondence between drought years and social and demographic changes in the Mississippian polities of the Savannah River drainage. In this chapter I have outlined a model, based on the projected influences of the AD 536 event, for the initiation of a dependence on lowland cultigens, with all the social changes that entailed. Especially interesting are the upland mast crop-lowland cultigen alternatives suggested by the implications of seasonal slippage in terms of precipitation and temperature. I believe that current knowledge is sufficient to inform a well designed program of investigation. However, much of it is still qualitative in nature and the analysis has not yet sufficiently advanced to verify specific linkages between social structures and subsistence means under varying environmental conditions.

ACKNOWLEDGMENTS: My interest in the AD 536 event, and its possible effect on the Early Woodland, was stimulated by discussions with Joel Gunn. My thinking on this topic was aided by thoughtful comments from Peter Weigl, Bruce Idol and Roger Kirchen.

References

Barnette, Karen L. (1978) Woodland Subsistence-Settlement Patterns in the Great Bend Area, Yadkin River Valley, North Carolina. MA thesis, Wake Forest University.

Anderson, David G., David W. Stahle and Malcolm K. Cleaveland (1995) Paleoclimate and the Potential Food Reserves of Mississippian Societies: A Case Study from the Savannah River Valley. *American Antiquity* 60:258-286.

Coe, Joffre L. (1964) Formative Cultures of the Carolina Piedmont. *Transactions of the American Philosophical Society* 54, part 5.

Crawford, Gary W. (1978) Report on the Plant Remains from the Parker Site, 31Dv4, North Carolina. *In* The Parker Site: A Woodland Site in Davidson County, North Carolina. Judith A. Newkirk, MA thesis, Wake Forest University.

Eastman, Jane M. (1994) The North Carolina Radiocarbon Date Study. *Southern Indian Studies* 42.

Idol, Bruce S., Stephen T. Trage and Roger W. Kirchen (1996) Report on Excavation at the Bethaabara 1754 Sleeping Hall Site, Forsyth County, North Carolina. Report prepared for Historic Bethabara Park, Winston-Salem, North Carolina.

Mikell, ? (1987) ?.

Mithen, Steven R. (1990) Thoughtful Foragers: A Study of Prehistoric Decision Making. Cambridge University Press.

Newkirk, Judith A. (1978) The Parker Site: A Woodland Site in Davidson County, North Carolina. MA thesis, Wake Forest University.

Sork, Victoria L, Judy Bramble and Owen Sexton (1993) Ecology of Mast-Fruiting in Three Species of North American Deciduous Oaks. *Ecology* 74(2):528-541.

Stahle, D. W., M. K. Cleaveland and J. G. Hehr (1988) North Carolina Climate Changes Reconstructed from Tree Rings: A.D. 372 to 1985. *Science* 240:1517-1519.

Webb, Paul A. and David S. Leigh (1995) Geomorphological and Archaeological Investigations of a Buried Site on the Yadkin River Floodplain. Southern Indian Studies 44:1-16.

Woodall, J. Ned (1990) Archeological Investigations in the Yadkin River Valley, 1984-1987. North Carolina Archeological Council Publication 25. Raleigh.

Woodall, J. Ned and Joel D. Gunn (1993) Population Dynamics on the Yadkin River Following the Medieval Optimum Global Climate Episode. Paper presented to the 50th annual meeting, Southeastern Archaeological Conference, Raleigh, North Carolina.

15. THE MIDDLE TO LATE WOODLAND TRANSITION IN THE SAVANNAH RIVER BASIN AND ADJACENT SUBREGIONS: REFINING TIME SCALES

David G. Anderson

Southeast Archaeological Center, National Park Service, 2035 East Paul Dirac Drive, Box 7, Tallahassee, Florida 32310

Global Climate Study and Archaeology in the Southeast

[Southeast; Tree rings, Global Climate Change, Teleconnections, El Niño; first millennium]

Global climate change is attracting tremendous interest and funding, and archaeology and history have much to contribute. There are very powerful climate-based data sets and models available for use in exploring relationships between climate and culture, that are attracting increasing attention by a wide range of scholars. I have no doubt that this will be an important area of research in the years to come.

The methods and concepts of global climate change can be brought to bear on the archaeological record. The articles in this volume document profound relationships between global climate and regional cultures in the mid-sixth century A.D., with the effects of the A.D. 536-545^{cal} event observed from a number of parts of the world. I say event, because I have no doubt that something did indeed happen at this time level, even if we do not fully understand what it was, or the global effects and teleconnections. The historical and archaeological data from western Europe, China, Africa, and the Maya area is pretty unambiguous. It is also clear that archaeological and historical records have important if not unique insights to offer those interested in the impacts of global change and teleconnected phenomena.

Archaeologists have begun exploring other global climate processes and signals in recent years. In particular, El Niño and El Viejo (also called "La Niña") signatures, corresponding to conditions of appreciably warmer and colder than normal water temperatures in the western Pacific and their attendant effects of global climate, are the subject of increasingly detailed research using both historical and archaeological data. Not surprisingly, given where the impacts of these processes were first noticed, historic and late prehistoric El Niño effects have been examined in particular detail in Peru and Chile in recent years. Examination of Spanish archival data has shown that a number of periods of major flooding, catastrophic erosion, and famioe over the past five centuries in that region are directly tied to El Niño/El Viejo effects (Quinn and Neal 1983, Quinn et al. 1987). An article by Sandweiss and colleagues (1996) explores much earlier, mid-Holocene El Niño effects in western coastal South America, using multiple independent lines of evidence, including zooarchaeological data, to reconstruct climatic conditions and their effect on landscape and culture.

In the Southeastern United States, El Niño/La Niña (or El Viejo) effects include extended periods of somewhat above and below average rainfall and temperature, and catastrophic storm frequency, particularly hurricanes and tornadoes (Diaz and Markgraf 1992, O'Brien et al. 1996, Green et al. 1997, Bove 1998). How fluctuations in climate effected human settlement in the region, particularly through their impact on crop yields, flooding patterns, and storm damage, will likely be increasingly explored by archaeological and historical research in the years to come. Climate research at a number of institutions around the country, including the Center for Ocean and Atmospheric Predictive Studies at Florida State University in Tallahassee, in the same building where I work, are exploring long and short term climatic effects on contemporary society, such as on crop yields, heating/cooling costs, devastating storm frequency and damages, and sea-level change (e.g., O'Brien et al. 1996, Green et al. 1997, Bove 1998). Much of this research includes analyses of recent historical records, and it is not a great step to move these studies much further back into the past.

As I will discuss below, I have examined tree ring data to understand, at annual time resolution, relationships social and environmental variables in the Savannah River valley and adjacent regions during the late prehistoric and early historic era (Anderson 1994; Anderson et al. 1995). Tree rings are increasingly seeing use in the exploration of climate and culture issues in other areas of the Southeast as well. Ned Woodall is using the longest east coast baldcypress tree ring chronology from the Black River in North Carolina (this volume). David Stahle, Malcolm Cleaveland, and Dennis Blanton (1998) are using the excellent and lengthy tree ring derived rainfall reconstruction available from southeast Virginia to explore developments in the early history of the

Jamestown and Roanoke colonies, work that is demonstrating how critical climate was to their success or failure.

Gordon Willey (1936) first realized the potential of dendrochronology in the 1930s when he was involved in the New Deal era excavations at Macon Plateau in central Georgia In many ways his study was far ahead of its time, and it is gratifying to see that this kind of research is now starting to come into its own across the Southeast. Climatological data will be increasingly used to examine major periods of cultural change in the region, such as the emergence of complex agricultural societies in the central Mississippi valley after ca. A.D. 1000[c14], and their subsequent rapid spread over much of the region. A dramatic consolidation of settlement and increase in organizational complexity occurred in the American Bottom near St. Louis around A.D. 1050[c14], in and near Cahokia (Milner 1998; Pauketat and Emerson 1997). The role climate could have played in changes such as these has received comparatively minimal attention, but this situation is changing (e.g., Lopinot 1994; Milner 1998).

Refining Prehistoric Time Resolution

[South Appalachian Region, South Atlantic Slope; Wiggle matching, AMS dating, Archaeomagnetism, Dendrochronology; Temporal Resolution, Points, Pottery/Ceramics, Design motifs; Woodland, Mississippian]

As all the articles in this volume acknowledged, prehistoric chronological resolution leaves much to be desired, and we need to move toward achieving temporal precision closer to that of history and some measures of paleoclimate. Our archaeological data gives us resolution on the one-to-two century scale at best for much of the Woodland period in the Southeast, yet to resolve fine-grained climatic events and impacts, we often need to have annual or at most decadal scale data. Major Holocene climate episodes, in contrast, may extend over hundreds or even thousands of years, whose relationship to cultural developments may or may not be well articulated (Wendland and Bryson 1974; Gunn 1994). In part because our temporal resolution is relatively imprecise, prehistoric archaeology within the region has been a largely episode, or phase, oriented study, focused more on long term changes and patterns than on the dynamics of rapid and short term environmental or social change. There are, fortunately, a number of ways in which we can refine prehistoric chronology locally.

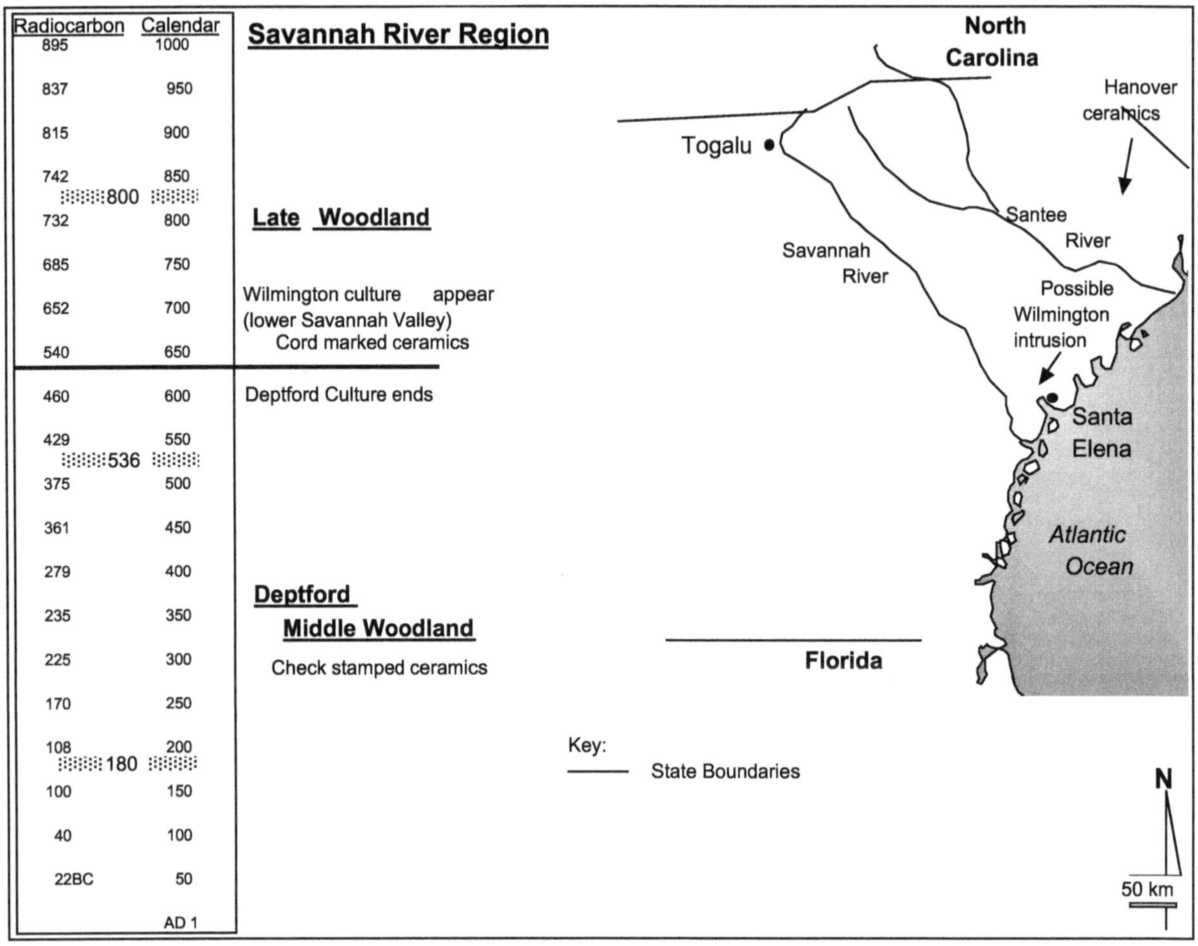

Figure 15.1. Map and Chronology of Piedmont and Coastal Georgia.

We of course need to work with traditional archaeological data categories to refine temporal resolution, and in the Southeast that certainly includes pottery and projectile points. Within Mississippian ceramic assemblages in the South Appalachian area, temporal resolution exists on the order of ±50 years for the period from approximately A.D. 1200-1700^{c14}, at least in some areas (e.g., Williams and Shapiro 1990). For the preceding Middle and Late Woodland era, Swift Creek complicated stamped pottery, which spans the interval from about A.D. 100 to 750^{c14}, appears to offer us comparable and perhaps even better temporal resolution. Research with Swift Creek ceramic design motifs, and particularly the occurrence of unique flaws (i.e., cracks, damaged areas) in the paddles that produced these vessels, is allowing us to recognize presumably contemporaneous vessels at a number of different Swift Creek sites (Snow and Stephenson 1998). As relative and absolute dating of these motifs becomes more precise, we should be able to produce increasingly sophisticated models of cultural development, interaction, and change for the terminal Middle Woodland/initial late Woodland period. Such information will be invaluable for determining the impact, if any, of the A.D. 536^{cal} event on this culture.

Research by paleoclimatologists should also help appreciably in refining archaeological time scales. Dendrochronological sequences or reconstructions dating back 1,000 or more years are now known from a number of parts of the region (Stahle and Cleaveland 1992, 1994; Stahle et al. 1985 a, 1985b, 1985c, 1988, 1991). When found in good archaeological context, and tied to secure reconstructions, extremely high precision dating should be possible in cases where wood or charcoal samples with 50 or more growth rings are found, the approximate minimum number needed for secure temporal placement. In areas where dendrochronological constructions are not available, or where only floating chronologies are available, it is still possible to achieve a resolution on the order of 10 to 20 years using radiocarbon dating on wood or charcoal samples where from ca. 50 to 100 annual growth rings are present. This can be done through the "wiggle-matching" technique, or the fitting of a series of high precision AMS radiocarbon dates taken from growth rings of known temporal separation to the secular radiocarbon production curve (Clark and Renfrew 1972, 1973, Renfrew 1996). Although this procedure has been known for some time, it is only now starting to see common application (e.g., Kuniholm et al. 1996). Another high precision absolute dating technique, archaeomagnetic dating, offers the potential of 10 to 20 year interval resolution in the Southeast (Wolfman 1984). To do this, however, the calibration curve for the region must be developed, which means archaeologists must make the investment in taking the requisite fired clay samples.

Tree Rings and Atlantic Slope Climate-Culture Interactions

[Savannah River, Southern Piedmont; Tree rings, Seasonal Slippage, Food Storage, paleoclimate, El Niño; Mississippian]

Thanks to the dendrochronological work of David Stahle, Malcolm Cleaveland, and their colleagues at the University of Arkansas, we also now have excellent year by year paleoclimate information from many areas of the Southeast (Stahle annd Cleaveland 1992, 1994, Stahle et al. 1985 a, 1985b, 1985c, 1988, 1991). Baldcypress ring widths provide a good proxy measure of spring/early summer rainfall, a relationship that has been demonstrated using historic rain gauge data with a very high degree of correlation. This type of data will be increasingly used in the years to come to explore the effects of annual and longer scale variations in climate on local cultures, both during the prehistoric and the historic eras.

I have used the tree ring derived paleoclimate data to calculate possible crop yields and stored food reserves each year during the Mississippian era, in the Savannah River valley and adjoining areas, where excellent long-term dendroclimatological reconstructions exist. These crop/storage estimates were then compared with the political history of the late prehistoric societies in this area, research reported at length elsewhere (Anderson 1994; Anderson et al. 1995). Periods of good and poor climate, as measured by extended periods of average and below average spring/early summer rainfall, were frequently found to be associated with periods of societal tranquillity and unrest, respectively. This relationship appears to reflect the importance of agricultural harvests in these societies, specifically the chiefly elites abilities to generate and then appropriate surplus production. However, as I have also taken care to point out, these relationships are by no means direct or unequivocal in every case, and are sometimes determined by a great many factors working in combination, not the least of which are regional political geography and the historical trajectories of the societies in question. I will return to these broader regional issues of the relationships between climate and culture later.

In a historic analog to these prehistoric investigations, tree ring derived rainfall measures were used to explore developments in the 1565-1587 Spanish colony of Santa Elena on the South Carolina coast. The Spaniards had the bad luck to settle Santa Elena during the worst decade of drought during the sixteenth century. The difficulties the colony faced no doubt played at least some role in the eventual decision to relocate the capitol of Spanish Florida to Saint Augustine in 1587, which occurred at the end of another severe climatic downtown, and probably not coincidentally was apparently the same year that the English "lost" colony to the north at Roanoke failed (Stahle et al. 1998). Given this, one could mount the argument that climatic fluctuations, as much as anything, explains why the people in the Southeastern United States speak English rather than Spanish. Early Spanish efforts at colonization in the region seem to have run into far more than their share of climatically-induced vicissitudes, from the hurricanes that destroyed the fleets of the 1528 Narvaez and 1559 Luna expeditions, to the multiyear long periods of droughty weather that characterized the beginning and end periods of settlement at Santa Elena (Figure 15.1).

As noted, examining the prehistoric Mississippian sequence in the Savannah River valley revealed similar correlations between climate and cultural phenomena. The same analysis, however, also indicated that the relationship was a highly complex one. The Savannah River Valley was largely abandoned by Mississippian groups, during a period of climatic downturn in the mid-fifteenth century, for example, but it also stayed abandoned during the century of favorable weather that followed. The rivalry and warfare between the powerful and expanding chiefdoms of Ocute and Cofitachequi in central Georgia and central South Carolina, with the Savannah chiefdoms caught in the middle, was likely of equal or greater importance in causing this abandonment (Anderson 1994, 1996a). Climatic induced stress—disrupting tribute flow and hence the position and authority of the chiefly elite, for example—could have helped push the Savannah River chiefdoms over the edge, but there were clearly other factors in play. The same was true in the early historic era, when the events at the Santa Elena and Roanoke colonies were played out against a much larger canvas, that of two powerful nation-states vying with one another for control of vast parts of the world.

Regional Scale Studies

[Atlantic Slope; Cultural Resources Archaeology, Regional Perspective; Woodland]

The Savannah River case indicates that adopting a regional perspective is likely to be just as critical to understanding culture change and interaction during the Woodland period, and particularly during the Middle-to-Late Woodland transition, the temporal focus of this volume, as it was in the later prehistoric era. Earlier generations of local archaeologists, particularly Joseph R. Caldwell in his classic 1958 monograph *Trend and Tradition in the Prehistory of the Eastern United States*, commonly proposed vast movements of prehistoric peoples over time and space, in maps filled with lines and arrows. While such a regional approach is something we should admire and emulate, we must also recognize that the sample sizes these scholars had to work with were extremely limited, based on work on a comparatively small number of sites. As a result, while many profound observations were made, many errors also occurred. Fortunately, we now have the data in hand to do a far better job.

Thanks largely to the rise of cultural resource management archaeology, there are now over 200,000 recorded archaeological sites in the Southeast, a dramatic increase from the ca. 10,000 or so recorded in 1970. The data in these records can and have been used to plot where people or at least sites were and were not on the landscape at various times in the past, notably over the course of the Paleoindian and Early, Middle, and Late Archaic in the case of my own work (Anderson 1991, 1996b). Unfortunately, such regional scale maps have yet to be prepared for the Woodland period, and we really do not have a very good idea about site distributions at the regional scale at the present. When such maps are ultimately produced for the Woodland period, I predict the results will tell us a great deal about population growth and change, and the rise and fall of cultural systems. The patterns that are revealed will generate new models and explanations, and will enable us to better evaluate the effects of climatic triggering events.

Cultural Change in the Coastal Plain

[Coastal Plain; Deptford/Linear check stamped, Wilmington/Cord-fabric impressed, Grog Temper, Immigration; Middle Woodland, Late Woodland]

What transpires in the terminal Middle Woodland and initial Late Woodland in the southern Atlantic Slope subregion? The 1,000 year long Deptford culture, characterized by linear check stamped pottery, comes to an end some time around A.D. 500^{c14}, and is replaced by Late Woodland complexes characterized by cord and fabric impressed pottery. This replacement is sometimes attributed to a movement of northern ideas and possibly people into the South Carolina and eastern Georgia Coastal Plain. In the northern part of the South Carolina Coastal Plain, this replacement is assumed to have been gradual. Indeed the stratigraphic columns from the few sites we have spanning the period, notably Minim Island and the Mattassee Lake sites from the lower Santee River, and the Blackwater Pond site cluster near the lower Wateree River, suggests a long cooccurrence for these finishes, with cord marked and fabric impressed finishes continuing long after linear check stamping drops out after ca. A.D. 500^{c14} (Anderson 1982; Cable 1997; Espenshade et al. 1989).

On the southern South Carolina and northern Georgia coast it has long been thought that grog tempered cord marked pottery of the Wilmington series arrived for the first time around A.D. 500 to 600^{c14}, when it (apparently) abruptly replaced earlier sand tempered Deptford materials (e.g., DePratter 1979, 1991). The traditional explanation has been that this reflected a movement of peoples into the region from the north (Caldwell 1952:316-317; Waring 1968:216, 221). Grog tempered ceramic manufacturing procedures had apparently been moving gradually southward for over a millennium, from antecedents in the Croaker Landing and Hanover series in coastal North Carolina, however, so a northern origin is not altogether unexpected (see Mathis this volume). Recent radiocarbon dates from sites in the Beaufort County, South Carolina area, furthermore, suggest that Wilmington may have appeared appreciably earlier (Espenshade 1994) and Deptford lasted appreciably later (e.g., Trinkley 1991, 1994) than ca. A.D. 500^{c14}.

Even given these caveats, the appearance of Wilmington ceramics in the vicinity of the mouth of the Savannah River does appear to be a more sudden and dramatic change than the slow spread (or long period of co-occurrence) suggested further to the north. While we have no unequivocal evidence in support of such an inference, it is tempting to speculate that what is sometimes called the 'Wilmington intrusion' (Waring 1968:221) reflects an actual movement of peoples from the north. In the context of this volume, one might speculate that northern peoples undergoing some type of climate-induced stress—Woodland era carpetbaggers if you will—forced their way into the lush southern South Carolina/northern Georgia Sea-Island area, where food may have been more plentiful. While this idea has never been seriously investigated, it should be, and at least now we have a possible causal mechanism to examine that is more intellectually satisfying, and possibly easier to test, than vague generalizations about migrations or the expansion of ceramic traditions. As Mathis (this volume) has shown, furthermore, what we are seeing at the mouth of the Savannah River could be the tail end of ripple effect involving groups farther to the north, a much larger Algonkian expansion whose causes could be related to climatic events.

Cultural Change in the South Appalachians, Piedmont, and North Georgia Subregions

[South Atlantic Slope; North Georgia; Regional Studies; Recognition, Decline, Absence, Pottery/Ceramics Designs/Finishes, Middle Woodland, Late Woodland 'problem'; Late Woodland]

Changes in settlement patterns appear to have occurred between the Middle and Late Woodland in some parts of the Southeast, but this does not seem to be associated with a marked population decline or organizational simplification over the region, or even within in the South Appalachian area. The long-distance Hopewellian-derived exchange network effectively ceases by ca. A.D. 300^{c14}, and with it the elaborate (i.e., log tomb, mound burial) mortuary treatment accorded presumably high status individuals in some societies. In some parts of the Southeast, however, notably along the Gulf Slope and in the Lower Mississippi Valley, the period following A.D. 500^{c14}, and particularly after ca. A.D. 750-800^{c14} there is evidence for increases in population, organizational complexity, warfare, and the use of agricultural domesticates. What used to be called the "good gray cultures" (*after* Williams 1963) of the Late Woodland period are now recognized as vibrant cultures in their own right (Nassaney and Cobb 1991). The Coles Creek, Weeden Island, and late Swift Creek cultures are classic examples of such societies, that in no way reflect a decline in organizational complexity from what came before.

In the South Appalachian area perhaps the most dramatic evidence for change at the Middle to Late Woodland transition is the so-called "Wilmington intrusion" along the coast of southern South Carolina and Georgia noted previously. While how abrupt the appearance of Wilmington pottery actually was has been questioned by several recent radiocarbon dates, I retain a healthy skepticism about the validity of at least some of these determinations, particularly those that push the boundaries of our previous dating of Deptford and Wilmington ceramics by several centuries. Some of these dates, frankly, are so divergent from previous determinations that I must await more supporting evidence before I can accept them. I personally believe the that the appearance of Wilmington ceramics in this area represents a major cultural disjunction, but readily grant that such an inference is at present unproven. Aside for this possible movement of peoples down the coast, I do not see evidence for a major cultural disjunction elsewhere in the South Carolina area at this time.

The same is true, I believe, in portions of the interior south Atlantic Slope, and here my conclusions differ somewhat from Lilly and Webb, who suggest a possible hiatus or decline in settlement may have occurred between the Middle and Late Woodland periods in northwest Georgia (this volume). At an increasing number of locations in the South Appalachian area the radiocarbon record, associations with other Late Woodland/initial Mississippian wares, as well as some stratigraphic evidence, however, clearly indicate that simple stamped Connestee and Cartersville-like assemblages, as well as related wares like Santee and Vining Simple Stamped, continue to occur well into the Late Woodland era (Anderson 1982, 1985, Anderson and Schuldenrein 1985, Anderson and Joseph 1988, Elliott and Wynn 1991, Pluckhahn 1997, Worth and Duke 1991, Worth 1996). Wetmore and her colleagues (this volume) in particular have amassed an exceptional amount of evidence in support of a continuation of Connestee from the Middle through the Late Woodland in western North Carolina. A similar case based on multiply dated materials in good context can be made in at least three other parts of the South Appalachian area, along the lower Santee River in South Carolina, the upper Savannah River in northern Georgia and South Carolina, and in the central Ocmulgee River of Georgia (Anderson 1982, Anderson and Schuldenrein 1985, Pluckhahn 1997; Worth's Raccoon Ridge site is likely to prove another such context). Across the South Appalachian area, furthermore, the Middle Woodland to Early Mississippian complicated stamped tradition follows what to my mind is a fairly even progression, from Deptford/Swift Creek to Napier, Woodstock, and Etowah.

Accordingly, what has been called the 'Late Woodland problem' in north Georgia (Rudolph 1991)—that a major population decline may have occurred following the end of the Middle Woodland—is not one of an absence of sites, but rather a recognition problem. Thanks to the work of a great many researchers in Georgia and the Carolinas over the past two decades, we now know that ceramic finishes like simple stamping have a long tenure and wide distribution on the South Atlantic slope, and are a critical (and previously unrecognized) indicator of Late Woodland/initial Mississippian period settlement.

Given the fact that large scale population relocations or 'abandonments' have been well documented in the late prehistoric Southeast (e.g., Anderson 1996a, Williams 1990), however, it is hard to argue that such events were not occurring well into the past as well. That is, regional political conditions (i.e., patterns of interaction between societies, encompassing such things as warfare, exchange, and alliance) were likely quite important in the spacing of societies across the Southeast far into past, probably throughout the Woodland and well into the Archaic period. Accordingly, Lilly and Webb's argument that the distributions of these wares are not uniform over the region, and that they may be absent from portions of northwest Georgia, is quite reasonable. In northwest Georgia there may well have been a disjunction, or abandonment of the area at the end of (or ending) Cartersville locally, and that this may well have happened some time in the fifth or sixth centuries. I believe this abandonment, if that is what occurred, was probably quite localized, however, and can probably be as readily explained by changes in regional political geography as to a sudden change in climate (although I will also acknowledge that the two factors can be linked). In other parts of the South Appalachian area, occupation continued uninterrupted from the Middle to the Late Woodland.

Subregional Modeling

[Southeast; Linking, Multiple Independent Evidence; ; Woodland]

At the locality or subregional scale, I particularly like the analysis and explanations that Woodall (this volume) has developed linking climate and settlement data in North Carolina. My investigations of tree rings have demonstrated similar processes of seasonal slippage that could account for some proportion of required cultural adaptations to climate change. Even so, a regional scale analysis is essential to understanding the subregional effects of linking mechanisms that themselves operate on the regional and global scale. Research with George Milner and Marvin Smith in compiling phase distribution maps for the late prehistoric Eastern Woodlands has shown, at least in the Mississippian Period, that abandonments in one area were often coupled with increases or at least appearances of population in other areas (Anderson 1991, Milner et al. 1998). It is often not possible to even comprehend developments in one area until one examines distributional and cultural historical data from over a larger scale. This study of A.D. 536^{cal} is particularly important in this regard, for it forces us to look at patterns and processes from a larger scale, and from different perspectives, than most of us are used to doing.

We need to examine what effects, if any, climate change could have had on what are, at least over much of the South Atlantic Slope region, still predominantly hunting and gathering peoples. There is little evidence for any kind of intensive agriculture, or even horticulture by the Woodland peoples in the South Appalachian area, particularly in the Coastal Plain (Yarnell and Black 1985, Smith 1992). Walker's research on estuarine clines (this volume) shows a means to address this question with site and subsistence data.

We also need to examine the paleoclimate data itself carefully for localized effects of events like that which occurred in A.D. 536^{cal}. As Gunn notes (this volume), there are few baldcypress trees that date prior to this time. Unfortunately, A.D. 536^{cal} is at the extreme limits of bald cypress life expectancy, so this could or could not be significant relative to the A.D. 536 event. The Black River, North Carolina dendrochronological reconstruction, for example, only extends back to A.D. 372^{cal}, and during its first few centuries is based on only a few trees and cores, reducing its utility (Stahle et al. 1988). As dendrochronologically-derived paleoclimatic reconstructions are pushed further back in time, with larger samples, we should have better information on this issue.

We need to develop models to explain why the changes we observe in the archaeological record could have occurred, and ideally test them with multiple independent lines of evidence. That is, we need to explore how climate and culture are linked, and the kinds of cultural responses that could have been available to groups experiencing climatic change. Some societies are more flexible than others, and I suggest that to the Woodland period peoples of many parts of the Southeast, while the events of the years A.D. 536^{cal} and after could have caused some noticeable effects, they were not of sufficient magnitude to produce dramatic cultural changes. The Weeden Island societies of the Gulf Coast were thriving at this time, Walker (this volume) documents a similar florescence further south in Florida, and while the Late Woodland Southeast apparently witnessed less long distance exchange and interaction when compared with the preceding Middle Woodland era, it was not a period of decreased cultural complexity in many areas.

Future Directions in Global Change and Subregional Cultures

[Globe, Southeast; ; Global Change Causes, Collapse; Future]

Archaeology and history, like paleontology in the decade following the asteroid impact hypothesis for the demise of the dinosaurs (Alvarez et al. 1980), this symposium has shown us, is coming to grips with the possibility of punctuated and dramatic change. The turmoil that continues to characterize the debate among paleontologists about the history of life on earth has shown that this process will not be gentle, nor will be it be universally accepted, at least not for a long time, and not without a lot of spilled ink.

Do I believe that dramatic climate change is possible, brought about by factors that as recently as 20 or 25 years ago might have had us considered serious candidates for a mental institution had we suggested them? Absolutely. Articles in *Science* and *Nature* appear almost every month addressing possible causal mechanisms for global climate change from sources such as volcanic eruptions, meteor impacts, and perturbations in the earth's orbit or the sun's movement through the galaxy. We have certainly had plenty of compelling reasons to accept such mechanisms in recent years, not the least of which was the impact of the Shoemaker-Levy comet train with Jupiter in 1994 and the subsequent massive disturbances observed in that planets' atmosphere (Harrington et al. 1994; Hammel et al. 1995; see also Shoemaker et al. 1990 for an assessment of the same thing happening to earth). Another sobering find has been the discovery of the Chicxulub impact crater in the northern Yucatan (Penfield and Camargo 1981; Hildebrand et al. 1991), verifying the central premise of the Alvarez hypothesis, that a giant meteor hit the earth about the same time that the dinosaurs went extinct. On a lesser scale, the last two decades have witnessed an increasing understanding of events such as the meteor explosion over Siberia in 1908 (e.g., Svetsov 1996), or the Santorini eruption and the disruption of Minoan civilization (Doumas 1978, 1980, 1983, Hardy 1990).

Where do we go from here? The organizer of this volume has focused attention on a major possible cause of culture change in the Woodland Southeast, global climate change. I look forward to seeing how climate change and cultural response is examined by archaeologists in the years to come.

References

Alvarez, Luis W., Walter Alvarez, Frank Asaro, and Helen V. Michel (1980) Extraterrestrial Cause for the Cretaceous-Tertiary Extinction. *Science* 208:1095-1108.

Anderson, David G. (1982) The Mattassee Lake Ceramic Artifact Assemblage. In *Mattassee Lake:*

Archaeological Investigations Along the lower Santee River in the Coastal Plain of South Carolina, by David G. Anderson, Charles E. Cantley, and A. Lee Novick, pp. 207-321. Interagency Archaeological Services Atlanta, Special Publication 1982.

Anderson, David G. (1985) Middle Woodland Societies on the Lower South Atlantic Slope: A View from Georgia and South Carolina. *Early Georgia* 13:29-66.

Anderson, David G. (1991) Examining Prehistoric Settlement Distribution in Eastern North America. *Archaeology of Eastern North America* 19:1-22.

Anderson, David G. (1994) The Savannah River Chiefdoms: Political Change in the Late Prehistoric Southeast. University of Alabama Press, Tuscaloosa.

Anderson, David G. (1996a) Chiefly Cycling and Large-Scale Abandonments as Viewed from the Savannah River Basin. In *Political Structure and Change in the Prehistoric Southeastern United States*, edited by John F. Scarry, pp. 150-191. University Presses of Florida. Gainesville.

Anderson, David G. (1996b) Approcches to Modeling Regional Settlement in the Archaic Period Southeast. In *The Archaeology of the Mid-Holocene Southeast*, edited by Kenneth E. Sassaman and David G. Anderson, pp. 157-176. University Presses of Florida. Gainesville.

Anderson, David G., and J. W. Joseph (1988) Prehistory and History along the Upper Savannah River: Technical Synthesis of Cultural Resource Investigations, Richard B. Russell Multiple Resource Area. Russell Papers. Russell Papers. National Park Service, Interagency Archeological Services. Atlanta, Georgia.

Anderson, David G., and Joseph Schuldenrein (editors) (1985) Prehistoric Human Ecology along the Upper Savannah River: Excavations at the Rucker's Bottom, Abbeville, and Bullard Site Groups. Russell Papers. National Park Service, Interagency Archeological Services. Atlanta, Georgia.

Anderson, David G., David W. Stahle, and Malcolm K. Cleaveland (1995) Paleoclimate and the Potential Food Reserves of Mississippian Societies: A Case Study from the Savannah River Valley. *American Antiquity* 60:258-286.

Bove, Mark C. (1998) Impacts of ENSO on United States Tornadic Activity. Ninth Symposium on Global Change Studies, Phoenix, Arizona.

Cable, John S. (1997) Ceramic Analysis. In Shaw Air Force Base Archeological Data Recovery at Sites 38SU45, 38SU133, and 38SU145, with Results of Test Excavations Conducted at Sites 38SU136, 38SU137, and 38SU141, by Charles E. Cantley, John S. Cable, Mark J. Brooks, Barbara E. Taylor, Peter A. Stone, Leonard R. Gardner, Leslie E. Raymer, and Lisa D. O'Steen, pp. 233-301. U.S. Air Combat Command Series Report of Investigations Number 7. Manuscript prepared by Geo-Marine, Inc. for the U.S. Army Corps of Engineers, Fort Worth District.

Caldwell, Joseph R. (1952) The Archeology of Eastern Georgia and South Carolina. In *Archeology of the Eastern United States*, edited by James B. Griffin, pp. 312–21. Chicago: University of Chicago Press

Caldwell, Joseph R. (1958) *Trend and Tradition in the Prehistory of the Eastern United States*. American Anthropological Association Memoir 88. Springfield, Illinois.

Clark, R. M., and C. Renfrew (1972) A Statistical Approach to the Calibration of Floating Tree-Ring Chronologies Using Radiocarbon Dates. *Archaeometry* 14:5-19.

Clark, R. M., and A. Sowray (1973) Further Statistical Methods for the Calibration of Floating Tree-Ring Chronologies. *Archaeometry* 15:255-266.

DePratter, Chester B. (1979) Ceramics. In *The Anthropology of St. Catherine's Island*, vol. 2: *The Refuge-Deptford Mortuary Complex*, edited by David H. Thomas and Clark S. Larsen., pp. 109–32. Anthropological Papers of the American Museum of Natural History, no. 56(1). New York.

DePratter, Chester B. (1979) *W.P.A. Archaeological Excavations in Chatham County, Georgia: 1937–1942*. University of Georgia, Laboratory of Archaeology Series 29. Athens.

Diaz, Henry F., and Vera Markgraf, eds. (1992) *El Niño: The Historical and Paleoclimatic Record of the Southern Oscillation*. Cambridge University Press, New York.

Doumas Christo G. (ed.) 1978. *Thera and the Aegean World I*. Thera Foundation, London.

Doumas Christo G. (ed.) 1980. *Thera and the Aegean World II*. Thera Foundation, London.

Doumas, Christo G. (1983) *Thera-Pompeii of the Ancient Aegean*. Thames and Hudson, London

Hardy, D. A. (1990) *Thera and the Aegean World* III. (3 vols.) Thera Foundation, London.

Elliott, Daniel T., and Jack T. Wynn (1991) The Vinning Revival: A Late Simple Stamped Phase in the Central Georgia Piedmont. *Early Georgia* 19:1–18.

Espenshade, Christopher T., and Paul Brockington, Jr. (1989) An Archaeological Study of the Minim Island Shell Midden: Early Woodland Dynamics in Coastal South Carolina. Brockington and Associates, Inc. Atlanta.

Espenshade, Christopher T., Linda Kennedy, and Bobby G. Southerlin (1994) What is a Shell Midden? Data Recovery Excavations of Thom's Creek and Deptford Shell Middens, 38BU2, Spring Island, South Carolina. Brockington and Associates, Inc., Atlanta. Submitted to Spring Island Plantation, Bluffton, South Carolina.

Green, Phaedra M., David M. Legler, Carlos J. Miranda V., and James J. O'Brien (1997) *The North American Climate Patterns Associated with the El Niño-Southern Oscillation.* Center for Ocean-Atmospheric Prediction Studies, COAPS Project Report Series 97-1. Florida State University, Tallahassee.

Gunn, Joel D. (1994) Introduction: A Perspective from the Humanities-Science Boundary. Global Climate-Human Life: Physical Contexts of Historical Ecology. *Human Ecology* 22:1-22, Special Issue.

Hammel, H. B., R. F. Beebe, A. P. Ingersoll, G. S. Orton, J. R. Mills, A. A. Simon, P. Chodas, J. T. Clarke, E. De Jong, T. E. Dowling, J. Harrington, L. F. Huber, E. Karkoschka, C. M. Santori, A. Toigo, D. Yeomans, R. A. West, 1995. HST Imaging of Atmospheric Phenomena Created by the Impact of Comet Shoemaker-Levy 9. *Science* 267, 1288-1296.

Harrington, J., R. P. LeBeau, K. A. Backes, and T. E. Dowling 1994. Dynamic response of Jupiter's atmosphere to the impact of comet P/Shoemaker-Levy 9. *Nature* 368, 525-527.

Hildebrand A. R., Penfield G. T., Kring D. A., Pilkington M., Camargo Z. A., Jacobsen S. B. and Boynton W. V. (1991) Chicxulub Crater: A Possible Cretaceous/Tertiary Boundary Impact Crater on the Yucatan Peninsula, Mexico, *Geology* 19:867-871.

Kuniholm, Peter I., Bernd Kromer, Sturt W. Manning, Maryanne Newton, Christine E. Latini, and Mary Jaye Bruce. Anatolian Tree Rings and the Absolute Chronology of the Eastern Mediterranean, 2220-718 B.C. *Nature* 381:780-783.

Lopinot, Neal H. (1994) A New Crop of Data on the Cahokian Polity. In *Agricultural Origins and Development in the Midcontinent*, edited by William Green, pp. 127-153. Office of the State Archaeologist, University of Iowa, Report 19. Iowa City.

Milner, George (1998) The *Cahokia Chiefdom: The Archaeology of a Mississippian Society*. Smithsonian Institution Press, Washington (In Press).

Nassaney, Michael S., and Charles R. Cobb (1991) *Stability, Transformation, and Variation: The Late Woodland Southeast*. Plenum Press, New York.

O'Brien, James J., Todd S. Richards, and Alan C. Davis (1996) The effect of El Niño on U.S. Landfalling Hurricanes. *Bulletin of the American Meteorological Society 77:773-774.*

Pauketat, Timothy R., and Thomas E. Emerson (1997) *Cahokia: Domination and Ideology in the Mississippian World*. University of Nebraska Press, Lincoln.

Penfield, G. T. and Camargo Z. A., (1981) Abstract. *Soc. Expl. Geophys. Tech. Progr*. 51:37.

Pluckhahn, Thomas J. (1997) Rethinking Early Mississippian Chronology and Cultural Contact in Central Georgia: The View from Tarver (9JO6). *Early Georgia* 25(1):21-54.

Quinn, William H., and Victor T. Neal (1983) Long-term variations in the Southern Oscillation, El Niño, and Chilean subtropical rainfall. *Fishery Bulletin*, 81:363-374.

Quinn, William H., Victor T. Neal, and Santiago E. Antunez de Mayolo (1987). El Niño Occurrences over the Past Four and-a-half Centuries. *Journal of Geophysical Research* 92 (C13):14,449-14,461.

Renfrew, Colin (1996) Kings, Tree Rings, and the Old World. *Nature* 381:733-734.

Rudolph, Teresa P. (1991) The Late Woodland "Problem" in North Georgia. In *Stability, Transformation, and Variation: The Late Woodland Southeast*, edited by Michael S. Nassaney and Charles R. Cobb, pp. 259–83. Plenum Press, New York.

Sandweiss, Daniel H., James B. Richardson III, Elizabeth J. Reitz, Harold B. Rollins, and Kirk A. Maasch (1996) Geoarchaeological Evidence from Peru for a 5000 Years B.P. Onset of El Niño. *Science* 273:1531-1533.

Shoemaker E. M., Wolfe R. F. and Shoemaker C. S. (1990) Asteroid and comet flux in the neighborhood of Earth, in *Global catastrophes in Earth history; an interdisciplinary conference on impact, volcanism and mass mortality*, edited by V. L. Sharpton, and P. D. Ward, pp. 155-170. Geological Society of America Special Paper 247.

Smith, Bruce D. (1992) *Rivers of Change: Essays on Early Agriculture in Eastern North America*. Smithsonian Institution Press. Washington, D.C.

Stahle, D. W., and M. K. Cleaveland (1992) Reconstruction and Analysis of Spring Rainfall Over the Southwestern USA for the Past 1000 Years. *Bulletin of the American Meteorological Society* 73:1947-1961.

Stahle, D. W., and M. K. Cleaveland (1994) Tree-Ring Reconstructed Rainfall over the Southeastern U.S.A. During the Medieval Warm Period and Little Ice Age. *Climatic Change* 26:199-212.

Stahle, D. W., M. K. Cleaveland, and J. G. Hehr (1985a) A 450-year Drought Reconstruction for Arkansas, United States. *Nature* 316:530-532.

Stahle, D. W., M. K. Cleaveland, and J. G. Hehr (1988) North Carolina Climate Changes Reconstructed from Tree Rings: 372 to 1985. *Science* 240:1517-1520.

Stahle, D. W., E. R. Cook, and J. W. C. White (1985b) Tree-Ring Dating of Bald cypress and the Potential for Millennia-Long Chronologies in the Southeast. *American Antiquity* 50:796-802.

Stahle, D. W., J. G. Hehr, G. G. Hawks, Jr., M. K. Cleaveland, and J. R. Baldwin (1985c) *Tree-Ring Chronologies for the Southcentral United States*. Tree Ring Laboratory and Office of the State Climatologist, Department of Geography, University of Arkansas. Fayetteville.

Stahle, D. W., M. K. Cleaveland, and R. S. Cerveny (1991) Tree-ring Reconstructed Sunshine Duration Over Central USA. *International Journal of Climatology* 11.

Stahle, David, Malcolm Cleaveland, and Dennis Blanton (1998) The Jamestown and Lost Colony Droughts. Manuscript cited with permission.

Svetsov, V. V. (1996) Total Ablation of the Debris from the 1908 Tunguska Explosion. *Nature* 383:697-699.

Trinkley, Michael B. (1991) Further Investigations of Prehistoric and Historic Lifeways on Callawassie and Spring Islands, Beaufort County, South Carolina. Chicora Foundation Research Series, No. 23. Columbia, South Carolina.

Trinkley, Michael B., and Natalie Adams (1994) *Middle and Late Woodland Life at Old House Creek, Hilton Head Island, South Carolina*. Chicora Foundation Research Series, No. 23. Columbia, South Carolina.

Waring, Antonio J., Jr. (1968) The Cultural Sequence at the Mouth of the Savannah River. In *The Waring Papers: The Collected Papers of Antonio J. Waring, Jr.*, edited by Stephen Williams, pp. 216–21. Papers of the Peabody Museum of American Archaeology and Ethnology, No. 58. Harvard University, Cambridge, Massachusetts.

Wendland, W. M., and Reid A. Bryson (1974) Dating Climatic Episodes of the Holocene. *Quaternary Research* 4:9-24.

Willey, Gordon R. (1936) Notes on Central Georgia Dendrochronology. *Tree-Ring Bulletin* 4:6-8.

Williams, J. Mark and Gary Shapiro (1990) *Lamar Archaeology: Mississippian Chiefdoms in the Deep South*. University of Alabama Press, Tuscaloosa.

Williams, Stephen (1963) The Eastern United States. In *Early Indian Farmers and Villages and Communities*, edited by William G. Haag, pp. 267-325. National Survey of Historic Sites and Buildings, National Park Service, U.S. Department of the Interior, Washington, D.C.

Williams, Stephen (1990) The Vacant Quarter and Other Events in the Lower Valley. In *Towns and Temples along the Mississippi*, edited by David H. Dye and Cheryl Anne Cox, pp. 170–80. University of Alabama Press. Tuscaloosa.

Wolfman, Daniel (1984) Geomagnetic Dating Methods in Archaeology. In *Advances in Archaeological Method and Theory* Vol. 7, edited by M. B. Schiffer, pp. 363-458. Academic Press, New York.

Worth, John (1996) Upland Lamar, Vinning, and Cartersville: An Interim Report from Raccoon Ridge. *Early Georgia* 24(1):34-83.

Worth, John, and W. Maxwell Duke (1991) Hogcrawl Creek: Early Mississippian Period Occupation in the Middle Flint River Floodplain. *Early Georgia* 19(1):19-42.

Yarnell, Richard A., and M. Jean Black (1987) Temporal Trends Indicated by a Survey of Archaic and Woodland Plant Food Remains from Southeastern North America. *Southeastern Archaeology* 4:93-107.

16. EPILOGUE

David G. Anderson

Southeast Archaeological Center, National Park Service, 2035 East Paul Dirac Drive, Box 7, Tallahassee, Florida 32310

The mid-sixth century appears to have been a period when human societies in many parts of the globe experienced appreciable stress. We do not know exactly what happened, but there is evidence to suggest that global climate changed somewhat in the years around and following ca. A.D. 536, growing slightly cooler. Tree ring records, sea level curves, and documentary accounts, as well as more indirect archaeological evidence, all indicate that something apparently happened to effect climate at a planetary scale.

What we have chosen to call the A.D. 536 event, if indeed a single event is ultimately implicated, and not a series of events and/or processes operating over a number of years, appears to have had long term effects on both climate and culture. The ensuing three centuries were somewhat cooler than normal, with the result that a number of cultural systems in widely differing parts of the world exhibit evidence for depopulation, changes in land use, large scale relocation or movement, or a reduction in organizational complexity.

Critically important lessons of this volume, accordingly, are that human cultural systems can be profoundly effected by comparatively sudden changes in global climate; that such changes can occur at any time; and that these changes do not need to be large or dramatic to have long-term and far reaching effects. Our global technological civilization, while unquestionably resilient, is not invulnerable to such changes, and we must beware of hubris when judging its ability to cope with what the universe can throw at us.

The changes observed in mid-sixth century cultures worldwide may eventually be linked to a specific physical mechanism, such as a volcanic eruption, meteor impact, or some other natural phenomenon. It is the belief of most of the authors of this volume that something significant, at least in terms of its effect on human cultures, did occur at this time. A fairly sudden change in global climate appears to be the most likely explanation for what has been observed. We ask our colleagues in the natural and physical sciences to seriously explore this possibility.

The extent to which global climate may have changed in the mid-sixth century, we must emphasize, was comparatively minor when compared to some of the truly dramatic events that have occurred in earth history, such as the extended periods of continental glaciation that characterize much of the Pleistocene epoch, or the cataclysmic asteroid impacts at the ends of the Permian and Cretaceous periods. Nonetheless, for human societies in many parts of the globe even a comparatively minor change in climate, like that inferred to have occurred at ca. A.D. 536, could have had profound effects. We must explore what happened in the mid-sixth century, and more importantly, determine whether similar events or processes occurred at other times in human history and prehistory, what their effects were on the cultures of the time, and assess the probability of similar events occurring in the years to come. Seriously addressing such questions, we believe, can profoundly shape how we as a civilization look at the world around us, and plan for the future.

As exciting and challenging as determining the physical causes of A.D. 536-like events is determining their effects on human cultural systems. Here archaeology and history have much to contribute, by providing detailed information on the short- and long-term consequences of global climate change. This potential is increasingly being recognized. In the mid-sixth century, as we have seen, different reactions to the A.D. 536 event occurred in societies at differing levels of organizational complexity. In general, those societies most dependent upon intensive agriculture, and at the same time least able to fall back on alternate resources, appear to have been the most vulnerable. These tended to be among the most complex civilizations of the time, as the case studies in this volume from Western Europe, China, the Mayan area, and eastern Africa indicate. Organizational change, typically a reduction in complexity or social unrest, as well as outright suffering (i.e., starvation, disease) appear to have been greatest in these societies. In contrast, less complex societies dependent on wild plant and animal resources, such as some of the native societies of the southeastern United States, seem to have come through this event more or less unscarred. Even in these societies, however, there are hints that all was not completely well, given evidence for changes in subsistence practices, material culture, and settlement patterning, as well as for occasional large scale population movements. That the extant regional civilizations did not completely collapse in most of the areas examined in this volume during the mid-sixth century, however, indicates that whatever happened to the global climate system was comparatively mild or short-lived. It also indicates that the state level cultural systems of the time were themselves quite resilient. A hallmark of complex organizational forms, particularly state level societies, of course, is the ability of their ruling classes or elites to marshal resources, and use them to advance their personal agendas. Survival and the maintenance of power are, of course, the most fundamental social agenda of such groups, and hence only the greatest of trauma could end these systems.

State-level societies in the sixth century, as indeed at any time, overcome periods of adversity by having the ability to buffer food shortfalls through the use of adequate appropriation and storage technologies, and having the physical force necessary to maintain control over these resources. While suffering may have been profound among commoners, and individual elites and their families may have perished in some areas, no doubt many elites and their supporters, and hence the organizational systems they formed, were able to continue, albeit in weakened or changed condition. It is thus critically important to explore relationships between climate change, food production (i.e., crop yields, fishing harvests), and cultural (i.e., socio-political or organizational) stability. Fortunately, a great deal of research along these lines is now being conducted, as climate modeling becomes more sophisticated, and the global impact of events like El Ni-o, or carbon dioxide and fluorocarbon emissions, is coming to be widely appreciated, far beyond the realm of a few knowledgeable researchers. Basic questions that are coming to be asked by governmental bodies as well as individual citizens in nations around the world include: (1) what kinds of changes might we expect to occur in global climate in the years to come? (2) how vulnerable are our current cultural and political systems to such changes? and, (3) how can we minimize or at least anticipate any possible negative consequences of such changes? As we have seen in this volume, the past offers a vast record for comparison, and archaeology and history can make major contributions to the answering of these questions. Such questions are fundamental to human existence and perhaps even to our survival. Just as the asteroid impact that killed off the dinosaurs sixty five million years ago gave our mammalian forebears the chance to spread over the planet and achieve dominance, a similar event could end our own species' reign on earth. As this volume has attempted to demonstrate, there appears to be little doubt that similar, albeit lesser events have also shaped the course of human development. We have seen through archaeology and history that whole civilizations have been profoundly effected by natural calamities, such as the eruption of Santorini on the island of Thera in the eastern Mediterranean in the mid-second millennium B.C., the eruption of Mount Tambora in 1815 (Stothers 1984), or the aftermath of the A.D. 536 event. The record of earth's history is being closely scrutinized for evidence of past trauma, and the incidence and effects of massive volcanic eruptions or the impacts of extra-terrestrial objects such as meteors, comets, or asteroids has moved from the realm of wild speculation to serious scientific examination in the recent years. We now know, for example, that large objects regularly hit the earth, with a probability of occurrence, or periodicity, that is related to the size of the impactor (Gehrels 1994). Thus, a major life-shattering event like the impact at (and creating) the Cretaceous-Tertiary, caused by a ca. 10 km diameter asteroid, occurs about once every 100 million years or so. It is fortunate that such events are extremely rare, because such an impact today would probably result in the extinction of our species and, as in past comparable events, most life on earth. Less dramatic impacts, by objects a few hundred meters to a kilometer or so in extent, occur more frequently, every few hundred thousand to few million years. Such an impact, while perhaps not fatal, would likely result in the collapse of our global technological civilization (Sagan 1994:311-313). Finally, every few hundred years, the earth is hit by an object a hundred meters or so in diameter, causing the devastation of the largest of nuclear weapons. Such an impact in a densely populated area would cause unimaginable suffering.

Whether the A.D. 536 event was caused by the impact of an object somewhere in the few hundred to few million year periodicity range, or resulted from some other cause, such as a massive volcanic eruption, is currently unknown. The physical signatures of such events should be ultimately discernible, although if an extraterrestrial source, such as a comet or asteroid is involved, these will likely differ appreciably depending on whether a land or ocean strike occurred, and on the composition of the impacting body.

How many unrecorded yet equally profound or even greater events have shaped the human past? We now know, for example, that a major asteroid impact occurred in the southern Pacific ocean approximately 2.15 million years ago, and that it apparently had significant and long-term effects on global climate (Gersonde et al. 1997). What role, if any, did this late Pliocene event play in the evolution of our own species? The emergence of the genus Homo occurs about this time, as well as a fairly dramatic upswing in stone tool use, and it is at least possible that these events are in some way linked. We know from the aftermath of Mount Pinatubo, Mount Tambora, and other examples dating back over the centuries that major volcanic eruptions can effect global climate, and have a major impact on human society (Simkin and Siebert 1994). The systematic examination of such events, through the cooperation of scholars in the various disciplines capable of exploring these kinds of questions, we believe, can greatly improve our understanding of the past.

Ultimately, what holds our interest as archaeologists and historians is how our species, and its varied cultures and individuals, have responded to perturbations in global climate such as those discussed here, and how they will likely respond in the future. The contents of this book suggest that we have much to contribute to the study of these questions.

References

Gehrels, T. (editor) (1994) *Hazards Due to Comets and Asteroids*. University of Arizona Press, Tucson.

Gersonde, R., F. T. Kyte, U. Bleil, B. Diekmann, J. A. Flores, K. Gohl, G. Grahl, R. Hagen, G. Kuhn, F. J. Sierro, D. Volker, A. Abelmann, and J. A. Bostwick (1997) Geological record and reconstruction of the late Pliocene impact of the Eltanin asteroid in the Southern Ocean. *Nature* 390:357-363.

Sagan, Carl (1994) Pale Blue Dot: A Vision of the Human Future in Space. Random House, New York.

Simkin, T., and Siebert, L., (1994) *Volcanoes of the World*. Geoscience Press, Tucson, Arizona.

Stothers, R.B. (1984) The great Tambora eruption of 1815 and its aftermath: *Science* 224:1191-1198.

www.ingramcontent.com/pod-product-compliance
Lightning Source LLC
LaVergne TN
LVHW071213210926
848482LV00015B/803